# Nanoscale Science and Engineering Forum



Presentations at the 2008 AIChE Annual Meeting

## Philadelphia, Pennsylvania
## 16 - 21 November 2008

ISBN: 978-1-61567-234-9

**Printed from e-media with permission by:**

Curran Associates, Inc.
57 Morehouse Lane
Red Hook, NY 12571
www.proceedings.com





Printed by Curran Associates, Inc. (2009)



AIChE
3 Park Avenue
New York, NY 10016-5991

Phone: (203) 702-7660
Fax: (203) 775-5177

www.aiche.org

# TABLE OF CONTENTS

# 16a $\alpha5\beta1$-Targeted Stealth Liposomes for Delivery of DNA to Cancer Cells

**Alison W. Tisdale**[1], Alice Hsu[2], R. Scott McIvor[2], and Efrosini Kokkoli[1]. (1) Department of Chemical Engineering and Materials Science, University of Minnesota, 421 Washington Ave. SE, Minneapolis, MN 55455, (2) Department of Genetics, Cell Biology and Development, University of Minnesota, 321 Church St. SE, Minneapolis, MN 55455

Successful gene therapy depends on efficient delivery of DNA to cells. Stealth liposomes are well characterized for the delivery of cancer drugs and have demonstrated numerous desirable attributes for a delivery system, including increased circulation time *in vivo*, ability to protect the encapsulated contents and successful delivery of the drug load to the cells of interest. In our lab we have previously demonstrated that delivery can be further improved by incorporating into the liposomes a peptide sequence which targets integrin $\alpha5\beta1$ expressed on colon cancer cells. Here we describe these targeted stealth liposomes as carriers for the delivery of poly-L-lysine-condensed plasmid DNA (pDNA). *In vivo* studies in mice with intrahepatic CT26 tumors showed successful delivery and expression of encapsulated pDNA. Further, when compared with polyethyleneimine (PEI)-mediated delivery and hydrodynamic injection of naked pDNA, these liposomes were substantially more efficient on a per $\mu$g basis. *In vitro* studies have confirmed that $\alpha5\beta1$ targeting improves delivery over traditional stealth liposomes. Taken together these results indicate that this system may be a viable method for specific, efficient *in vivo* delivery of DNA to cancer cells.

# 16b ABC Triblock Copolymer Micelleplexes for Potent Gene Silencing and In Vivo Tumor Targeting

**Dana J. Gary**, Rahul Sharma, and You-Yeon Won. School of Chemical Engineering, Purdue University, 480 Stadium Mall Drive, West Lafayette, IN 47907

Polymers, particularly polycations, are very popular non-viral gene carriers because they can spontaneously associate with the negatively-charged phosphate backbone of nucleic acids to form protective "polyplexes". However, among the drawbacks of these systems is that under physiological salt conditions, where repulsive forces between like-charged particles are screened and/or nonspecific interactions with negatively-charged blood components may be present, aggregation of cationic polyplexes can result. Conjugation of poly(ethylene glycol), or PEG, to cationic polymers has been shown to confer several advantageous properties including colloidal stability and reduced aggregation of complexes due to steric shielding and increased hydrophilicity/water solubility. Both polycation (poly(2-(dimethylamino)ethyl methacrylate) or PDMAEMA) and PEGylated polycation (PEG-PDMAEMA) formulations were employed in this study as siRNA delivery vehicles. In addition, we are introducing a novel chemistry for triblock copolymer micelles made from poly(ethylene glycol)-poly(n-butyl acrylate)-poly(2-(dimethylamino)ethyl methacrylate) (PEG-PnBA-PDMAEMA). This micelle-based approach is different from the more traditional polymeric carriers being studied in that micelles form gene silencing complexes ("micelleplexes") based on particles rather than individual polymer chains which consequently make them larger in relative size. We expect this may have important implications in the ultimate efficiency of the systemic delivery process.

In this study, we examined fundamental in vitro properties such as binding strength, enzymatic resistance, cytotoxicity, protein aggregation and gene silencing efficiency across three types of polymeric carrier systems. By comparing PDMAEMA with PEG-PDMAEMA and PEG-PDMAEMA with PEG-PnBA-PDMAEMA, we sought to elucidate the role of extent of PEGylation and size/ architecture, respectively, on performance and hypothesize how this may ultimately translate into in vivo efficiency. A series of in vitro/pre-in vivo experiments showed that each of the three systems tested were stably complexed with siRNA, enzyme-resistant, fairly non-cytotoxic and resisted aggregation in the presence of serum levels of protein. Most importantly, the in vitro silencing levels of the exogenous luciferase gene were about 85% for each system, and were on par with a commercial transfection reagent, Lipofectamine2000, which was also tested. Encouraging preliminary in vivo biodistribution studies with fluorescently-tagged PEG-PDMAEMA and PEG-PnBA-PDMAEMA complexes in athymic mice possessing solid tumors indicate that both complexes do accumulate and get retained in tumor tissue. This is a promising result which suggests that our carriers are good candidates for exploiting the Enhanced Permeability and Retention (EPR) effect, responsible for the nonspecific uptake of favorably-sized nanoparticles in tumors. With some further testing and optimization, we believe our systems will prove to be viable options for in vivo tumor-targeted therapies.

# 16c Rgd-Targeting of PAMAM Dendrimers for Efficient Delivery of siRNA to Malignant Glioblastoma Cells

**Carolyn L. Waite**[1], Sarah M. Sparks[2], Kathryn E. Uhrich[2], and Charles M. Roth[3]. (1) Chemical and Biochemical Engineering, Rutgers University, 98 Brett Road, Piscataway, NJ 08854, (2) Chemistry and Chemical Biology, Rutgers University, 98 Brett Road, Piscataway, NJ 08854, (3) Rutgers University, 599 Taylor Road, Piscataway, NJ 08854

RNA interference utilizes short-interfering RNAs (siRNAs) to selectively silence a gene and, when targeted towards an oncogene, has potential for use as an anticancer therapy. The advancement of RNAi-based therapies is limited by the lack of effective siRNA delivery vectors. The use of cationic polymers as siRNA delivery vectors is promising, as synthetic polymers can be easily functionalized or modified for a particular application. Rational design of polymeric carriers, though, has been complicated by the fact that most chemical modifications affect multiple aspects of the delivery process.

In this work, polyamidoamine (PAMAM) dendrimers are being studied as vectors for the targeted delivery of siRNAs to U87 malignant glioma cells. The highly organized, branched structure of PAMAM makes it an attractive polymer for targeted drug delivery due to favorable ligand presentation and the presence of many reactive amine groups. The extent of primary amine acetylation of generation 5 poly(amidoamine) (PAMAM) dendrimers was studied as a modification for the delivery of siRNA to U87 malignant glioma cells. A linear decrease in cytotoxicity was observed upon primary amine acetylation. To evaluate the siRNA delivery by acetylated PAMAM and subsequent gene silencing dynamics, U87 cells stably expressing d1EGFP were analyzed using flow cytometry. We observed that a modest fraction (approximately 20%) of primary amines can be modified while maintaining the GFP silencing ability comparable to unmodified PAMAM, but higher degrees of amine neutralization notably reduced the gene silencing efficiency of PAMAM-siRNA polyplexes. This trend might be explained by a marked reduction in endosomal buffering capacity of dendrimers upon amine acetylation, which counteracted the increase in siRNA unpackaging.

To improve the siRNA delivery of dendrimers to tumor cells, $\alpha_v\beta_3$ integrin-targeting RGD peptides have been conjugated to PAMAM dendrimers using a sulfo-LC-SPDP crosslinker. An increase in intracellular delivery of fluorescently-labeled siRNA to U87 cells that highly express the $\alpha_v\beta_3$ integrin receptor was observed upon incorporation of RGD targeting peptides into the delivery system. Enhanced delivery was not observed, however, when RGD-targeted dendrimers were delivered to A172 malignant glioma cells that do not express $\alpha_v\beta_3$ integrin receptors. This indicates that the uptake of RGD-conjugated dendrimers occurs at least in part through the $\alpha_v\beta_3$ receptor mediated mechanism.

Currently, we are evaluating the effect of varying extents of RGD-conjugation to PAMAM dendrimers on cellular delivery of siRNA. We expect that increasing the extent of RGD conjugation will have a synergistic effect on cellular binding and subsequent siRNA delivery due to multivalent interactions between targeted dendrimers and cell-surface $\alpha_v\beta_3$ receptors. However, a dramatic change of the dendrimer chemistry might alter its intracellular trafficking and have a detrimental effect on its ability to deliver siRNA. Therefore, surface plasmon resonance will be used to quantatively study the interactions between targeted dendrimers with $\alpha_v\beta_3$ integrin proteins, and the dendrimer chemistry (extent of acetylation and degree of peptide conjugation) will be optimized to achieve a maximum biological response.

# 16d Drug Release Kinetics from Dendritic Nanodevices for Sustained Delivery

**Yunus Emre Kurtoglu** and Rangaramanujam M. Kannan. Chemical Engineering, Wayne State University, 5050 Anthony Wayne Drive, Detroit, MI 48202

Recent research has shown that Polyamidoamine (PAMAM) dendrimers have great prospect as targeted drug delivery nanodevices. PAMAM dendrimers are highly hyper-branched synthetic polymers with well defined spherical structure and size. The surface of the spherical structure consists of functional groups (cationic, anionic or neutral) that make them very hydrophilic and highly water soluble, thus offer enhancements in solubility for hydrophobic drugs upon attachment. In addition, PAMAM dendrimers were shown to enhance cell entry on drug resistant cells by changing cellular uptake pathway. It was previously demonstrated that PAMAM dendrimers can accumulate in the tumor tissue via enhanced permeation retention effect (EPR) and receptor mediated tumor targeting. One key challenge success of such systems hang on is to design systems that release the drug payload form dendritic nanodevice in a predetermined fashion in the targeted site while staying intact elsewhere in the body. The goal of this work was to engineer a model PAMAM dendrimer based nanodevice for sustained drug delivery applications. This model device is used to evaluate the applicability of ester and amide linkages in preparation, determine the corresponding drug release rates, and to understand the nanoscale structural effects on drug release mechanism.

Ibuprofen, a poorly water-soluble anti-inflammatory drug, was used as a model drug to study the release kinetics and mechanism of -NH2 and -OH terminated G4-PAMAM dendrimers (M.W ~14kDa) containing amide and ester linkages respectively. mPEG (Mw=5000) was also conjugated to Ibuprofen via an ester linkage for comparison with dendrimer conjugates. The payloads of G4-OH and G4-NH2 and mPEG conjugates were 8.9, 5.34 and 3 percent by weight respectively. The hydrolytic and enzymatic stability of the ester and amide linked conjugates were analyzed in various pH buffers and in diluted human plasma. Amide conjugates were stable in all buffers for 10 days and did not release significant amount of the drug. Ester linked G4-OH-Ibu conjugate showed pH dependent hydrolytic cleavage and sustained drug release for more than 10 days. Neither amide nor ester bonded PAMAM conjugates showed enzymatic hydrolysis in diluted human plasma within 12 hours. On the other hand, ester bonded mPEG conjugate was cleaved enzymatically within the same amount of time. The stability of G4-OH-Ibu against enzymatic hydrolysis suggests that the conjugate structure blocks esterase enzyme activity to cleave the ester linkage and release the drug. Nanometer scale spherical structure of G4 PAMAM dendrimer conjugate provided a unique way to protect the ester linkage from enzymatic cleavage in plasma. In this work a dendritic nanodevice that can provide sustained Ibuprofen release intracellularly while maintaining stability in plasma is described.

## 16e Design of a Multivalent Binding Platform for Enhanced Non-Viral Gene Transfer to $\alpha_v\beta_3$ Expressing Cells

**Quinn Ng** and Tatiana Segura. Chemical and Biomolecular Engineering, University of California, Los Angeles, 420 Westwood Plaza, 5531 Boelter Hall, Los Angeles, CA 90095

Non-viral gene delivery has been widely investigated over the past decade as a means to guide tissue regeneration, treat disease and study gene function. However, low efficiencies of gene transfer and inability to target desired cell populations have limited the use of this approach. Although current non-viral delivery approaches have utilized receptors that are uniquely expressed at the cell surface for targeting and to enhance gene transfer, strategies that target the density of receptors at the cell surface have not been explored. We believe that by targeting multiple ligands simultaneously (multivalent binding), we can enhance targeting of the desired cell type and enhance gene transfer through the engagement of biological pathways that are unique to clustered receptors. Interestingly adenovirus type 2 and 12 utilize this approach to enhance cellular internalization through proteins at their surface that contain spatially constrained integrin binding peptides that can interact with multiple integrin receptors simultaneously. In our present study, we investigated the effect spatially constrained Arg-Gly-Asp (RGD) peptides on the surface of DNA/polyethylene imine (PEI) polyplexes have on the efficiency of gene transfer to cells that have different densities of $\alpha_v\beta_3$ integrin receptors. RGD peptides were spatially constrained through immobilization to nanoparticles (nano-RGD) and the resulting nano-RGD nanoparticles were used to decorate the surface of DNA/PEI polyplexes. 5-nm nano-RGDs were synthesized with a theoretical density of 15 RGD peptides per particle and were shown to require 5 orders of magnitude lower concentration of RGD to prevent HeLa cell adhesion to vitronectin coated plates compared to free RGD peptide. This indicates that multivalent integrin binding is more effective in preventing cell adhesion than monomeric binding. The effect of nano-RGD on the ability of DNA/PEI polyplexes to transfect HeLa cells was studied for nano-RGD nanoparticles that were either covalently or electrostatically immobilized to the polyplex. Further, the effect on non-viral gene transfer efficiency of $\alpha_v\beta_3$ integrin density on the cell surface was studied using HeLa cells that were trypsin treated (low integrin density) or not (high integrin density). Polyplexes that were modified with nano-RGD resulted in a 5.4-fold or 37-fold increase in non-viral gene transfer efficiency over unmodified polyplexes for low or high integrin density cells respectively. This indicates that the presentation of the RGD peptides on the surface of DNA/PEI polyplexes affects cells with a high density of $\alpha_v\beta_3$ more than those that have low $\alpha_v\beta_3$ density. Further, nano-RGD was able to enhance gene transfer only when the nanoparticles were covalently bound to the DNA/PEI polyplexes. Polyplexes modified with nano-RGD through electrostatic interactions resulted in little to no enhancement of transgene expression compared to unmodified DNA/PEI polyplexes. Interestingly, experiments show that the degree of internalization for nano-RGD modified polyplexes and unmodified polyplexes are similar despite the increase in gene expression, suggesting that the clustered RGD has an effect on cellular trafficking rather than internalization alone. Experiments exploring the optimal concentration of nano-RGDs to enhance gene transfer resulted in an optimal concentration of 10 x $10^{13}$ nano-RGD particles, indicating that the density of nano-RGD clusters on the surface of the polyplex affects gene transfer. The use of platforms that can engage multiple receptors at the cell surface may not only enhance the targeting but also activate intracellular pathways that are unique in clustered receptors and result in enhanced gene transfer.

**Title: Preparation and Stabilization of Biodegradable Polyelectrolyte Complexes for Targeted Drug Delivery**

**Session:** 22B02 Bionanotechnology for Gene and Drug Delivery I

**Abstract:** Use of nanotechnology in pharmaceutical research is receiving a significant attention by academia as well as industries for targeted drug delivery. Water-soluble, polymeric nanoparticles have evolved as an alternative to other available systems mainly because of their non-toxic and biodegradable nature. This work focuses on preparation of such nanoparticulate polyelectrolyte complexes (PECs) for targeted drug delivery applications. PECs result mainly from electrostatic interactions between polycations and polyanions and have a core-shell structure with free amine groups available on the shell surface for conjugation with antibodies, genes, toxins, and drug molecules. Core-shell structured PEC nanoparticles have been precipitated using polycations namely, poly (methylene-co-guanidine) hydrochloride (PMCG), and Spermine tetrahydrochloride (ST), and polyanions namely, sodium alginate (SA) and Chondroitin Sulfate (CS). Effect of various process parameters such as positive to negative charge ratio of polyions, pH, stabilizer concentration, mixing speed, time for addition of polyanion in polycation solution on, PEC formation, stability and morphology has been studied. PECs with size ranging from 100 nm to 100 μm have been precipitated with variation in process parameters. It is found that increase in positive to negative charge ratio, a pH of 4.2, increase in mixing speed and decrease in addition time decreases the PEC size (to around 100 nm) and increases the stability of particle size. Separation of PECs from the polyionic solution has been attempted using centrifugation which enhances agglomeration among PECs and increases PEC size from 100 nm (before centrifugation) to around 400 nm (after centrifugation).

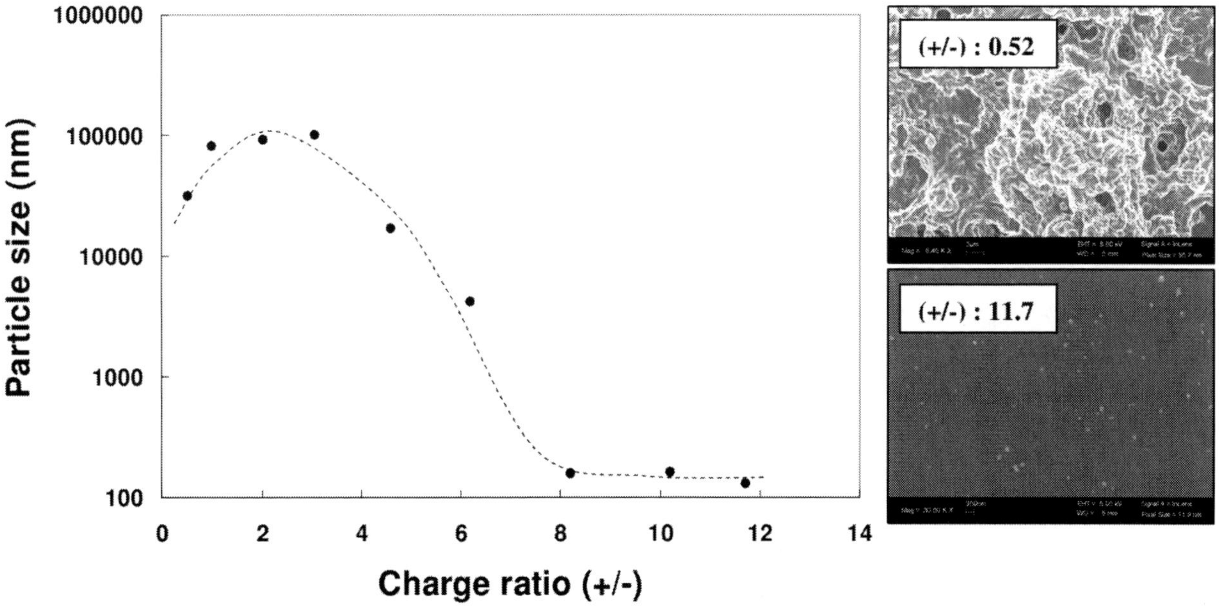


**Fig. 1. Effect of charge ratio (+/-) on size and morphology of PECs**

## 16g Prodrug Conjugates to Improve the Delivery Capacity of Nanoparticles for Protein Therapeutics

**Marian E. Gindy**, Chemical Engineering, Princeton Unversity, E-Quad, Olden Street, Princeton, NJ 08544 and Robert K. Prudhomme, Princeton University, Dept of Chemical Eng Olden St, Princeton, NJ 08544.

Hydrophilic peptides and protein macromolecules represent an important class of therapeutic compounds, although wide clinical application is hindered by low oral bioavailability, short circulation half-life, and in vivo instability due to enzymatic degradation. Polymeric nanoparticles are ideally suited to circumvent these limitations, as they can be made to extend circulation times in vivo and protect encapsulated molecules from degradation. Nonetheless, the capacity of polymeric particles to encapsulate hydrophilic compounds at sufficient concentrations is generally limited.

The approach developed in this work relies on the synthesis of prodrugs to modulate the supersaturation of hydrophilic macromolecules during the particle formation process. Prodrugs are prepared through chemical conjugation of the protein to hydrophobic anchor molecules via labile linkages. Depending on the properties of the anchor molecule and degree of protein modification, the prodrug solubility could be appropriately tuned. Using this strategy, polymer-stabilized protein particles were prepared via Flash NanoPrecipitation, a controlled precipitation process which utilizes amphiphilic block copolymers to direct particle self-assembly. Particles of nanometer size, narrow size distributions, and protein loadings in excess of 35wt% were prepared. Release of the prodrug from the particle is anticipated based on enzymatic degradation of the polymer, whereupon the prodrug is instantaneously cleaved into the active form. The prodrug conjugation strategy, coupled with Flash NanoPrecipitation technology, thus present an attractive system which permit the nanoparticle formulation of a wide range of active solutes irrespective of their intrinsic water solubilities.

## 17a The Design of Functional Nanostructured Materials

**Ravi S. Kane**, Chemical and Biological Engineering, Rensselaer Polytechnic Institute, 4005 Center for Biotechnology, Troy, NY 12180

Interfacing nanoscale materials with bioactive molecules is important for the design of therapeutics, sensors, and functional composites. We have been investigating the structure and function of biological molecules immobilized onto nanoscale scaffolds such as carbon nanotubes. While it is well appreciated that interfacing biological molecules with nanomaterials can influence nanomaterial properties, we will describe how the nanoscale environment can in turn influence the activity and stability of adsorbed proteins. Furthermore, we will discuss the ability to remotely control the activity of proteins adsorbed onto carbon nanotubes. We will discuss the mechanistic basis of this phenomenon, and its use to design self-cleaning nanotube-based coatings as well as nanotube conjugates that selectively destroy a target toxin from a mixture of proteins. Functionalization with biomolecules also provides a method for controlling the assembly of nanomaterials. We will also discuss these and other approaches for organizing nanomaterials into macroscopic structures.

## 17b Getting around the Innate Immune Response - Nanoscale Insights into Shape, Flexibility, and the 'marker of Self' System

**Dennis E. Discher**, Chemical and Biomolecular Engineering, University of Pennsylvania, Room 129 Towne Building, 220 South 33rd Street, Philadelphia, PA 19104-6393

The Innate Immune System comprises the cells, soluble factors, and related mechanisms that defend the host from 'Foreign', in a non-specific manner. Foreign includes invading microbes, as has been typical for eons, and also implanted or injected material systems. We have been exploring means to evade or delay clearance by the innate immune system, particularly shape, flexibility, and also ligand-specific mechanisms. Shape effects of drug delivery vehicles are largely unexplored, especially in vivo, but we have recently shown that flexible cylinders persist in vivo longer than spheres of identical composition (1). Flexible filomicelles increase both dosage and tumor-selective effects in vivo relative to spheres and thus appear promising as anticancer drug delivery systems. However, Macrophages – as the main cells of the Innate Immune System – still clear the injected carriers and limit dose; while at the same time, Macrophages somehow choose to leave our own 'Self' cells alone. The Foreign vs Self decision certainly does NOT reside in steric repulsion by the glycocalyx or synthetic polymer 'mimics' of it such as PEG. We have sought to address how Macrophages specifically recognize Self and to define its physicochemical limits, and we have focused on the cell-surface protein CD47 found on all of our own cells. We demonstrate a two-step procedure for recognizing intruders that helps avoid misdirected attacks. In the first step, which is well known, Macrophages adhere and begin engulfing objects studded with antibodies or plasma complement proteins that bind interlopers and will also bind to the body's own cells. But before a macrophage engulfs its target, it also checks for the second form of identification on all self cells, CD47. The ~100 aa extracellular domain of CD47 proves sufficient to induce a macrophage to disengage a cell from the same species or even a synthetic particle decorated with this domain (2). We detail the divergence in Foreign vs Self adhesive signaling mechanisms, the dependence on protein densities, and the particle size dependence for this fundamental facet of immunocompatability. REFERENCES: (1) Y. Geng, P. Dalhaimer, S. Cai, R. Tsai, M. Tewari, T. Minko, and D.E. Discher. 2007. Shape effects of filaments versus spherical particles in flow and drug delivery. Nature Nanotechnology 2: 249-255. (2) R. Tsai and D.E. Discher. 2008. Inhibition of 'Self' Engulfment through deactivation of Myosin-II at the Phagocytic Synapse between Human Cells Journal of Cell Biology 180: 989-1003.

## 17c Drug Delivery and Imaging Using Nanoparticles Produced by Block-Copolymer Directed Flash Nanoprecipitation

**Robert K. Prud'homme**, Chemical Engineering, Princeton University, A-217; Engineering Quadrangle, Olden Street, Princeton, NJ 08544

Nanoparticle formulations of hydrophobic drugs present unique opportunities for treatment of solid tumor cancers, for delivery of drugs by aerosol administration, and as a route to novel vaccine adjuvants. The common requirements of these applications are precise control of particle size and surface functionality. For cancer therapy particles in the size range of 100-200 nm passively pass through defects in the vasculature in tumors and deposit by "enhanced permeation retention". In addition to delivery, the ability to monitor the fate of the nanoparticles is also of important since anti-cancer agents are invariably toxic to healthy tissue. Our process --Flash NanoPrecipitation – a controlled precipitation process that produces stable nanoparticles at high concentrations using amphiphilic diblock copolymers to direct self-assembly enables the production of composite nanoparticles that enable simultaneous imaging and delivery. Uniform particles with tunable sizes from 50-500 nm can be prepared in an economical and scalable manner. The key to the process is the control of time scales for micromixing, polymer self-assembly, and particle nucleation and growth. The PEG protective layer creates long-circulating particles and the inclusion of PEG chains with terminal ligands allows drug targeting. The incorporation of gold nanoparticles, magnetic nanoparticles, or fluorophores into the composite particle enables imaging by x-ray, MRI, or confocal microscopy, respectively. The incorporation of up converting phosphor crystals into the composite nanoparticles enables a highly efficient form of photodynamic therapy.

# Synthesis of aligned carbon nanotubes on double-sided metallic substrate by chemical vapor deposition

*Huan Wang, Jiyun Feng, Xijun Hu and Ka Ming Ng**

Department of Chemical Engineering

Hong Kong University of Science and Technology

Clear Water Bay, Hong Kong

**Abstract**

Aligned multi-walled carbon nanotubes (MWCNTs) were grown on both sides of a metallic or metal-coated substrate by water vapor-assisted chemical vapor deposition (CVD). Aligned CNT films of thickness ranging from 1 μm to over 100 μm were obtained. By manipulating various operating factors – position of substrate in the reactor, amount of water vapor, amount of catalyst, reactor temperature, and growth time, the morphology and thickness of these carbon nanotube films could be adjusted.

*Keywords:* Carbon nanotubes; Chemical vapor deposition; Thermal interface material

## 1. Introduction

With the continually diminishing size of electronic devices, the heat generated by such devices can cause structural damage due to over-heating.[1] It is highly desirable to design and fabricate thermal interface materials (TIMs) with exceptionally high thermal conductivity for transporting heat efficiently from electronic components to a heat sink.[2, 3]

One of such materials is carbon nanotube (CNT) which has shown unique and attractive mechanical,[4] electrical,[5] and thermal properties[6] since its discovery by Iijima.[7] Several studies have revealed that CNTs have unusually high thermal conductivity in their axial direction. For example, molecular dynamic simulations of a single-walled nanotube (SWCNT) by Berber et al. indicated that the thermal conductivity of the SWCNT can be as high as 6600 $Wm^{-1}K^{-1}$ at room temperature.[1] Kim et al.[8] measured the thermal conductivity of a single multi-walled carbon nanotube (MWCNT) using a micro-fabricated suspended device, and their measurement showed that the thermal conductivity was larger than 3000 $Wm^{-1}K^{-1}$ at room temperature. Realizing that CNTs are good TIM candidates, efforts have been made on the use of dispersed CNTs as thermal conducting fillers in polymer composites. For example, Liu et al.[9] reported a thermal conductivity ranging from 1.1 to 1.9 $Wm^{-1}K^{-1}$ as the CNT loading in a polymer composite increased from nil to 3.8 wt%. The below expectation enhancement could be attributed to the random orientation of CNTs in the polymer matrix and the existence of interface thermal resistances between the actual heat source/ sink and CNTs. To avoid these problems, Huang et al.[10] grew aligned CNTs on a silicon substrate and then fabricated a polymer composite film with such protruded, aligned CNTs running from one side of the TIM device to the other. An enhancement in thermal conductivity from 0.56 $Wm^{-1}K^{-1}$ for the pure elastomer matrix to 1.21 $Wm^{-1}K^{-1}$ for the same polymer embedded with a 0.4 vol % aligned CNT array was obtained. Despite the fact that aligned CNTs should have formed ideal thermal conducting paths through the composite structure, this enhancement was still far below expectation.

In this study, the use of aligned CNTs as TIM was further explored in order to take advantage of the exceptional thermal conductivity of CNTs. Instead of growing the CNTs on silicon, they were grown on a metal surface by water vapor-assisted chemical vapor deposition (CVD). Figure 1 shows the proposed multilayer composite structure, with two films of aligned CNTs grown on the opposite faces of a suspended substrate. Provided that the substrate is sufficiently thin and highly conductive, this double-sided design for a TIM device has the distinct advantage that it can fit between the heat source and heat sink to minimize the resistance to heat flow due to the presence of an air gap.

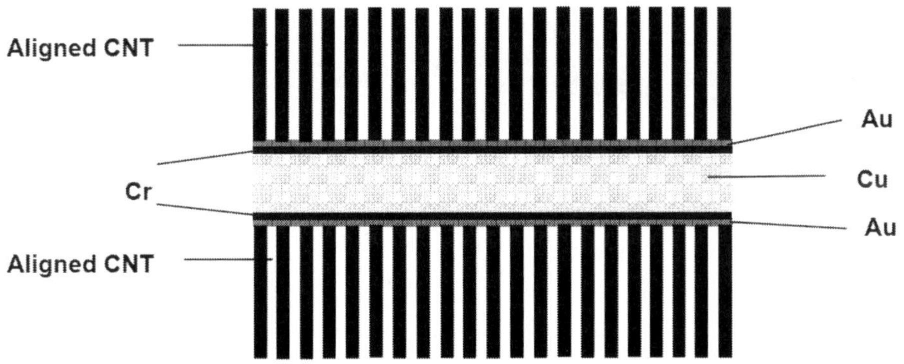


Figure 1. The proposed structure of thermal interface material.

In addition, operating factors that can significantly influence the growth of CNTs on such a double-sided metal substrate were investigated in detail. It was very difficult to characterize samples with ductile substrate such as thin metal plate or foil by scanning electron microscopy (SEM). Therefore, optimization of the operating factors was accomplished by using silicon wafer as substrate. The effect of water vapor was included because it was known to significantly influence the growth of aligned CNTs. For example, Hata et al.[11] synthesized aligned single-walled carbon nanotubes on silicon wafer with a height of up to 2.5 mm in the presence of water vapor.

## 2. Experimental

### *Materials*

The carbon source was industrial grade ethylene and the gaseous phase was made up of industrial grade argon and hydrogen for the synthesis of CNTs. The catalyst used was ferrocene purchased from Aldrich. Double-sided polished P-type silicon wafer with a thickness of 400 μm, copper plate with a thickness of 0.3 mm, and copper foil with a thickness of 125 μm and 7 μm were used as substrates for aligned CNT growth.

### *CVD reactor operations*

CVD synthesis of aligned CNTs was performed in a horizontal quartz tube (2.5 cm in diameter and 70 cm in length) housed in a single-zone furnace (Carbolite, UK) as shown in Figure 2. A quartz boat (2.2 cm wide and 10 cm long) was placed in the middle of the quartz tube, where the temperature was the highest along the whole tube. The substrates used in this study were placed in the middle of the boat except otherwise specified. It has been known that ferrocene decomposition forms iron particles as the catalysts for aligned CNTs growth. With a sublimation temperature of about 175 °C, it was stationed at the inlet of the quartz tube where the temperature was above the sublimation temperature of ferrocene (about 200 °C).

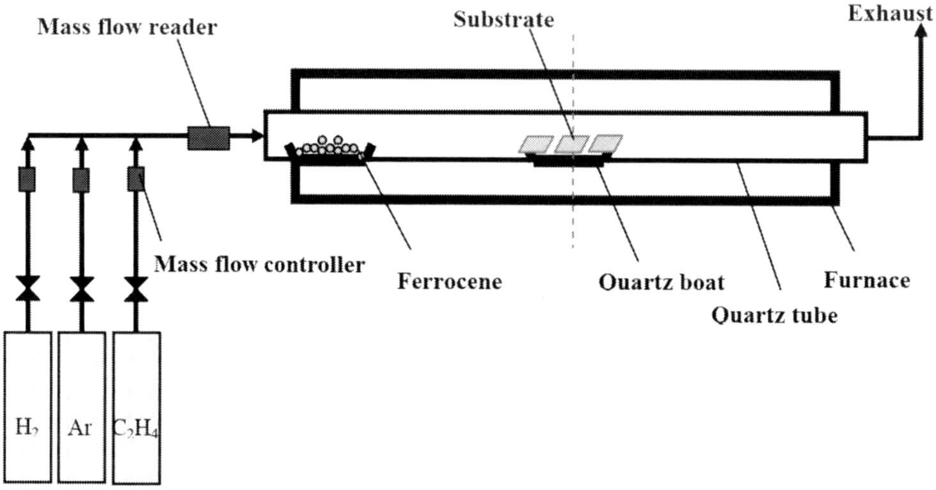


Figure 2. Experimental setup for aligned CNT synthesis by chemical vapor deposition.

Each reaction run started with the CVD reactor at room temperature. A specified amount of double deionized (DDI) water was injected using a micro-syringe under the quartz boat holding the ferrocene. Then, flow of argon was initiated while heating up the CVD reactor. Typically 20 min was required to reach the desirable reactor temperature and at that point ethylene and hydrogen were introduced. CVD growth of aligned CNTs was carried out at about 725~ 800 °C with the flow rates of ethylene, hydrogen and argon set at 40, 16 and 200 sccm, respectively. The flow of ethylene/hydrogen/argon mixture was maintained for the entire growth period ranging from 10 to 60 min. After the reaction, the furnace was allowed to cool down to room temperature while maintaining the flow of argon.

***Preparation of substrates***

Different metals including silver, copper, and gold were used as substrates for aligned CNTs growth. Gold was found to be the best substrate. However, copper is much cheaper and has a higher thermal conductivity (388 $Wm^{-1}K^{-1}$) than that of gold (315 $Wm^{-1}K^{-1}$). For these reasons, a thin film of gold was sputtered on the surface of the copper substrate. It is known that the adhesion of gold on copper is not sufficiently strong. By trial and error, it was found that good adhesion could be achieved with chromium between gold and copper. Thus, chromium films of thickness 12 nm and gold films of thickness 20 nm were deposited on both sides of the basis substrates using an ARC-12M sputtering machine.

***Characterization of aligned CNT on metal-coated substrate***

The aligned CNTs synthesized by CVD were characterized by SEM (JEOL-JSM 6700F operating at an accelerating voltage of 5 kV). For the ductility of copper, attempts to prepare SEM samples with copper substrates by cutting were not successful. Silicon wafer, which can easily be cut by a quartz knife, was used to replace the copper substrate to observe cross section of the aligned CNT samples by SEM. In addition, the detailed structure of the CNTs was determined by high-resolution transmission electron microscopy (HRTEM) using JEOL-2010 operating at an accelerating voltage of 200 kV. TEM samples were prepared by dispersing the CNTs in ethanol with mild sonication for 30 min, and then air-drying a drop of the solution on a copper grid coated with a holey carbon film. X-ray diffraction patterns of samples were recorded with a new powder x-ray diffraction system (Model X'pert, Panalytical). This provides a simple way to characterize the alignment of as-grown CNTs without damaging their original orientation. The XRD pattern was shown in Figure 3. The well resolved XRD peaks of aligned CNTs sample which can be assigned to (002), (100), and (110) diffractions, indicating that such a material has well alignment. Raman spectra were recorded with a RM3000 (Renishaw) in backscattering mode using an Ar+ laser excitation (514.5 nm, 5mW; resolution 1 $cm^{-1}$) in Figure 4. The Raman spectra, obtained in the range 800- 2000 $cm^{-1}$, shows a strong characteristic peak is located at 1590 $cm^{-1}$, corresponding to the C-C optical mode vibration along the graphite sheet. The peak in at 1343 $cm^{-1}$ is a Raman scattering peak due to defects in the mixture of amorphous carbon and CNTs. In this work, Raman spectroscopy was mainly used to verify the multi-walled nature of the CNTs and the overall homogeneity of the sample.

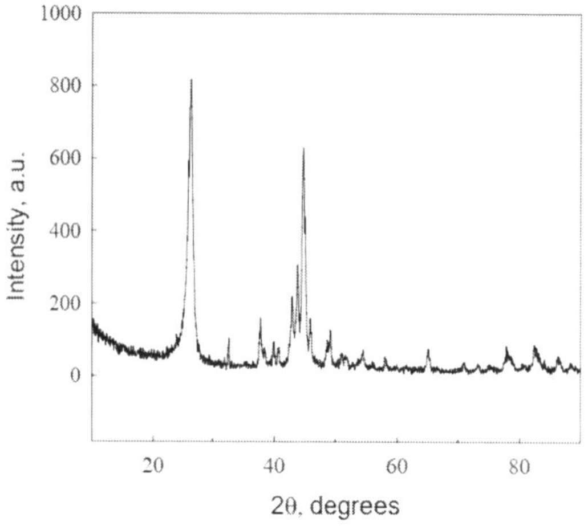


Figure 3. XRD pattern of aligned CNTs.

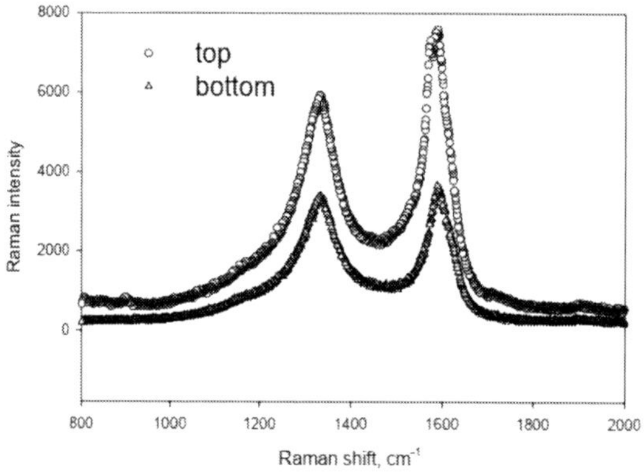


Figure 4. Raman spectra of aligned CNTs.

***Measurement of thermal resistance***

The thermal resistance of the TIM sample with metal substrate was evaluated using a steady state measurement system which was designed in accordance with ASTM D5470. The sample pressed between two copper blocks was subject to a constant heat flow rate of $Q$ generated by resistive cartridges heaters embedded in one copper block while the counter block was cooled with circulating water. A load cell was used to control the sample contact pressure to 0.3 MPa. Class A platinum resistance temperature detectors (RTD) with an accuracy of 0.3 $^{o}$C were used to measure the temperature at various points along the two copper blocks. Details of the design, operations, and calculations can be found in Zhang et al.[12]

**Results and discussion**

***Effect of position of the substrate in the quartz boat***

The first issue addressed was the effect of the position of a substrate in the quartz boat. Note that the substrate was simply laid flat on the boat in all cases. Figures 5(a)-(b), (c)-(d) and (e)-(f) show the photomicrographs of the top and bottom of the substrates placed in three different positions – front, middle and rear of the quartz boat, respectively. When the substrate was placed in the front of the quartz boat, amorphous carbon was found on top of the aligned CNTs layer (Figure 5(a)). Normally, the water vapor in the CVD system which could have removed amorphous carbon produced by pyrolysis was not effective when the substrate was in the front of the quartz boat. With the substrate in the middle, the morphology of aligned CNTs was good (Figures 5(c)-(d)). The CNT layer thickness was measured at three positions as indicated by A, B, and C in Figure 5(c) and an average thickness was obtained from these three values. Under these conditions, the thicknesses of the CNT films were almost the same on both sides of the substrate at about 50 μm. When the substrate was placed on the rear of the quartz boat, aligned CNTs could grow on both sides of the substrate. However because the substrate was far away from the catalyst, the aligned CNT layer on the bottom face of substrate (about 10 μm) was thinner than that on the top (about 25 μm) (Figures 5(e)-(f)). Based on these results, all substrates were placed in the middle of the quartz boat in the subsequent experiments.

**a) Position: front (top)**

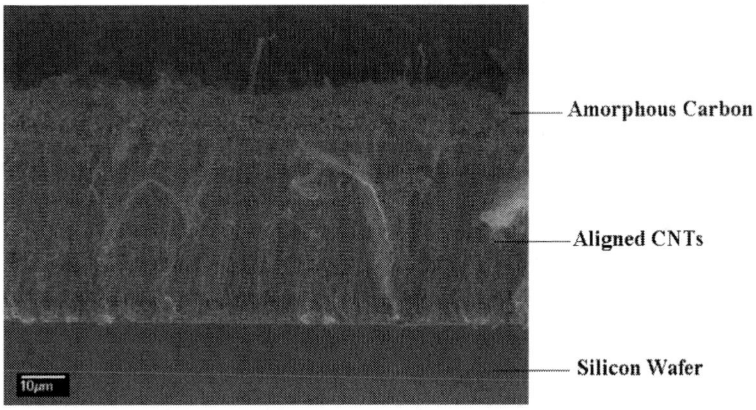

**b) Position: front (bottom)**

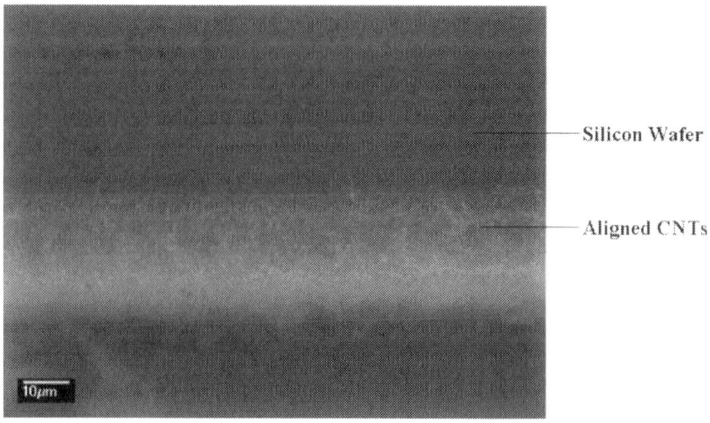


**c) Position: middle (top)**

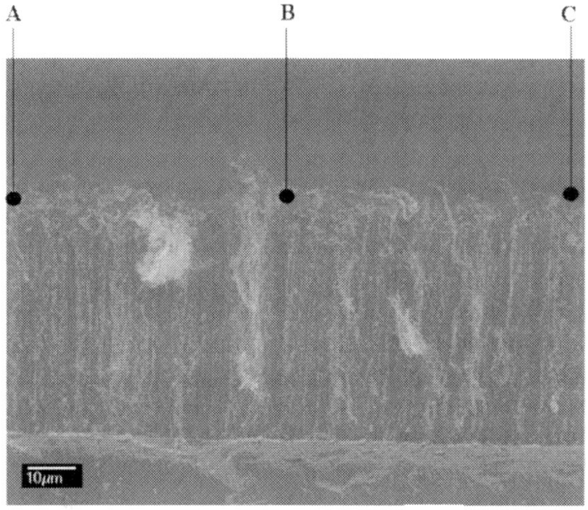


**d) Position: middle (bottom)**

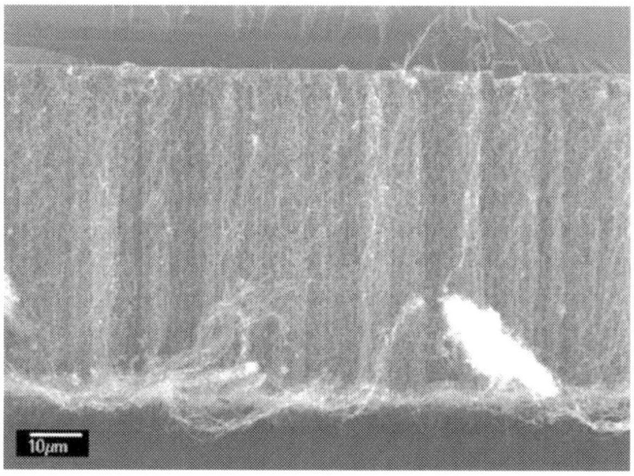


**e) Position: rear (top)**

**f) Position: rear (bottom)**

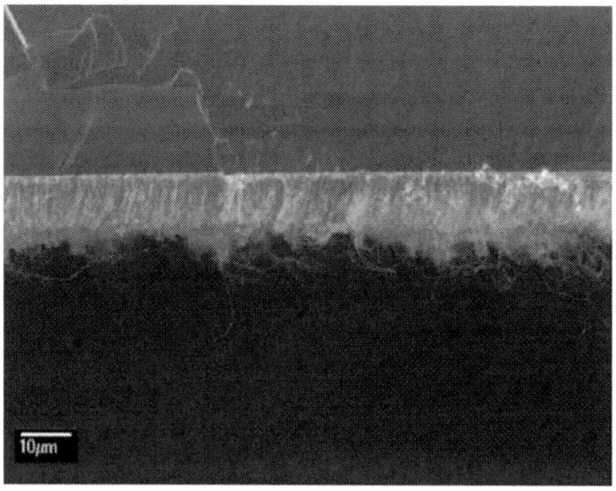


Figure 5. SEM images of aligned CNT layer with the substrate placed at various positions of the quartz boat: (a)-(b) front; (c)-(d) middle; (e)-(f) rear of the quartz boat. (Volume of water injected: 0.3 mL, catalyst: 100 mg ferrocene, growth temperature: 750 °C, growth time: 20 min)

*Effect of water vapor*

It is well known that water vapor can significantly influence the growth of aligned CNTs on substrates[13]. Figures 6(a)-(l) show CNT layers on the top and bottom surfaces of the substrate when different volumes of water were injected into the quartz tube. Figures 6(a) and (b) show that relatively thin CNT layers (top: 1 μm, bottom: 1 μm) were synthesized when no water was injected into the quartz tube. Much longer and denser aligned CNTs were obtained with the injection of water. The relationship between the thickness of CNTs film and the increasing volume of water is shown in Figure 7. Figures 6(c)-(l) are the corresponding photomicrographs.

**a) Volume of water: 0.1 ml (top)**

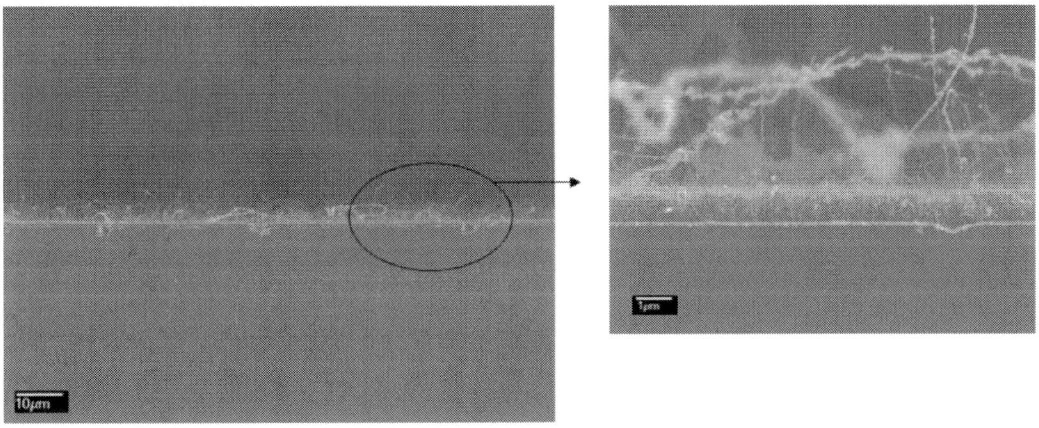


**b) Volume of water: 0.1 ml (bottom)**

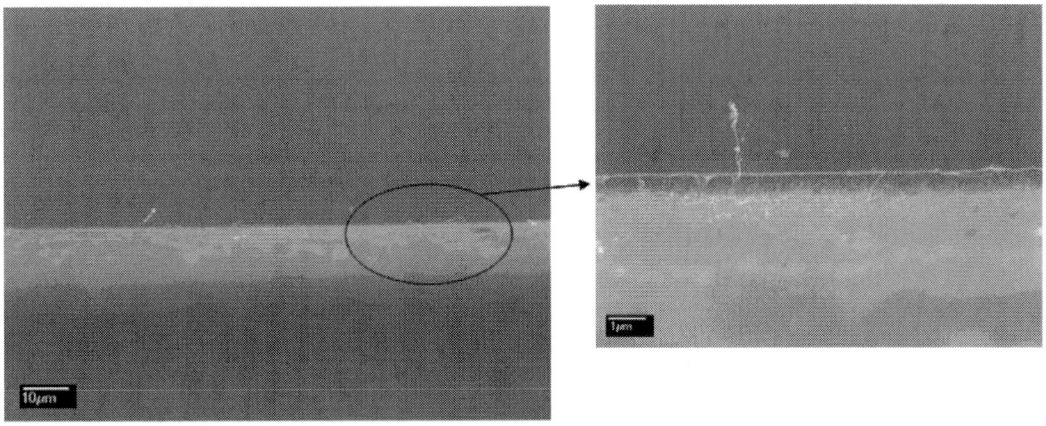

**c) Volume of water: 0.2 ml (top)**

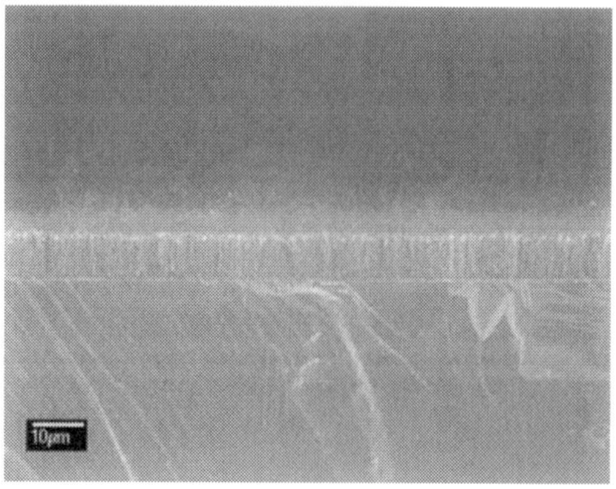


**d) Volume of water: 0.2 ml (bottom)**

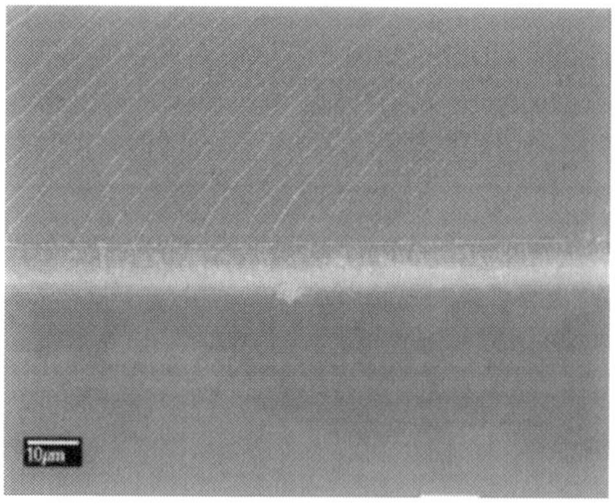


**e) Volume of water: 0.3 ml (top)**

**f) Volume of water: 0.3 ml (bottom)**

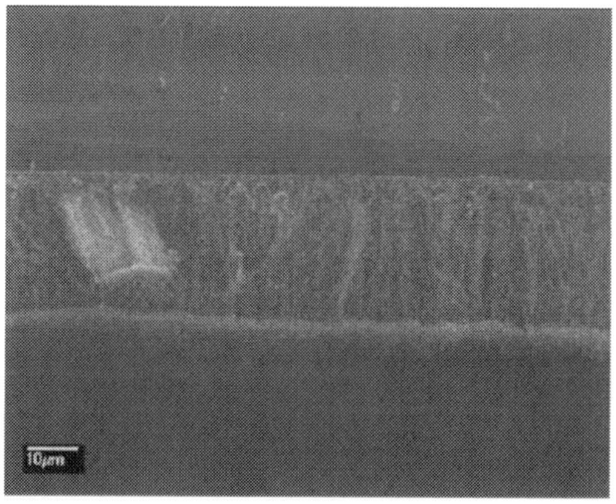


**g) Volume of water: 0.4 ml (top)**

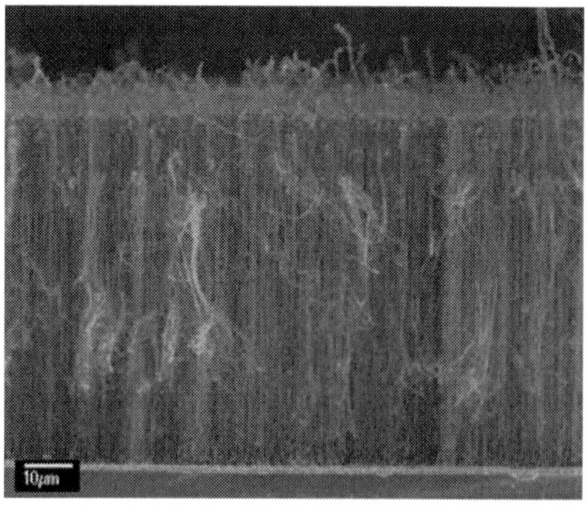


**h) Volume of water: 0.4 ml (bottom)**

**i) Volume of water: 0.6 ml (top)**

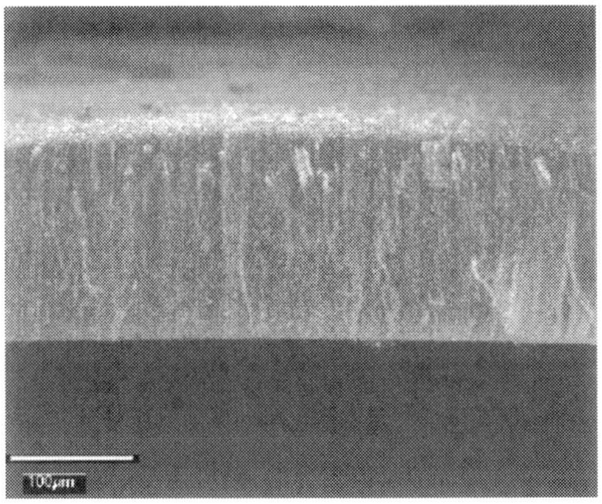

**j) Volume of water: 0.6 ml (bottom)**

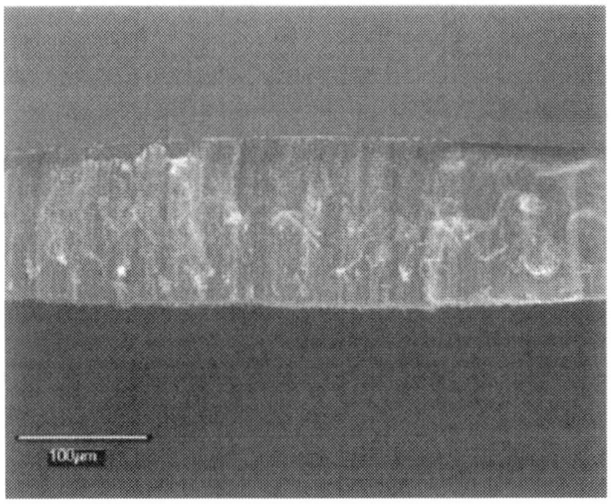

**k) Volume of water: 0.8 ml (top)**

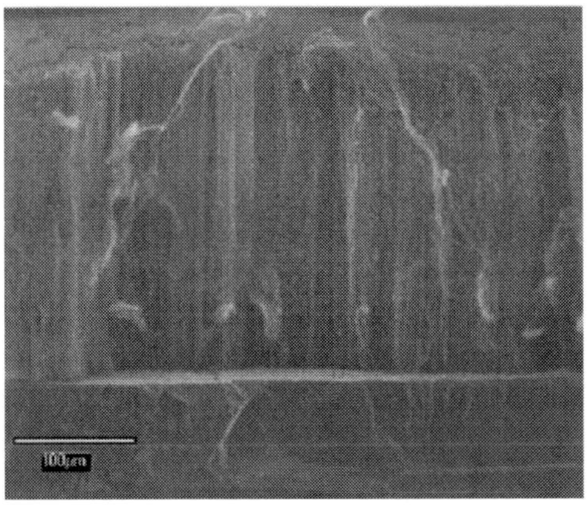

**l) Volume of water: 0.8 ml (bottom)**

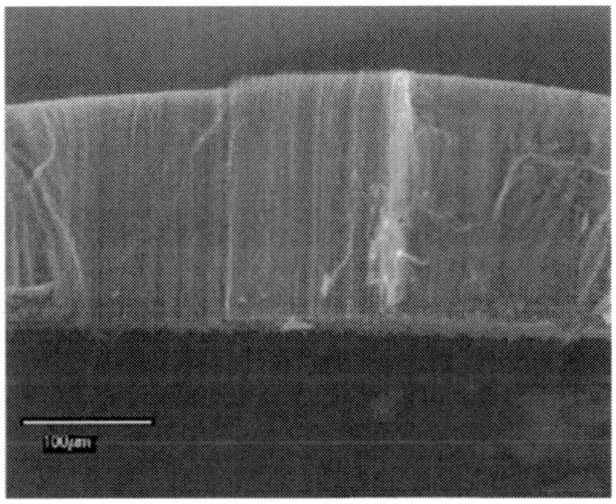

Figure 6. SEM images of aligned CNT layer with various amounts of water injected: (a)-(b) nil; (c)-(d) 0.1 mL; (e)-(f) 0.2 mL; (g)-(h) 0.3 mL; (i)-(j) 0.4 mL; (k)-(l) 0.6 mL. (Catalyst: 100 mg ferrocene, growth temperature: 750 °C, growth time: 20 min)

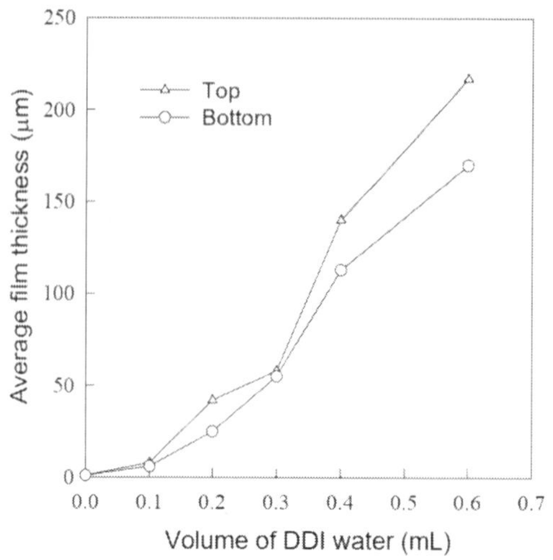


Figure 7. Effect of volume of water injected on the CNT film thickness.

*Effect of the weight of ferrocene catalyst*

Apart from the position of substrate in quartz boat and the amount of water injected into the quartz tube, the amount of ferrocene catalyst can also render a significant impact on the morphology and thickness of the CNT layer. Figures 8(a)-(d) depict SEM images of CNTs on the top face of substrate produced with different amounts of catalyst ranging from 30 to 200 mg. Figure 8(a) shows that few aligned CNTs could be found on the substrate when only 30 mg of ferrocene was used. Figure 9 shows the dependence of CNT layer thickness on the weight of ferrocene. Thicker CNT films could be synthesized by using a larger amount of ferrocene.

**a) m= 30 mg**

**b) m= 100 mg**

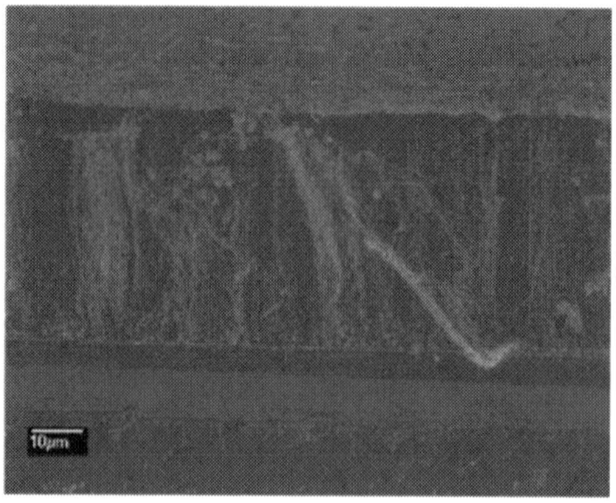

**c) m= 150 mg** **d) m= 200mg**

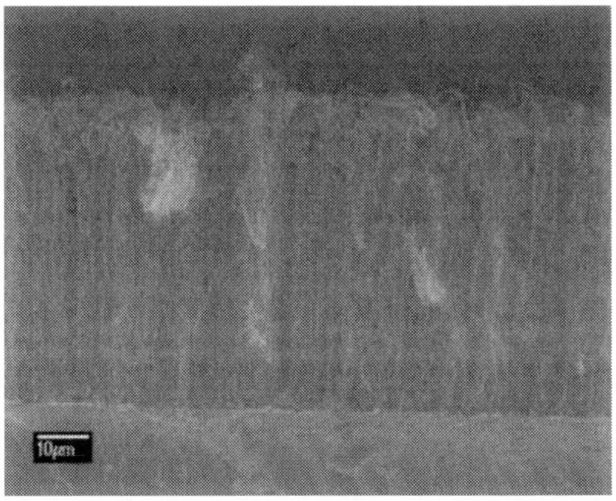


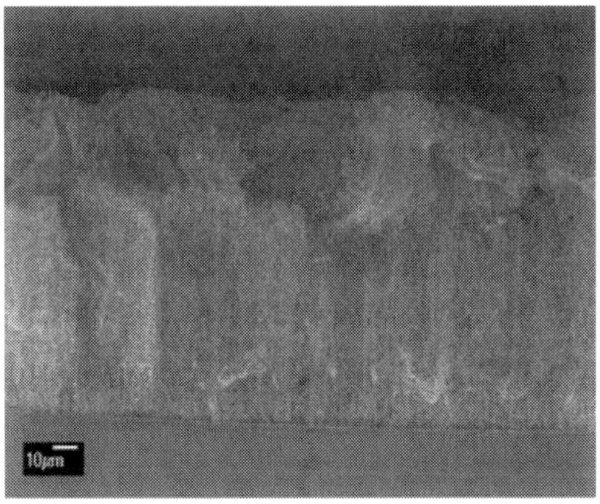


Figure 8. SEM images of aligned CNT layer on top of the substrate for various amounts of ferrocene: (a) 30 mg; (b) 100 mg; (c) 150 mg; (d) 200 mg. (Volume of water injected: 0.2 mL, growth temperature: 750 °C, growth time: 20 min)

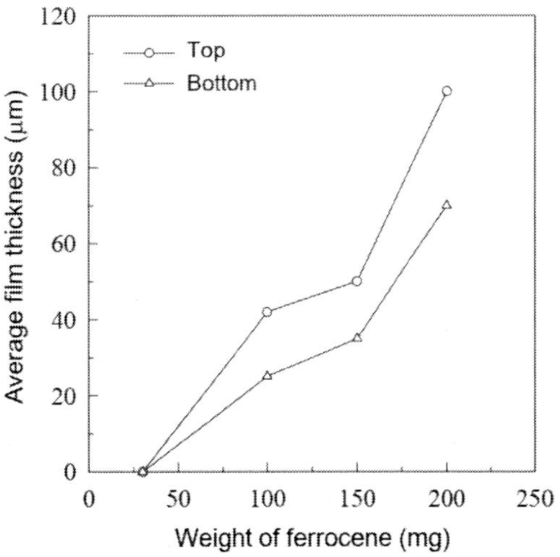


Figure 9. Effect of ferrocene amount on the CNT film thickness.

*Effect of growth temperature*

The SEM images of CNTs grown on the top face of the substrate at increasing growth temperatures are shown in Figure 10. The relationship between the thickness of CNT layer and these growth temperatures is shown in Figure 11. At 725 °C, the temperature was not sufficiently high for having catalyst deposited on the substrate and no aligned CNTs were found (Figure 10(a)). The decomposition of ferrocene to Fe catalyst particles was possible at a higher temperature and an aligned CNT layer could be synthesized at 750 °C. This experimental result is in agreement with those of Hata et al.[11] who reported water-assisted synthesis of single-walled carbon nanotubes on silicon wafer at 750 °C by CVD. At a temperature of 800 °C, the aligned CNT film was 220 μm and 125 μm on the top and bottom surfaces of the substrate, respectively. However, the CNT film was actually made up of two layers with one on top of the other at the higher growth temperatures (Figures 10(c), (d)). Zhang et al.[14] and Zhu et al.[15] reported similar observations. The mechanism for the formation of these multiple layer stacks was discussed by Zhu et al.[15] They suggested that the interfaces between the aligned CNT and catalyst particles on the substrate were etched away by water vapor. In the presence of ethylene, new CNTs grew from the original catalyst particles to form a new layer of CNTs. It is unclear whether the boundary between two layers might present additional thermal resistance because one layer could be readily peeled away from the other.[15]

**a) T= 725 °C**

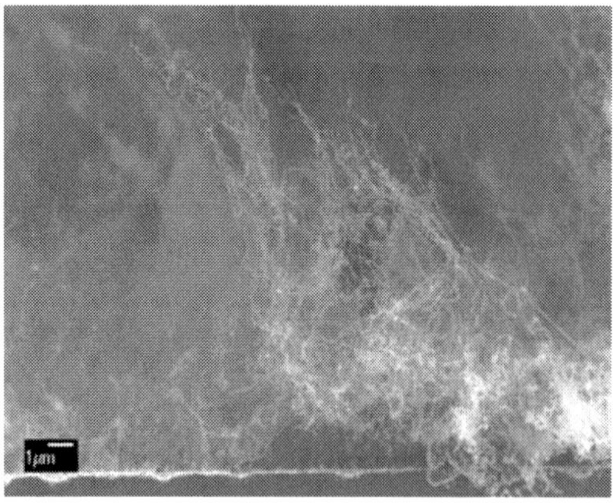


**b) T= 750 °C**

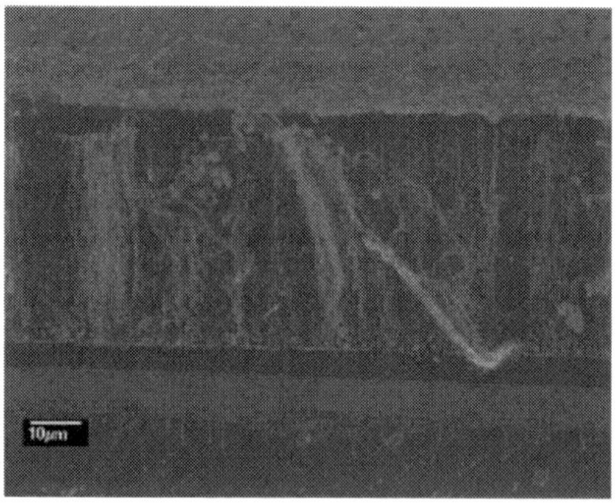

c) T= 775 °C                                d) T= 800 °C

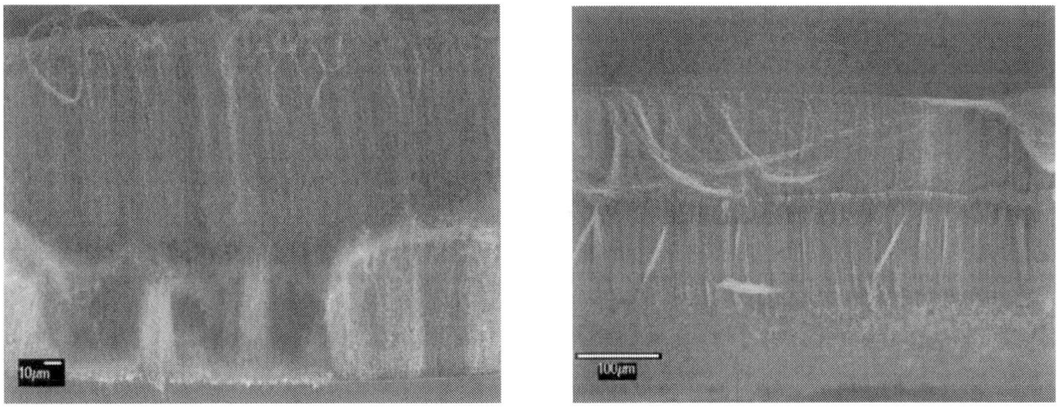


Figure 10. SEM images of aligned CNT layer on top of the substrate grown at different temperatures: (a) 725 °C; (b) 750 °C; (c) 775 °C; (d) 800 °C. (Volume of water injected: 0.2 ml, catalyst: 100 mg, growth time: 20 min)

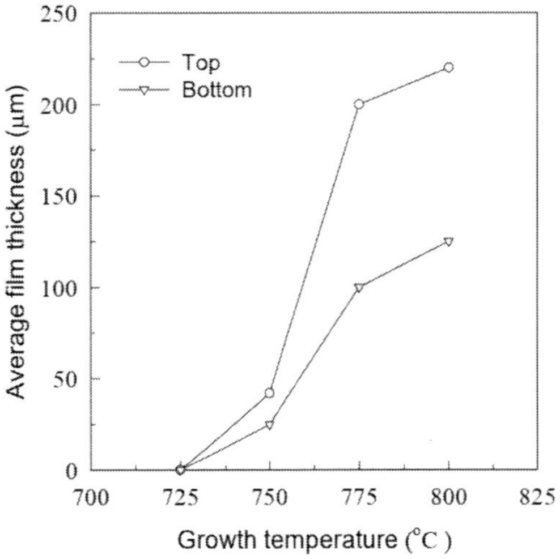


Figure 11. Effect of growth temperature on the thickness of aligned CNT film.

***Effect of growth time***

Figure 12 shows the SEM images of CNTs grown on the top face of substrate at different growth times. The relationship between the thickness of aligned CNT film and growth time at 750 °C is shown in Figure 13. The thickness of the CNT films on both sides of the substrate stayed relatively constant from 30 min to 60 min. This decreased growth rate could be attributed to the fixed amount of ferrocene and water in the reaction system.

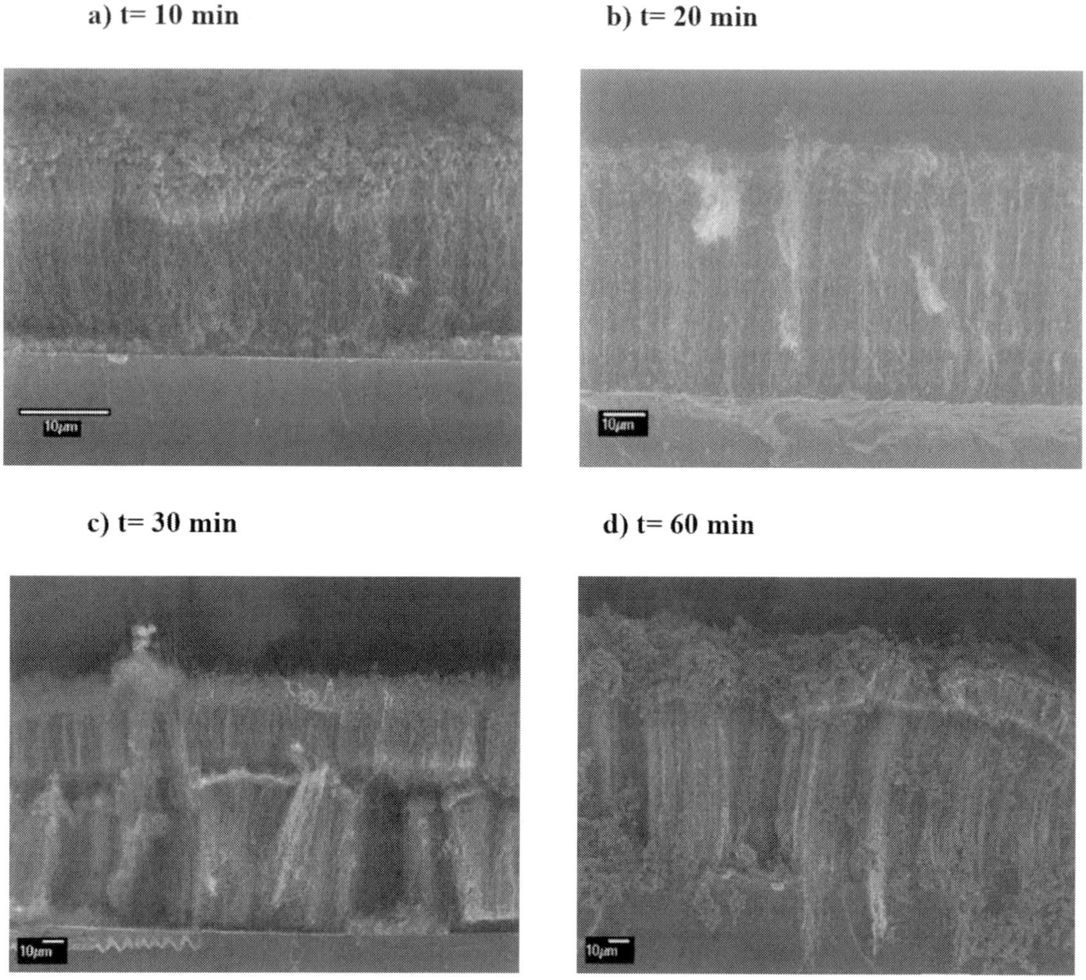


Figure 12. SEM images of aligned CNT layer on top of the substrate for various growth time: (a) 10 min; (b) 20 min; (c) 30 min; (d) 60 min. (Volume of water injected: 0.2 mL, catalyst: 100 mg ferrocene, growth temperature: 750 °C)

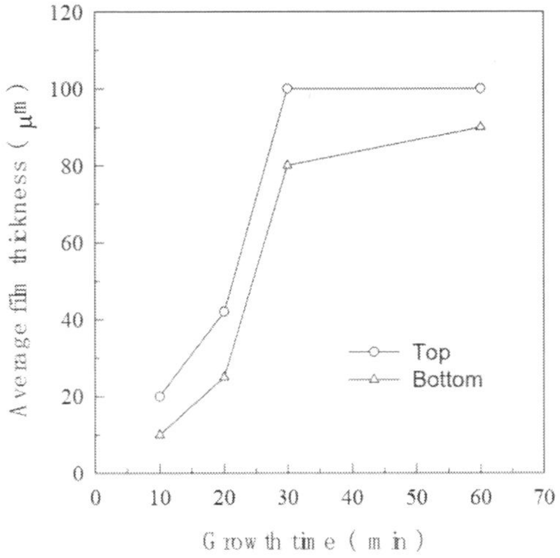


Figure 13. Effect of growth time on thickness of aligned CNTs film.

***Detailed structure of aligned CNTs by TEM***

The detailed structure of the CNTs was studied by HRTEM (Figure 14). We found that the CNTs were of different sizes but had a similar structure. Figure 14(a) shows a bundle of synthesized aligned CNTs, comprising mainly small carbon nanotubes. Figure 14(b) shows a CNT with trapped Fe catalyst particles. Most of the CNTs had an end closed with a metal particle at the tip (Figures 14(c)). Similar results were reported by Zhang et al.[16] Figure 14(d) shows a particular nanotube with an inner and outer diameter of approximately 8 and 20 nm, respectively, and contained about 25 graphitic layers in each sidewall.

**a)**

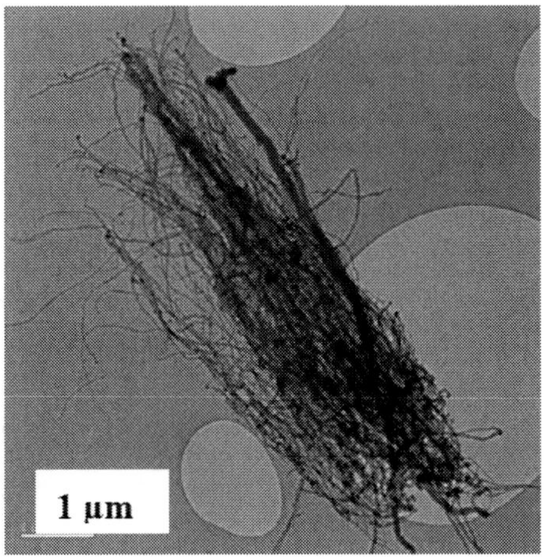


**b)**

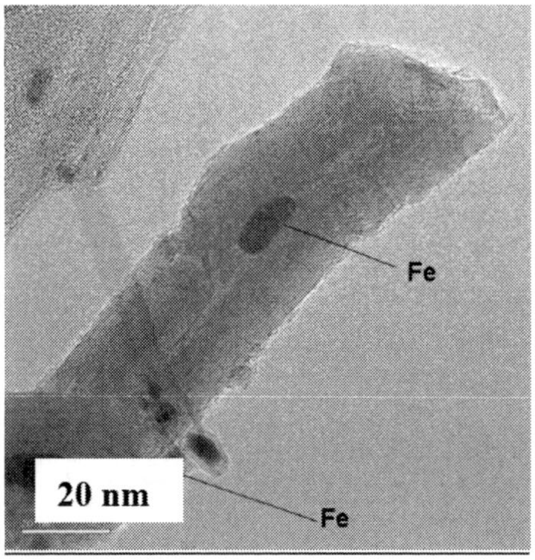

c) d)

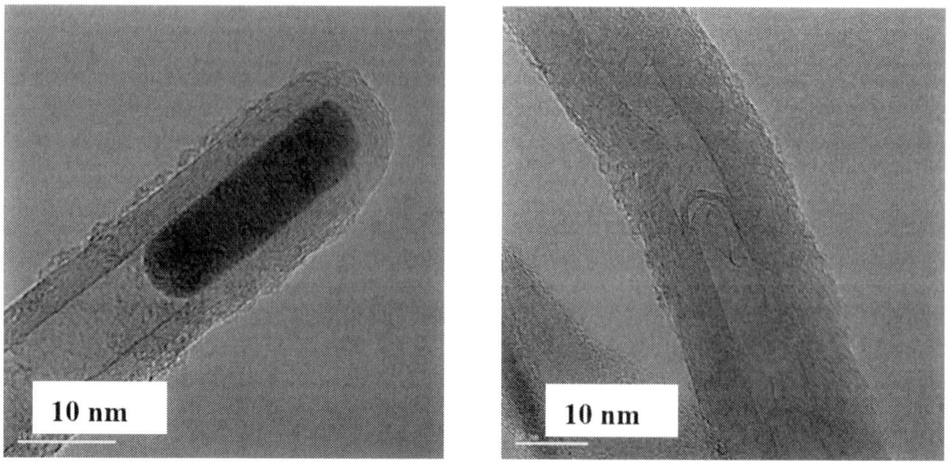


Figure 14. Detailed structure of aligned CNTs by TEM. (a) a bundle of CNTs; (b) a CNT with trapped Fe catalyst particles (indicated by arrows) (c) a CNT with a metal particle at its tip and (d) a typical CNT with bamboo structure. (Volume of water injected: 0.3 mL. catalyst: 100 mg ferrocene, growth temperature: 750 °C, growth time: 20 min)

***Aligned CNTs grown on copper plate and copper foil as TIM***

Based on experiments on silicon substrate, the following conditions were used to synthesize 50 μm thick CNT films on both sides of a copper substrate sputtered with Cr and Au: volume of water injected: 0.3 mL, catalyst: 100 mg ferrocene, growth temperature: 750 °C, growth time: 20 min. Figure 15 shows the photographs of copper plate (12.7 mm diameter; 0.3 mm thickness) and copper foil (12.7 mm diameter, 0.007 mm thick; 25.4 mm by 25.4 mm, 0.0125 mm thick) with aligned CNTs grown on both sides. Recall that the average film thickness was about 50 μm under these conditions (Figure 7).

a)

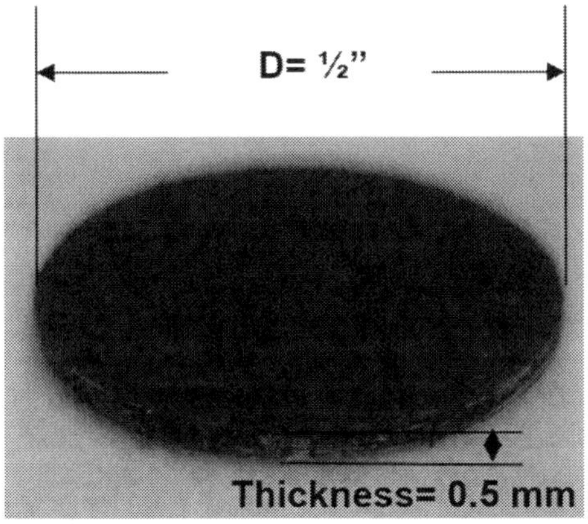


b)

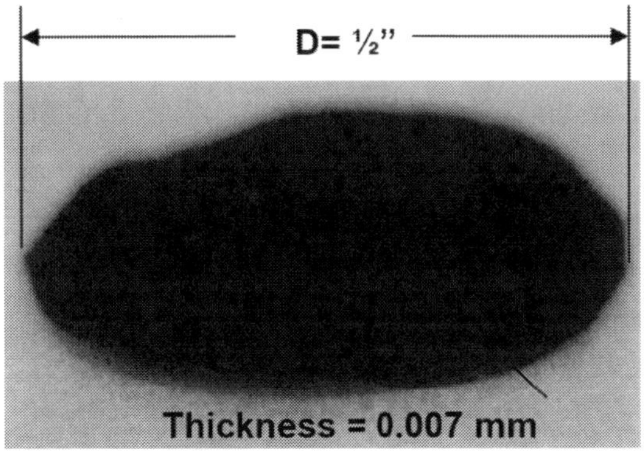


c)

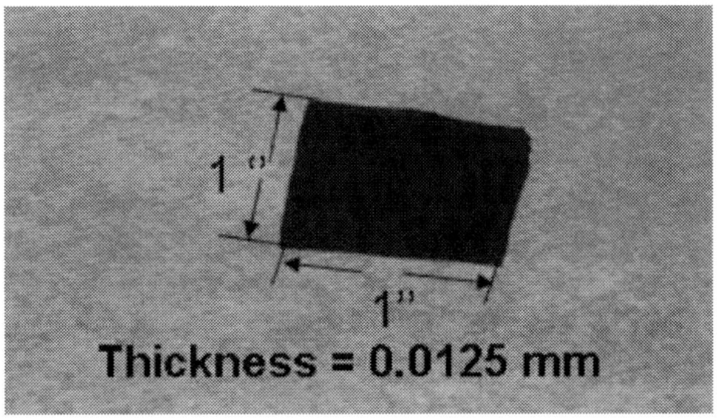


Figure 15. Photos of the aligned CNT TIM: (a) 0.5 mm thick copper plate as substrate, (b) 0.007 mm thick copper foil as substrate, (c) 0.0125 mm thick copper foil as substrate. (Volume of water injected: 0.3 mL, catalyst: 100 mg, growth temperature: 750 °C, growth time: 20 min)

***Measurement of thermal resistance***

Using the system and calculation procedure developed by Zhang et al.,[12] the thermal resistance of the 12.5 μm copper foil (or a total TIM sample thickness of 112.5 μm) was determined to be 12 $mm^2$ K/W under an applied contact pressure of 0.3 MPa. This is lower than those reported in the literature[10, 12, 17, 18] (Table 1). For example, the thermal resistance of the aligned CNT composite film synthesized by Huang et al.[10] ranged from 100 to 450 $mm^2$ K/W as the sample thickness increased from 100 to 500 μm. The thermal resistance of the CNT film synthesized by Xu and Fisher[17] using plasma enhanced CVD was 100 $mm^2$ K/W under a contact pressure of 0.16 MPa and

23 mm$^2$ K/W under 0.445 MPa. Our results are comparable to those of Zhang et al.,[12] who reported a thermal resistance of 15 mm$^2$ K/W with a film thickness Ranged from 30 to 70 μm.

**TABLE 1: Thermal performance of different TIMs in the literature**

| TIM device | Authors | Pressure (MPa) | Thermal resistance (mm$^2$ K/W) |
|---|---|---|---|
| Aligned carbon nanotube composite film | Huang et al. | 3 | 100~450 |
| CNT by PECVD | Xu et al. | 0.16 | 100 |
| | | 0.445 | 23 |
| Aligned CNTs by CVD | Zhang et al. | 0.1 | 15 |
| Aligned CNTs on double-sided copper foil | This study | 0.3 | 12 |

## 4. Conclusions

A TIM consisting of aligned CNTs on double-sided metal substrate was designed and synthesized by water vapor assisted CVD. The morphology and thickness of CNT films were strongly influenced by the position of substrate in the quartz boat, amount of water injected, amount of catalyst, growth temperature, and growth time. Thicker CNT films could be synthesized by higher growth temperature, a larger amount of ferrocene, or longer growth time.

This study can be extended in different ways. The thermal conductivity of the aligned CNT TIM sample can be determined by varying the film thickness. At present, the film thickness could not be directly measured but we can get around this problem by using the injection molding method of Huang et al.[10] Instead of fixing the amount of water, the humidity level in the reactor can be controlled. Clearly, thermal resistance smaller than 12 mm$^2$ K/W can be realized by using thinner copper foils and shorter aligned CNTs. The performance of this double-sided TIM device for minimizing the well-known air gap thermal resistance should be directly investigated. Efforts in these directions are now underway.

**Acknowledgments.**

We thank K. Zhang and M. Yuen for their assistance in measuring the thermal resistance.

**References**


(1) Berber, S.; Kwon, Y. K.; Tomanek, D. *Phys. Rev. Lett.* **2000**, *84*, 4613.

(2) Gwinn, J. P.; Webb, R. L. *Microelectronics J.* **2003,** *34*, 215.

(3) Mahajan, R.; Chiu, C. P.; Prasher, R. *Electronics Cooling* **2004**, 10.

(4) Treacy, M. M.; Ebbesen, T. W.; Gibson, J. M. *Nature* **1996**, *381*, 678.

(5) Dai, H. J.; Wong, E. W.; Lieber, C. M. *Science* **1996**, *272*, 523.

(6) Hone, J.; Whitney, M.; Piskoti, C.; Zettl, A. *Phys. Rev. B* **1999**, *59*, R2514.

(7) Lijima, S. *Nature* **1991**, *354*, 56.

(8) Kim, P.; Shi, L.; Majumdar, A.; McEuen, P. L. *Phys. Rev. Lett.* **2001**, *87*, 215502-1.

(9) Liu, C. H.; Huang, H.; Wu. Y.; Fan, S. S. *Appl. Phys. Lett.* **2004**, *84*, 4248.

(10) Huang, H.; Liu, C. H.; Wu, Y.; Fan, S. S. *Adv. Mater.* **2005**, *17*, 1652.

(11) Hata, K.; Futaba, D. N.; Mizuno, K.; Namai, T.; Yumura, M.; Lijima, S. *Science* **2004**, *306*, 1362.

(12) Zhang, K.; Yuan, M. F.; Wang, N.; Miao, J. Y.; Xiao, G. W.; Fan, H. B. *2006 Electronic components and Technology Conference*, 177.

(13) Zhao, Z. B.; Qu, J. Y.; Qiu. J. S.; Wang, X. Z.; Wang, Z. Y. *Chem. Commun.* **2006**, 594.

(14) Zhang, H. R.; Liang, E. J.; Ding, P.; Chao, M. J. *Phys. B* **2003**, *337*, 10.

(15) Zhu, L. B.; Xiu, Y. H.; Hess, D. W.; Wong, C. P. *Nano. Lett.* **2005**, *5*, 2641.

(16) Zhang, X. F.; Cao, A. Y.; Wei, B. Q.; Li, Y. H.; Wei, J. Q.; Xu, C. L.; We, D. H. *Chem. Phys. Letts.* **2002**, *362*, 285.

(17) Xu, J.; Fisher, T. S. *Proc 9$^{th}$ intersociety conference on thermal and thermomechanical phenomena in electronic systems* **2004**, *34*, 549.

(18) Liu, C. H.; Huang, H.; Wu, Y.; Fan, S. S. *Appl. Phys. Lett.* **2004**, *84*, 4248.

## 20b Vapor Grown Carbon Fiber - Microfibrous Matrix Composites

**Amogh N. Karwa**[1], Bruce Tatarchuk[2], and Virginia A. Davis[1]. (1) Department of Chemical Engineering, Auburn University, 213 Ross Hall, Auburn, AL 36849-5127, (2) Chemical Engineering, Auburn University, 230 Ross Hall, Auburn, AL 36849

The Center for Microfibrous Materials Manufacturing at Auburn University uses high speed, wet-lay paper-making processes to manufacture low cost, high performance three-dimensional microfibrous materials for filtration and catalysis applications. The incorporation of Vapor Grown Carbon Fibers (VGCFs) into these advanced materials creates new opportunities for high performance applications. VGCFs are being added directly to the sheet forming process with different microfibrous materials. The physical, pressure drop, thermal stability and electrical characteristics of the resulting sheets will be presented.

## 20c Catalyzed Growth of Nascent Caps of Single-Wall Carbon Nanotubes of Various Chiralities

**Diego Armando Gomez-Gualdron** and Perla B Balbuena. Chemical Engineering, Texas A&M University, 2701 Longmire Dr. Apt107, College Station, TX 77845

Ab-initio calculations are performed on single-wall carbon nanotube (SWCNT) caps of different chiralities. The nascent caps are the minimum structures for which the chirality can be determined; we hypothesize that their early formation is related to the abundance of certain types of chiralities during the catalytic of growth of nanotubes, in particular on cobalt nanoparticles. Small differences are found in the stability of certain nascent caps, the most interesting difference is between same diameter (9,1) and (6,5) nanotubes, whose chiralities are uniquely defined, with the (6,5) being more abundant than the (9,1) in some synthesis processes. The interaction energies of the most stable caps for the chiralities (6,5), (9,1), (7,5), (6,6) and (10,2) with an optimized 9-atom cobalt cluster are calculated as well. This interaction energy is expected to be favorable based on the analysis of the cluster-cap electrostatic potential map found in our calculations. The most favorable interaction is found for the (6,5) cap, which has an adsorption energy 2eV more stable than that of the (9,1) cap. The way the cluster interacts with the cap depends on the chirality, the polarization of the composite systems is found to correlate with the SWCNT chiral angle. The reorganization of the geometry of the cluster also changes for different caps, revealing some interesting features of the active sites of the caps and about the variation of the strength of the carbon-cobalt interaction as the chiral angle changes. Further, the growth of the nanotubes out of the nascent cap is studied by progressively adding $C_2$, this reaction mechanism is chosen based upon the compatibility of the two unpaired electrons of the cap with the two free electrons offered by the radical. It is found that the energetic of the successive addition favors the (6,5) chirality. In addition, it is observed that the closer the cap is to the armchair chirality, the higher the number of favorable sites for a successful $C_2$ addition leading to the completion of one hexagon. This fact would explain why the most abundant chiralities in the production of carbon nanotubes for most CVD processes are those lying to close to armchair chiralities. All the aforementioned factors show the complexity of the process leading to the selection of a specific chirality. It seems that complex and synergistic interactions of kinetic and thermodynamic factors determine the selectivity of the process of nanotube growth.

## 20d Single-Walled Carbon Nanotube (SWNT) Diameter Tuning Using C10 Co-Mcm-41

**Nan Li**, Xiaoming Wang, Gary L. Haller, and Lisa D. Pfefferle. Chemical Engineering, Yale University, 9 Hillhouse Ave, New Haven, CT 06520

Selectivity of SWNT diameter has become crucial to its applications. Small (<0.7 nm) and uniform diameter SWNT of diameter exhibits very special electronic properties because of its high radius of curvature; SWNT with a large (>1.6nm) diameter tends to be very de-bundled and even exists individually which makes its application much easier without any further functionalization. Cobalt incorporated MCM-41 by carbon monoxide (CO) disproportionation has been proven to be an efficient catalyst for the synthesis of SWNT with narrow diameter distribution at a wide range of mean diameters (0.6-1.7nm). We have shown using in-situ X-ray Absorption Spectroscopy (XAS), CO significantly enhances the mobility of the cobalt particles on the MCM-41 surface. In this study, we have found that by operating at low/high reaction temperature compared with those normally used in CO disproportionation (750°C), the mean diameter of SWNT varies greatly while all maintaining a narrow distribution. Extended X-Ray Absorption Fine Structure (EXAFS) studies together with Transmission Electron Microscopy (TEM) images indicate that cobalt particles which act as catalyst are growing larger along with the increase of reaction temperature from 450°C to 950°C, resulting in the systematic shift of mean SWNT diameter from 0.6nm to 1.7nm confirmed by Raman spectroscopy, TEM and fluorescence spectroscopy. As we reported earlier, metal content of produced SWNT after purification is 0.7 wt% and the surface area of the purified SWNT without any further functionalization is as high as 1200 $m^2/g$. The total carbon yield in this study varies from 10wt% to 50wt% (weight per catalyst weight) depending on the SWNT diameter, and the SWNT selectivity is all above 95%. All these features and the choice of SWNT diameter suggest many promising applications of SWNT in electronics, bio-medical engineering, environmental engineering, new catalytic support, etc.

## 20e Ostwald Ripening-Induced Termination of Carbon Nanotube Growth

**Placidus B. Amama**[1], Cary L. Pint[2], Laura McJilton[2], Terry P. Murray[3], Robert H. Hauge[4], and Benji Maruyama[5]. (1) Materials & Manufacturing Directorate, UTC/ Wright Patterson Air Force Research Laboratory, 2941 Hobson Way, Dayton, OH 45433, (2) Carbon Nanotechnology Laboratory, Rice University, 6100 Main St, MS 100, Houston, TX 77005, (3) Materials Engineering, University of Dayton, Kettering Lab - Room 165, Dayton, OH 45469, (4) Smalley Institute, Rice University, 6100 Main Street MS 362, Houston, TX 77005, (5) Materials & Manufacturing Directorate, Wright Patterson Air Force Research Laboratory, 2941 Hobson Way, Dayton, OH 45433

One of the most challenging problems in carbon nanotube synthesis today is understanding the termination of growth. Many mechanisms have been proposed for the termination of carpet growth. In particular, the impact of growth environment and catalyst/support interaction are known to be critical to growth, but have not been investigated explicitly in the context of Ostwald ripening. Here, we explore the role Ostwald ripening plays in inducing the termination of carpet SWNT growth by studying the ripening rate of the catalyst as a function of the annealing ambient and the catalyst support interaction. Coarsening rates of iron catalysts on alumina supports deposited by three different methods are investigated: atomic layer deposition, magnetron sputtering and e-beam evaporation. The Ostwald ripening rates of each alumina/Fe combination also is studied as a function of thermal treatment in H2, atomic H, H2 and H2O, and inert ambients.

## 20f A Mechanochemical Model of Growth Termination In Vertical Carbon Nanotube Forests and Experimental Approaches toward Infinite Film Synthesis

**Jae-Hee Han**[1], Charles Robert Welch[2], Charles P. Marsh[3], Thomas A. Carlson[3], and Michael S. Strano[4]. (1) Chemical Engineering, Massachusetts Institute of Technology, 77 Massachusetts Avenue, NE47-577, Cambridge, MA 02139, (2) Information Technology Laboratory, U.S. Army Engineer Research and Development Center, 3909 Halls Ferry Road, Vicksburg, MS 39180, (3) Construction Engineering Research Laboratory (CERL), U.S. Army Engineer Research and Development Center, 2902 Newmark Drive, Champaign, IL 61822, (4) 66-566 Department of Chemical Engineering, Massachusetts Institute of Technology, 77 Massachusetts Avenue, Cambridge, MA 02139

Understanding the mechanisms by which vertical carbon nanotubes (CNTs) array terminate their growth may lead to the production of aligned materials of infinite length. Both single- and doubled-walled carbon nanotubes (SWNTs and DWNTs) films demonstrate characteristic yet unexplained deflections of the top surface near the edges and corners of the film. We show that this upturn in the surface can be explained by assuming a mechanical coupling between neighboring nanotubes.[1] A Monte Carlo simulation of film growth is able to qualitatively reproduce the shape by assuming that the coupling is limited by the enthalpy of the carbon forming reaction. The shape of the surface is approximately conic with hyperbolic cross sections that allow for the calculation of a threshold force ($F_{max}$ = 34 to 51 nN for SWNTs, 25 to 27 nN for DWNTs) and elastic constant ($k$ = 384 to 547 N/m for SWNTs and 157 to 167 N/m for DWNTs) from the images of experimentally synthesized films. Despite differences in nanotube type and precursor chemistry, the values appear consistent. The origin of the mechanical coupling will be discussed. With this theoretical hypothesis, up to now, we have experimentally synthesized the millimeter-thick vertical CNTs array via water injection route during synthesis to circumvent the growth termination of the film in chemical vapor deposition system. Notably we observed such characteristic deflections of the top surface of the films grown and were able to control the degree of both deflections and straightness, possibly originated from the mechanical coupling between arrays of CNTs in the forests. For the further detailed quantitative analysis, small angle X-ray scattering (SAXS) studies on different forests grown under various water injection modes will be presented.

Reference

[1] Han, J. H.; Graff, R. A.; Welch, B.; Marsh, C. P.; Franks, R.; Strano, M. S. *ACS Nano* **2008**, *2*, 53-60.

## 71a Targeting Single-Walled Carbon Nanotubes for the Treatment of Breast Cancer

**Roger G. Harrison**[1], Luis F. F. Neves[1], Yahya A. Lazrak[1], David E. Martyn[1], Peter S. McFetridge[1], Daniel E. Resasco[2], and Kenneth E. Bartels[3]. (1) Chemical, Biological and Materials Engineering, University of Oklahoma, 100 E. Boyd St., T-335, Norman, OK 73019, (2) Cbme, University of Oklahoma, 100 East Boyd, Room T-335, Norman, OK 73019, (3) Veterinary Clinical Sciences, Oklahoma State University, Veterinary Medical Teaching Hospital, Stillwater, OK 74078-2041

Single-walled carbon nanotubes (SWNTs) are unique in that they strongly absorb near-infrared (NIR) light, while biological systems have very low levels of absorption of NIR light. This project focuses on developing a means for targeting the SWNTs to the tumor vasculature. It is proposed to target the SWNTs by conjugation with the human annexin V protein. Human annexin V (molecular weight of 36,000) is a monomeric protein that binds with high affinity to phophatidylserine (PS) in phospholipid bilayers. PS is the most abundant anionic phospholipid of the plasma membrane and is tightly segregated to the internal side of the plasma membrane in most cell types. Anionic phospholipids are largely absent from the external surfaces of resting mammalian cells under normal conditions. Recently, it has been found that PS is expressed on the external surface of endothelial cells that line the blood vessels in tumors but is not expressed on the outside surface of the vascular endothelium in normal organs. Thus, annexin V can be used to specifically target the endothelial cells of the tumor vasculature. The underlying hypothesis is that a SWNT-annexin V conjugate injected into the bloodstream will localize in the tumor vasculature and that exposure of the tumor to NIR radiation will cause the tumor to be selectively killed by a combination of heating and cutting off its blood supply.

In this study, we have covalently coupled the human protein annexin V to carboxymethylcellulose adsorbed on single-walled carbon nanotubes (SWNTs) by using a zero-length crosslinking agent that couples carboxyl groups to primary amines (1-ethyl-3-[3-dimethylaminopropyl]carbodiimide hydrochloride, or EDC). The binding of annexin V (biotinylated) to phosphatidylserine allowed for the determination of the dissociation constant (Kd). The dissociation constant indicates a relatively high affinity of the receptor (phosphatidylserine) for the ligand (annexin V). The UV-vis-NIR absorption spectra of the SWNT-CMC-biotinylated annexin V complex was determined.

The SWNT-CMC-annexin V complex bound to human endothelial cells in vitro was subjected to laser irradiation at 980 nm on a 24-well plate in order to determine the effect of the energy adsorption by the SWNTs on cell viability. The cell viability was evaluated by performing the Alamar Blue assay, cell counting and cell observation with a microscope.

## 71b Polymersomes as Carriers for Genetic and Protein Therapeutics

**David A. Christian**[1], Shenshen Cai[1], Diana M. Bowen[1], Younghoon Kim[1], J. David Pajerowski[2], and Dennis E. Discher[1]. (1) Chemical and Biomolecular Engineering, University of Pennsylvania, Room 129 Towne Building, 220 South 33rd Street, Philadelphia, PA 19104-6393, (2) Bioengineering, University of Pennsylvania, Room 129 Towne Building, 220 South 33rd Street, Philadelphia, PA 19104-6393

Polymersomes are polymer-based vesicular shells that form upon hydration of PEG-based amphiphilic block copolymers. These high molecular weight amphiphiles impart physicochemical properties that allow polymersomes to stably encapsulate or integrate a broad range of active molecules as well as to circulate for several hours in vivo (t1/2 ~ 20 hours). This robustness together with recently described mechanisms for controlled breakdown of degradable polymersomes as well as escape from endolysosomes suggests that polymersomes might be usefully viewed as having structure/property/function relationships somewhere between lipid vesicles and viral capsids. Here we describe the assembly and development of polymersomes to encapsulate and deliver genetic (i.e. siRNA and AON) and protein therapeutics.

## 71c Immune-Evasion of Viruses and Nanoparticles as a Result of 'Self' Mimicry

**Richard Tsai**, University of Pennsylvania, Room 129 Towne Building, 220 South 33rd Street, Philadelphia, PA 19104-6393 and Dennis E. Discher, Chemical and Biomolecular Engineering, University of Pennsylvania, Room 129 Towne Building, 220 South 33rd Street, Philadelphia, PA 19104-6393.

Interactions of viruses and nano-carrier systems have been study extensively both in vitro and in vivo, but a major hurdle has been the inability of these foreign nanoparticles to evade recognition and rapid clearance by macrophages of the innate immune system. The longest circulating cells in the human body are red blood cells which are able to circulate without triggering such rapid clearance. Macrophages differentiate foreign vs. self through a 'marker of self' found on the surface of red cells and other cells. We show that re-engineering of CD47 to be coupled either to a synthetic surface or expressed on viral particles can inhibit recognition by macrophages. Our results indicate that CD47 significantly reduces phagocyotsis by pathways that also apply with microparticles. Viral particles expressing the 'self' protein have a lower level of transduction to macrophages leading to more specific targeting for gene delivery. Exploiting the body's own system of distinguishing foreign from 'self' paves the way for avoiding toxic limits set by clearance and delivering more drug and genetic material to intended targets.

## 71d Novel Architecture for Dendritic Cell-Targeted Vaccines

**Jordi Mata-Fink** and K. Dane Wittrup. Chemical Engineering, Massachusetts Institute of Technology, 400 Main St., E19-563, Cambridge, MA 02142

Peptide vaccines are an attractive new tool for cancer therapy. Delivery of tumor antigenic peptides to dendritic cells can stimulate potent tumor-specific T cells. Work over the last decade has shown that particulate vaccines activate more T cells *in vitro* than soluble protein, and that peptides targeted to dendritic cells are more effective immune stimulants than untargeted peptides *in vivo*. We describe a new nanostructured, modular peptide vaccine particle that provides exquisite control over composition to a degree unseen in current particles. Our architecture combines the benefits of a high antigen dose and dendritic cell targeting with the capability to easily add new functionalities. We describe the construction of the vaccine particles and, using ovalbumin as a model antigen, demonstrate their effectiveness in an *in vitro* system.

## 71e Remotely Triggered Liposome Release by near Infrared Light Absorption Via Hollow Gold Nanoshells

**Guohui Wu**, Department of Chemical Engineering, University of California, Santa Barbara, Santa Barbara, CA 93106, Alexander Mikhailovsky, Chemistry, University of California, Santa Barbara, CA 93106, Htet Khant, Chemical Engineering, University of California, Santa Barbara, CA 93106, Wah Chiu, Baylor College of Medicine, Houston, TX 77030, and Joesph A. Zasadzinski, Department of Chemical Engineering, University of California, Santa Barbara, CA 93106.

A major challenge for future drug delivery is to release encapsulated drugs with both spatial and temporal control. Liposomes show great promise as drug delivery vehicles, but it is often necessary to compromise liposome stability for rapid drug release. Targeting liposomes to specific tissues requires developing highly specific ligands with strong affinities to receptors over-expressed on diseased cells. This high binding affinity can also lead to "binding-site barriers" in which the tightly bound nanocarriers prevent drug penetration into the tissue. To address the joint challenges of controlled release and specific targeting, we present a novel photo-activated nanoshell/liposome composite by which near-complete liposome contents release can be initiated within seconds by irradiating hollow gold nanoshells (HGN) with a near infrared (NIR) pulsed laser. Targeting the drug release is limited by the dimensions of the laser beam focus; no specific ligands or antibodies are required. The HGNs were encapsulated within liposomes or tethered to the outer surface of liposomes. Electron cryo-microscopy was used to confirm the encapsulation or tethering. We found free HGNs outside liposomes can also trigger release, but with lower efficiency. Absorbing pulsed NIR light causes the HGNs to rapidly increase in temperature leading to the formation of microbubbles, whose collapse causes transient membrane rupture and content release. LC-MS and quantitative PCR have been applied to confirm that majority of drug molecules were not damaged. This has potential applications in using optical triggered nanoshells to assist macro biomolecules to enter cells for gene delivery.

# 71f Surface Modification of Hydrophobic Nanoparticles and Its Biological Application

**Rui Shen** and Hong Yang. Chemical Engineering, University of Rochester, Gavett Hall 206, Department of Chemical Engineering, University of Rochester, Rochester, NY 14627

A silane-based ligand has been used for surface modification of three different types of nanoparticles synthesized from nonhydrolytically solvents. These modified particles can be easily transferred from hydrophobic solvents to various aqueous solutions by reacting with this surfactant. After surface modification, silica can be further deposited on the surfaces of these nanoparticles. The amounts of silane ligand, ammonium hydroxide and silica precursors were adjusted to change structures of these core-shell particles. This method worked well for all three nanoparticle systems including iron oxide, silver and Pt@iron oxide, indicating the method should be generic for surfactant capped nanoparticles. Cores of some of these nanoparticles could be removed to form hollow structures. Both core-shell and hollow particles could be loaded with penicillin and tested for drug delivery. Study on inhibition zones show that hollow particles have larger loading capability than core-shell particles.

## 71g Synthesis and Evaluation of Cationic Beta-Cyclodextrin Nanogels for Drug Delivery across the Blood Brain Barrier

**Linfeng Wu**, Departments of Surgery, Pennsylvania State University, 500 University Drive, Hershey, PA 17033 and Tao L. Lowe, Departments of Surgery, Bioengineering, and Materials Science and Engineering, Pennsylvania State University, 500 University Drive, Hershey, PA 17033.

The blood brain barrier (BBB) poses a big challenge for cerebral drug delivery since it is composed of a tightly packed capillary endothelial cell monolayer, preventing the passage of many substances. Recently, we developed a biodegradable nanogel using the Michael addition reaction. Beta-cyclodextrin was integrated in the nanogels in order to increase the payload of hydrophobic drugs. Permanently charged quaternary ammonium groups were also integrated in the nanogels. We used bovine brain microvascular endothelial cells (BBMVECs) monolayer as an in vitro BBB model to evaluate the permeability of the nanogels towards the BBB. In vitro cytotoxicity was also conducted on these nanogels.

## 72a Biomolecules Versus Plain Adhesion Molecules for Larger-Scale Targeting: An Unconventional Perspective

**Maria M. Santore**, Polymer Science and Engineering, University of Massachusetts, Conte Polymer Laboratory, 120 Governor's Drive, Amherst, MA 01003

This talk aims to raise some questions concerning the use of biomolecules as means to affect specific binding of particles and cells in the "bio-nano" field, with applications in materials assembly, sensors, diagnostics, and drug delivery. DNA pairing and antibody binding are widely used and highly effective, as long as nonspecific interactions are adequately suppressed: The strategy is that molecular-level specificity translates directly to specificity at larger length scales. Nature, however, does not typically rely on a single molecular pair to mate each type of cell pairs: multiple types of adhesion molecules are involved; the individual molecules may not be entirely specific, and outcomes are both highly selective and dynamic. This raises questions concerning our strategies for selective adhesion beyond the molecular scale: (1) How can multiple surface features be engineered for synergy (2) How can the same target molecules be employed different scenarios to achieve appropriate selectivity in each? And (3) How can synthetic interfacial constructs be engineered for dynamic selectivity and adhesion? The presentation shows our naïve efforts to design interfaces with targeted interactions that do not rely on the specificity of immobilized biomolecules.

## 72b Polymersomes for Delivery, Release and Cell Tracking

**Daniel A. Hammer**, Bioengineering, University of Pennsylvania, 240 Skirkanich Hall, 210 South 33rd Street, Philadelphia, PA 19104

*Text Not Available.*

## 72c Optical Modulation of Single Walled Carbon Nanotubes for Life Science and Biomedical Applications

**Michael S. Strano**, 66-566 Department of Chemical Engineering, Massachusetts Institute of Technology, 77 Massachusetts Avenue, Cambridge, MA 02139

Nanoscale sensing elements offer promise for single molecule analyte detection in physically or

biologically constrained environments. Molecular adsorption can be amplified via modulation of sharp singularities in the electronic density of states that arise from 1D quantum confinement1. Single-walled carbon nanotubes (SWNT), as single molecule optical sensors2,3, offer unique advantages such as photostable near-infrared (n-IR) emission for prolonged detection through biological media, single-molecule sensitivity and, nearly orthogonal optical modes for signal transduction that can be used to identify distinct classes of analytes. Selective binding to the SWNT surface is difficult to engineer4. In this lecture, we will briefly review the immerging field of fluorescent diagnostics using band gap emission from SWNT. In recent work, we demonstrate that even a single pair of SWNT provides at least four optical modes that can be modulated to uniquely fingerprint chemical agents by the degree to which they alter either the emission band intensity or wavelength. We validate this identification method in vitro by demonstrating detection and identification of six genotoxic analytes, including chemotherapeutic drugs and reactive oxygen species (ROS), which are spectroscopically differentiated into four distinct classes. We also demonstrate single-molecule sensitivity in detecting hydrogen peroxide, one of the most common genotoxins and an important cellular signal. Finally, we employ our sensing and fingerprinting method of these analytes in real time within live 3T3 cells, demonstrating the first multiplexed optical detection from a nanoscale biosensor and the first label-free tool to optically discriminate between genotoxins. We will also discuss our recent efforts to fabricate biomedical sensors for real time detection of glucose and other important physiologically relevant analytes in-vivo. The response of embedded SWNT in a swellable hydrogel construct to osmotic pressure gradients will be discussed, as well as its potential as a unique transduction mechanism for a new class of implantable sensors.

1. Saito, R., Dresselhaus, G. & Dresselhaus, M. S. Physical Properties of Carbon Nanotubes (Imperial College Press, London, 1998).

2. Barone, P. W., Baik, S., Heller, D. A. & Strano, M. S. Near-Infrared Optical Sensors Based on Single-Walled Carbon Nanotubes. Nature Materials 4, 86-92 (2005).

3. Jeng, E. S., Moll, A. E., Roy, A. C., Gastala, J. B. & Strano, M. S. Detection of DNA hybridization using the near infrared band-gap fluorescence of single-walled carbon nanotubes. Nano Letters 6, 371-375 (2006).

4. Heller, D. A. et al. Optical detection of DNA conformational polymorphism on single-walled carbon nanotubes. Science 311, 508-511 (2006).

## 73a Characterization of Length Fractionated Single Walled Carbon Nanotubes

**Jeffrey A. Fagan**, Matthew L. Becker, Jeffrey R. Simpson, Jaehun Chun, Barry J. Bauer, Angela R. Hight Walker, and Erik K. Hobbie. Polymers Division, National Institute of Standards and Technology, 100 Bureau Dr., Gaithersburg, MD 20899-8542

The optical properties of length separated single walled carbon nanotubes (SWCNTs) separated by both size exclusion chromatography and centrifugation in a high density medium will be presented. SWCNTs were characterized by UV-visible-near infrared absorption and fluorescence spectroscopy, dynamic light scattering, Raman scattering, small angle neutron scattering, and atomic force microscopy. Comparisons will be made between the exhibited behavior of the same length SWCNT populations dispersed and separated via different procedures. The longest SWCNTs populations become colored without any chirality separation and display excellent optical properties. Length dependence of the absorbance, fluorescence and Raman spectra are observed in both systems; differences between the measured values for DNA and deoxycholate dispersed SWCNTs fractions are attributable to the dispersing agent.

## 73b Solution Phase Processing of Carbon Nanotubes with Optoelectronic Tweezers

**Peter Pauzauskie**[1], Arash Jamshidi[2], Joe Zaug[1], Justin Valley[2], Joe Satcher Jr.[1], and Ming Wu[2]. (1) Chemical Sciences Division, CMELS, Lawrence Livermore National Laboratory, 7000 East Ave., Livermore, CA 94551, (2) Electrical Engineering, University of California, Berkeley, Cory Hall, UC Berkeley, Berkeley, CA 94720

Carbon nanotube (CNT) processing will require a massively parallel technique to enable high-throughput patterning and separations. Optoelectronic tweezers (OET) have emerged in recent years as a powerful tool for highly parallel manipulation of micron scale particles, including solid-state nanowires [1]. In this presentation we demonstrate that aqueous dispersions of both carboxylic-functionalized and individual sodium dodecyl benzene sulfonate (SDBS) coated multiwalled CNTs are addressable with forces generated by optoelectronic tweezers (OET). Due to the large aspect ratio of carbon nanotubes, OET is capable of transporting CNTs in solutions with conductivites ranging from 10 – 100 mS/m while using 100,000x less optical power density than optical tweezers. AFM images are used to characterize nanotube diameters and in situ Raman spectroscopy is presented as a way to monitor nanotube ensembles trapped within individual OET potential wells. Digital video microscopy is used to quantify inter-tube repulsion dynamics resulting from dipole-dipole interactions. Potential applications involving nanotube separations and optically-patterned nanotube arrays will be discussed.

References:

[1] Jamshidi, A.; Pauzauskie, P. J.; Schuck, P. J.; Ohta, A. T.; Chiou, P. Y.; Chou, J.; Yang, P. D.; Wu, M. C. Nature Photonics 2008, 2, 85-89.

## 73c Improving the Effectiveness of Interfacial Trapping

**Randy K. Wang**[1], Wei-Chiang Chen[1], and Kirk J. Ziegler[2]. (1) Chemical Engineering, University of Florida, 512 Nuclear Science Building, Gainesville, FL 32611, (2) Department of Chemical Engineering, University of Florida, Room 237 CHE, P.O. Box 116005, Gainesville, FL 32611

Single-walled carbon nanotubes can be coated with surfactants or polymers to aid dispersion but bundled nanotubes remain due to large van der Waals attractions. Recently, we showed that liquid-liquid interfaces could be used to remove nanotube bundles from aqueous suspensions. A simple model has been developed to characterize the free energy changes upon movement of nanotubes from the aqueous phase to the interface. This model predicts that bundled nanotubes preferentially adsorb at the interface because of changes in free energy. Further improvements to the dispersion quality can be obtained by introducing different solvent ratios, a second extraction step, and other organic solvents. After the improvements, the interfacial trapping process has dispersion characteristics comparable to and, by some measures, better than ultracentrifugation. Interestingly, in some systems, mixing an organic solvent with the aqueous suspension results in solvatochromic shifts in the fluorescence and absorbance spectra. These shifts indicate a change in the nanotube environment which is attributed to swelling of the surfactant micelle. These shifts can be measured for a wide variety of solvents which may provide new information about the surfactant structure surrounding carbon nanotubes and their photophysical properties.

## 73d Viscoelasticity and Stability of Single Walled Carbon Nanotube – Unsaturated Polyester Resin Dispersions

**Matthew J. Kayatin** and Virginia A. Davis. Department of Chemical Engineering, Auburn University, 212 Ross Hall, Auburn, AL 36849

We have investigated the mechanical property enhancement resulting from the high shear dispersion of Single-Walled Carbon Nanotubes (SWNTs) into a viscous unsaturated polyester resin matrix. Specifically, the complex structure, dispersion state, and flow behavior of this material was characterized using bulk rheological properties. Under controlled strain oscillatory shear viscoelastic behavior was observed from the lower semi-dilute to isotropic concentrated regime spanning the percolation threshold. Network elasticity was found to scale with increasing SWNT concentration and identifiable by the development of frequency independent linear viscoelastic moduli. The concentration dependence was removed using colloidal scaling for weakly attractive particles resulting in universal loss and storage moduli for the system. Transmission electron microscopy was performed to probe the dispersion state. As predicted by the rotational Peclet number, non-Brownian behavior was observed for SWNT bundles in a viscous medium.

In the dilute regime the application of low strain rates, over long time scales, resulted the formation of SWNT aggregates identifiable by a sharp increase in apparent viscosity. The low shear aggregation was reversible, to some extent, by the application of a shear stress great enough to overcome the aggregate attractive forces. To identify the dominant phenomena controlling the aggregation process dispersions were studied with varying SWNT purity and production type (HiPCO, laser ablation). Thermogravimetric Analysis, X-Ray Photoelectron Spectroscopy, and Raman Spectroscopy were performed to characterize the various SWNTs. Additionally, Electrostatic Force Microscopy was used to determine the magnitude of the various SWNT surface charges and to correlate aggregation behavior with the dominant Van der Waals attractive forces.

## 73e Covalent Functionalization of Single Walled Carbon Nanotubes Alters Their Densities Allowing Electronic and Other Types of Separation

**Woo-Jae Kim**[1], Nitish Nair[1], Chang Young Lee[2], and Michael S. Strano[3]. (1) Department of Chemical Engineering, Massachusetts Institute of Technology, 77 Massachusetts Avenue, building 66-580, Cambridge, MA 02139, (2) Chemical Engineering, Massachusetts Institute of Technology, 77 Massachusetts Avenue, Building 66-366, Cambridge, MA 02139, (3) 66-566 Department of Chemical Engineering, Massachusetts Institute of Technology, 77 Massachusetts Avenue, Cambridge, MA 02139

We show that covalently attached functional groups can alter the densities of individual SWNT in a predictable and highly controllable manner. A volume additivity model based upon molecular group contributions is able to estimate the density difference between 4-hydroxy phenyl functionalized and non-functionalized HiPco SWNT as approximately 98.3 kg/m3, compared with 97.9 kg/m3 measured from density gradient centrifugation. Conversely, the density difference between the (6,5) (0.75 nm diameter) and (9,8) (1.17 nm diameter) SWNT is smaller at 23.4 kg/m3. We conclude that covalent functionalization can provide an effective handle to separate out particular SWNT from typical diameter distribution. We show that SWNT mixtures where metallic SWNT have been selectively reacted produce two distinct density fractions corresponding to functionalized metallic and pure semiconducting SWNT. The results are confirmed by Raman spectroscopy where the higher density fractions contain an increased disorder mode with a corresponding decrease in intensity for the low density fraction. This method also allows for the first independent measure of (n,m) SWNT having different chemical conversion with functional groups and this will allow rigorous analysis of SWNT chemistry than is possible with un-calibrated spectroscopies such as Raman or photoluminescence.

# 73f Antenna Chemistry with Metallic Single-Walled Carbon Nanotubes

**Juan G. Duque**[1], Matteo Pasquali[1], and Howard K. Schmidt[2]. (1) Department of Chemical and Biomolecular Engineering, Rice University, 6100 Main Street, MS-100, Houston, TX 77005, (2) Chemical and Biomolecular Engineering, Rice University, 6100 Main Street, Houston, TX 77005

Single-Wall Carbon Nanotubes (SWNTs) have remarkable mechanical and electrical properties [1]. Individual metallic SWNT are ballistic conductors with conductivity that support high DC current densities [2]. As a consequence, metallic SWNT have a high axial dielectric constant [3], and rapidly polarize in response to externally applied electric fields. Therefore, metallic SWNT can be considered nano-scale antennae, while their high aspect ratio , ~ 1000, provides a localized apparent field amplification factor, β, at their tips. SWNT networks have been shown to efficiently absorb and thermalize electromagnetic radiation generally across the RF [4], microwave (MW) [5] and optical [6] frequency regimes. We demonstrate that these electrodynamic effects can be exploited under electromagnetic radiation to drive electrochemical reactions with aqueous suspensions of individualized nanotubes. Experimental results provide evidence of selective activation of metallic SWNT. In the presence of high oxidation potential transition metal salts to reduce metallic nanoparticles at the ends of SWNT, yielding novel composite nanostructures that encompass dumbbells and rings. Nanotube antenna chemistry could impact nanomedicine, energy harvesting and synthesis. 1. Saito, R., M. Fujita, et al., Applied Physics Letters, 1992. 60(18): p. 2204-2206. 2. McEuen, P.L., M.S. Fuhrer, et al., IEEE Transactions on Nanotechnology, 2002. 1(1): p. 78-85. 3. Benedict, L.X., S.G. Louie, et al., Physical Review B, 1995. 52(11): p. 8541-8549. 4. Gannon, C.J., P. Cherukuri, et al., Cancer, Published online 24 October, 2007. 5. Imholt, T.J., C.A. Dyke, et al., Chemistry of Materials, 2003. 15(21): p. 3969-3970. 6. Ajayan, P.M., M. Terrones, et al., Science, 2002. 296(5568): p. 705-705.

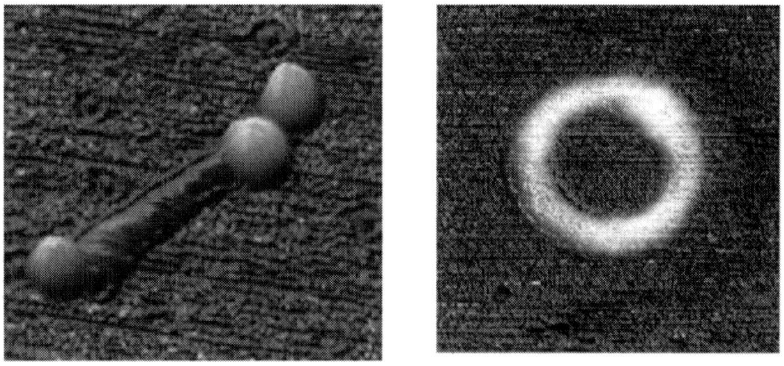

# 122a Synthesis of High Mechanical Strength Silica Films In Supercritical $CO_2$, Utilizing Blends of POSS-Paa Copolymers and PEO-B-Ppo-B-PEO Triblock Copolymers as Templates

**Alvin H. Romang**, Department of Chemical Engineering, University of Massachusetts-Amherst, 159 Goessman lab, 686 N. Pleasant st, Amherst, MA 01003 and James J. Watkins, Department of Polymer Science and Engineering, University of Massachusetts, 120 Governors Dr., Amherst, MA 01003.

Reduction of the feature size in microelectronics requires a reduction of the dielectric constant for the low-k barrier layers. Reduction of the dielectric constant is accomplished through two methods: varying the chemistry, and the creation of pores in the silica matrix. However, the porosity of the silica matrix is limited by the required mechanical strength of the films.

Here we show the synthesis of mesoporous organosilicate low dielectric constant films with increased hardness and modulus through the addition of polyhedral oligomeric silsesquioxane (POSS) in the silica matrix utilizing supercritical $CO_2$ as the reaction medium. The mesoporous organosilicate films are synthesized by selectively depositing organosilicate precursors inside sacrificial Pluronic block copolymer templates (PEO-b-PPO-b-PEO) with the addition of POSS molecules. The block copolymer templates are blended with copolymers of polyacrylic acid (PAA) and POSS functionalized acrylates to disperse the POSS uniformly in the template. The strong interaction between PAA and PEO facilitates the dispersion of POSS molecules, which tend to aggregate in the polymer matrix. The resulting films exhibit up to three times the hardness of the control films synthesized without the addition of POSS molecules for given porosities.

# 122b Three Dimensional Inorganic Nanoarrays Created by Electrodeposition through a Crystalline Protein Monolayer

Daniel B. Allred[1], Anchi Cheng[2], Mehmet Sarakaya[3], Francois Baneyx[1], and **Daniel T. Schwartz**[1]. (1) Department of Chemical Engineering, University of Washington, Box 351750, Seattle, WA 98195-1750, (2) Department of Cell Biology, Scripps Research Institute, La Jolla, CA 92037, (3) Department of Materials Science and Engineering, University of Washington, Box 351750, Seattle, WA 98195-1750

Three dimensional transmission electron microscopy (3D-TEM) is used to characterize the structure of materials electrodeposited through a monolayer of the hexagonally-packed intermediate (HPI) layer protein from D. radiodurans. The crystalline protein monolayers were prepared on unique TEM-compatible electrodes using methods and materials described previously.[1-5] Moderate pH electrodeposition baths ($4 \leq pH \leq 11$) were necessary to preserve the crystalline protein structure. The TEM tilt series data and 3D reconstruction was performed at the National Resource for Automated Molecular Microscopy at Scripps. The abstract figure shows 3D-TEM reconstruction data for cuprous oxide electrodeposited through the crystalline protein monolayer. The top image shows electron density contours that represent the inorganic deposit (red) and the protein template (white). Here the inorganic material is electrodeposited cuprous oxide. The electrodeposited material grows through the pores in the protein as a hexagonal nanoarray, with the larger cuprous oxide particle organized with 3-fold symmetry and smaller particles on top of the 6-fold protein sites. The bottom image shows the 3D architecture of the electrodeposited material by itself, with the protein removed. One sees that each of the large cuprous oxide particles sits on a thin, fully interconnected hexagonal lattice on the substrate. The protein itself is largely invisible to the electron beam, so its structure is revealed as a negative replica of the electron-dense inorganic materials. The fact that the protein structure we observe closely matches previous studies, demonstrates that the 3D pore structure in HPI-layer protein is fully accessible to electrodeposition. Moreover, we show that electrodeposition of complex inteconnected structures is possible below 10 nm, augering well for continued down-scaling of the complex 3D interconnect structures in VLSI. [1]. D.B. Allred, M. Sarikaya, F. Baneyx, and D.T. Schwartz, Electrochemical Nanofabrication using crystalline protein masks, Nano Lett. 5, 609-613 (2005). [2]. D.B. Allred, M. Sarikaya, F. Baneyx, and D.T. Schwartz, Bacterial surface layer proteins for electrochemical nanofabrication, Electrochim Acta 53, 193 - 199 (2007). [3]. D.B. Allred, M. Zin, H. Ma, M. Sarikaya, F. Baneyx A.K-y. Jen, and D.T. Schwartz, Direct nanofabrication and TEM characterization on a suite of utrathin film substrates, Thin Sol. Films 515 5341–5347 (2007). [4]. A. Presenda, D.B. Allred F. Baneyx, D.T. Schwartz, M. Sarikaya, Stability of s-layer proteins for electrochemical nanofabrication, Coll Surf B 57 256-261(2007). [5]. D.B. Allred, A. Cheng, M. Sarikaya, F. Baneyx, and D.T. Schwartz, Three-dimensional architecture of inorganic nanoarrays electrodeposited through a surface-layer protein mask, Nano Letters DOI: 10:1021/nl0803444 (2008).

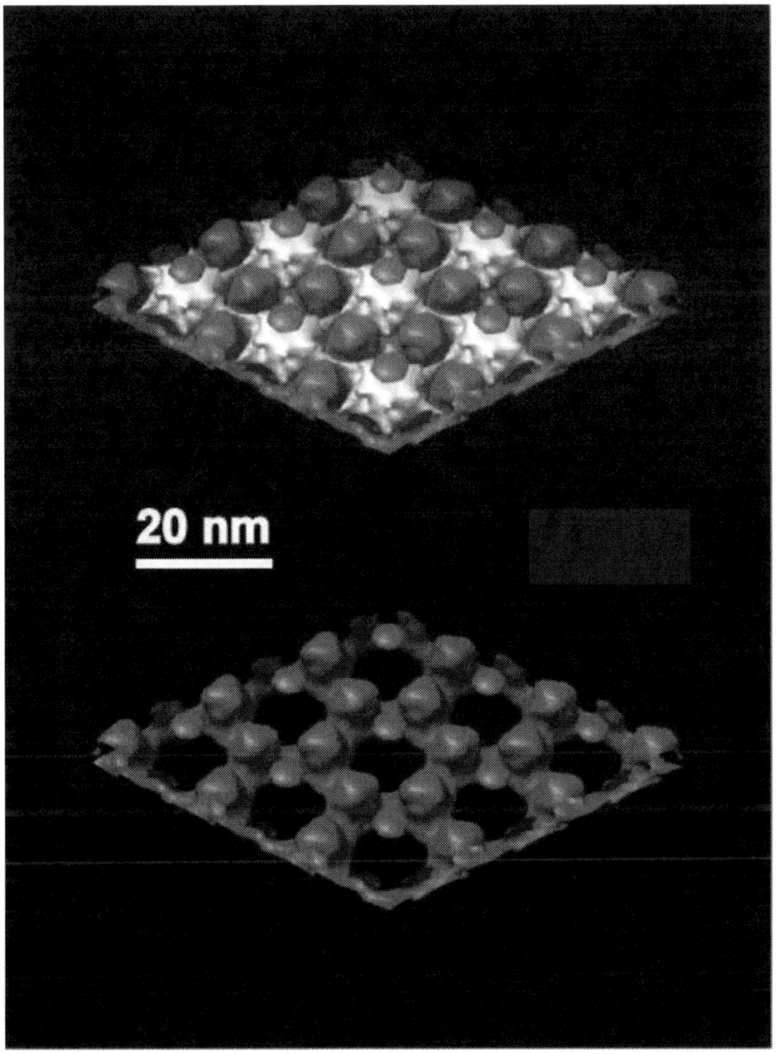
20 nm

## 122c Peptide-Mediated Deposition of Nanostructured $TiO_2$ into the Periodic Structure of Diatom Biosilica

Clayton Jeffryes[1], Timothy Gutu[2], Jun Jiao[2], and **Gregory L. Rorrer**[1]. (1) Oregon State University, Department of Chemical Engineering, Corvallis, OR 97331, (2) Department of Physics, Portland State University, Portland, OR 97207

Nanostructured titanium dioxide ($TiO_2$) semiconductor materials have significant applications in photocatalysis and photovoltaics. In this study, a biomimetic approach was developed to fabricate nanostructured $TiO_2$ that possessed hierarchical structure using diatom shells as a template. Diatoms are single-celled algae that make silica shells called frustules that possess periodic structures ordered at the micro- and nanoscale. Biosilica frustules were isolated from cell cultured Pinnularia diatom by hydrogen peroxide treatment of cell biomass. Poly-L-lysine (PLL) was conformally adsorbed onto surface of the frustule biosilica. The condensation of soluble Ti-BALDH to nanostructured $TiO_2$ by PLL-adsorbed diatom biosilica loaded 0.8 g Ti / g Si onto the frustule. $TiO_2$ nanoparticles preferentially deposited into the 200 nm frustule pores and also coated the frustule outer surface. The periodic pore array templated these concentrated pockets of $TiO_2$ nanoparticles into an ordered, lattice-like structure. Thermal annealing at 680 $^{o}$C converted the as-deposited $TiO_2$ to its anatase form with an average nanocrystal size of 19 nm, as verified by XRD, electron diffraction, and SEM/TEM. This is the first reported study of directing the peptide-mediated deposition of $TiO_2$ into a hierarchical nanostructure using a biologically fabricated template.

## 122d Tud-C: A Tunable, Hierarchically Structured Mesoporous Zeolite Composite

Jia Wang, Pcmt/dct, Delft University of Technology, Julianalaan 136, Delft, 2628BL, Netherlands and **Marc-Olivier Coppens**, Howard P. Isermann Department of Chemical Engineering, Rensselaer Polytechnic Institute, 110 8th Street, Troy, NY 12180.

In many catalytic applications of zeolites, diffusion limitations and pore blockage are serious limitations to their efficient use. A well-connected, hierarchical system of micro- and meso-/macropores may solve those problems. Thus, meso-structured zeolites have attracted significant attention. In recent years, due to the development of supra-molecular templating techniques (such as using the tri-block copolymer P123), promising methods were developed to introduce mesopores into zeolites. However, the practical implementation of such materials is limited, since the templates are often expensive or the methods tend to be complicated.

Here, we report a facile one-pot method to synthesize a composite, denoted as TUD-C[1], which consists of fairly uniform ZSM-5 nanocrystals embedded in a mesoporous matrix. This method uses only one organic templating/scaffolding molecule, namely TPAOH, the molecule already used to template ZSM-5. Micro- and mesopores form separately under different conditions, allowing optimized control over the micro- and meso-porosity. The hierarchical micro-mesoporous texture of TUD-C was confirmed by high-resolution transmission electron microscopy (TEM). 27Al MAS NMR and FTIR demonstrate the presence of Brønsted acid sites. The Si/Al ratio and crystallization temperature are important parameters to influence the material texture and composition. TUD-C shows a remarkable hydrothermal stability. Our synthesis method distinguishes itself by its simplicity, low cost, and flexibility, and provides an alternative to another method discussed last year to synthesize the hierarchical composite TUD-M[2].

Reference List

[1] J.Wang, J.C.Groen, W.Yue, W.Zhou, M.-O.Coppens, Chem.Commun. 2007, 14, 4653-4655.

[2] J.Wang, J.C.Groen, W.Yue, W.Zhou, M.-O.Coppens, J.Mater.Chem. 2008, 18, 468-474.

# 122e Confined Synthesis of Silicalite-1 Nanocrystals In Threedimensionally Ordered Mesoporous Carbon

**Wei Fan**[1], Sandeep Kumar[1], Yoo Won Choel[2], Andreas Stein[2], R. Lee Penn[2], and Michael Tsapatsis[1]. (1) Department of Chemical Engineering and Materials Science, Institute of Technology,University of Minnesota, 421 Washington Avenue SE, Minneapolis, MN 55455, (2) Department of Chemistry, Institute of Technology,University of Minnesota, 207 Pleasant Street SE, Minneapolis, MN 55455

The sustained interest in the synthesis of zeolite nanocrystals is justified because they can serve as catalysts, seeds for thin film growth and for preparation of polymer-zeolite nanocomposite membranes. In this study, silicalite-1 was synthesized inside three-dimensionally-ordered mesoporous carbon with a uniform pore size that can be tuned from 20 to 40 nm. The three-dimensionally-ordered mesoporous (3DOm) carbon was prepared using ordered arrays of silica nanoparticles (Lys-Sil) as a template[1,2]. Electron Microscopy, X-ray diffraction and $N_2$ adsorption and desorption reveal that the silicalite-1 nanocrystals can reliably replicate the 3DOm porosity, resulting in the first highly ordered mesopororous/ microporous material with inverse opal mesoporosity and MFI zeolite microporosity. Applications in catalysis and membranes will be discussed along with implications in understanding zeolite nucleation and growth.

1. Yokoi, T.; Sakamoto, Y.; Terasaki, O.; Kubota, Y.; Okubo, T.; Tatsumi, T. *J. Am. Chem. Soc.* ***2006,*** *128,* 13664-13665.

2. Davis, T. M.; Snyder, M. A.; Krohn, J. E.; Tsapatsis, M. *Chem. Mater.* ***2006,*** *18*, 5814-5816.

# 122f Microporous Silicate Nucleation and Growth: Mechanistic Investigations toward Rational Design of Nanocrystalline Materials

**Jeffrey D. Rimer**, Molecular Design Institute, Department of Chemistry, New York University, 100 Washington Square East, New York, NY 10003, Alexandra Navrotsky, Peter A. Rock Thermochemistry Laboratory and NEAT ORU, University of California at Davis, Davis, CA 95616, Dion Vlachos, Director of Center for Catalytic Science and Technology (CCST), University of Delaware, Newark, DE 19716, and Raul F. Lobo, Center for Catalytic Science and Technology, Department of Chemical Engineering, Univesity of Delaware, 150 Academy St., Newark, DE 19716.

Developing nanostructured materials for emerging applications in areas ranging from energy to pharmaceuticals involves the rational design of synthetic pathways, largely through self-assembly, that ideally can be tuned to attain desired material properties, such as particle size and morphology. The ability to a priori control these properties is desirable and advantageous on many levels, but requires rather detailed knowledge regarding the interfacial interactions, kinetics, thermodynamics, and mechanistic aspects governing growth. To this end, our research focuses on improving the existing, but not well-developed, molecular-level understanding of microporous silicate (i.e., zeolite) crystallization. Microporous silicates have received much attention over the past decade due to increased interest in their implementation in expanding technologies (e.g., optical electronics [1a], selective separations [1b-d]) – all of which entail adapting the economies of scale approach to industrial zeolite production to selectively tailor materials at much smaller length scales. We employ a hierarchical approach using combined experiments and modeling to study the all-silica zeolite, silicalite-1, which has served as a prototype for mechanistic studies of microporous growth.

Investigations largely focus on silica nanoparticles (2-6 nm) that are precursors, and possible building units, in the synthesis of silicalite-1. We have shown that silica nanoparticles evolve by Ostwald ripening during the induction period, leading to a compositional shift from an initially amorphous-like particle to one more akin to silicalite-1 [2a]. Similarly, both the kinetic rates and enthalpies of dissolution reveal that the molecular structure exhibits an internal reorientation from an initially disordered to a more zeolite-like arrangement [2b]. The results of these studies, aside from providing relevant time scales involved in silicalite-1 crystallization, offer insight into the early stages of nucleation, suggesting that solution-mediated processes may be the dominant mechanism.

The kinetics and thermodynamics of nanoparticle formation were examined through a combination of small-angle scattering and microcalorimetry measurements, identifying particle self-assembly at a specific silica concentration and pH [3]. We developed a model coupling silica speciation and electrostatic double layer theory to predict the self-assembly, phase behavior, and surface charge (i.e., colloidal stability) of silica nanoparticles. This model provides an explanation for thermodynamic phenomena associated with silicalite-1 crystallization [4], and it has served as a basis for establishing a predictive silicalite-1 growth model, capable of quantitatively capturing growth rates over a wide range of reaction conditions. We will present these collective findings within the context of mechanisms proposed in the literature, highlighting advancements made toward understanding silica nanoparticle self-assembly, composition, structure, and its roles in both nucleation and crystallization of silicalite-1. In a broader scope, these techniques offer a general methodology that can be applied to other microporous, mesoporous [4a], and/or biomineral silica-based materials to develop predictive growth models, thereby serving as a framework toward rational syntheses of novel nanomaterials.

1. (a) Tosheva and Vltchev, Chem. Mater. 2005, 17, 2494, (b) Lai et al., Science 2003, 300, 456, (c) Lee et al., Science 2003, 301, 818, (d) Lai et al., Adv. Func. Mater. 2004, 14, 716.

2. (a) Rimer et al., J. Phys. Chem. B 2005, 109, 12762, (b) Rimer et al., Chem. Mater. 2007, 19, 4189.

3. (a) Rimer et al., Langmuir 2005, 21, 8960, (b) Fedeyko et al., J. Phys. Chem. B 2004, 108, 12271, (c) Rimer et al., J. Phys. Chem. C, 2008, submitted.

4. (a) Rimer et al., Chem. Eur. J. 2006, 12, 2926, (b) Yang et al., Chem. Mater. 2002, 14, 2803.

# 122g Investigation of the Effect of Synthesis Sol Composition on Silicalite-1 Crystallization Using ATR / FTIR Spectroscopy

**Vladimiros Nikolakis**, Agnelos Patis, and Vassilios Dracopoulos. FORTH/ ICEHT, Stadiou Str, Platani, PO Box1414, Patras, 26504, Greece

The elucidation of Silicalite-1 crystallization mechanism has been the subject of numerous investigations [1-10]. This zeolite is usually crystallized from alkaline sols containing siliceous species and tetrapropylammonium cations, and its nucleation and growth take place in the presence of a suspension of nanoparticles having sizes less than ~10 nm. The nanoparticle structure and role in crystallization are fundamental problems that have also received a lot of attention. These nanoparticles are pre-organized inorganic/organic composites and their structure is most likely different than that of silicalite-1 crystals.

We have investigated the crystallization of silicalite-1 crystals in aqueous sols using Attenuated Total Reflection / Fourier Transform Infrared (ATR/FTIR) spectroscopy[10]. The effect of SiO2, and tetrapropylammonium hydroxide (TPAOH) concentration on the structure of the intermediate species as a function of time has been studied and will be presented. The compositions studied have been selected because they enable the acquisition of spectra with high signal/noise ratio and because they have been studied the past.The existence of oligomeric siliceous species has been verified for all compositions after the end of tetraethoxysilane (TEOS) hydrolysis. On the other hand, more condensed and highly crosslinked siliceous structures have been observed only when the SiO2 concentration was above a certain limit. These structures have been considered to be the "primary" nanoparticles reported by other researchers, and their IR spectra revealed that their structure is different than that of the silicalite-1 crystals. It has also been found that TPAOH content affects the structure and connectivity of both oligomeric and nanoparticle siliceous species. Increasing the TPAOH content resulted in a decrease of the SiO2 connectivity of the siliceous nanoparticles as well as to a broader distribution of the oligomeric siliceous species.

The evolution nanoparticle structure as a function of crystallization time has also been monitored. It has been found that upon heating of the synthesis sols, the "primary" nanoparticles transform to a new form that has an increased number of SiOSi bonds. These "intermediate" particles do not have the structure of silicalite-1 and they are present for the most part of crystallization. Finally once the silicalite-1 nuclei are formed they grow in a suspension of these "intermediate" particles. This transformation was more pronounced in the diluted sols than in the sols with low water content. Finally, our results will be analyzed in the context of the crystallization mechanisms proposed in the literature.

References

[1] T.M. Davis, T.O. Drews, H. Ramanan, C. He, J. Dong, M.A. Katsoulakis, E. Kokkoli, A.V. McCormick, R.L. Penn, M. Tsapatsis, Mechanistic principles of nanoparticle evolution to zeolite crystals, Nature Materials 5 (2006) 400-408

[2] J.M. Fedeyko, J.D. Rimer, R.F. Lobo, D.G. Vlachos, Spontaneous formation of silica nanoparticles in basic solutions of small tetraalkylammonium cations, Journal of Physical Chemistry B 108 (2004) 12271-12275

[3] V.Nikolakis, E. Kokkoli, M. Tirrell, M. Tsapatsis, D.G. Vlachos, Zeolite growth by addition of subcolloidal particles: Modeling and experimental validation, Chemistry of Materials 12 (2000) 845-853

[4] S.Y Yang, and.A. Navrotsky, Study on synthesis of TPA-silicalite-1 from initially clear solutions of various base concentrations by in situ calorimetry, potentiometry and SAXS, Chemistry of Materials 16 (2004) 3682-3687

[5] S.L. Burkett, and M.E. Davis, Mechanisms of structure direction in the synthesis of TPA/Si-ZSM-5, Journal of Physical Chemistry 98 (1994) 4647-4653

[6] C.E.A. Kirschhock, R. Ravishankar, L. Van Looveren, P.A. Jacobs, J.A. Martens, Mechanism of transformation of precursors into nanoslabs in the early stages of MFI and MEL zeolite formation from TPAOH-TEOS-H2O and TBAOH-TEOS-H2O mixtures, J. Phys. Chem B. 103 (1999) 4972-4978

[7] J..M. Fedeyko, D.G. Vlachos, and R.F. Rimer, Langmuir 21 (2005) 5197-5206

[8] B..J. Shöeman, Analysis of the nucleation and growth of TPA-silicalite-1 at elevated temperatures with the emphasis on colloidal stability, Microporous and Mesoporous Materials 22 (1998) 9-22

[9] B.J. Shöeman, A high temperature in situ laser light-scattering study of the initial stage in the crystallization of TPA-silicalite-1, Zeolites 18 (1996) 97-105

[10]A. Patis, V. Dracopoulos, and V. Nikolakis, Investigation of Silicalite-1 Crystallization Using Attenuated Total Reflection/Fourier Transform Infrared Spectroscopy, J. Phys. Chem C, 111 (2007) 17478 - 17484

# 122h Molecular Interactions Between Structure-Directing Organic Molecules, Fluoride Species, and Siliceous Zeolite Frameworks Established by Two-Dimensional NMR

**Ramzy M. Shayib**[1], Sylvian Cadars[1], Stacey I. Zones[2], Allen Burton[2], and Bradley F. Chmelka[1]. (1) Chemical Engineering, University of California Santa Barbara, Santa Barbara, CA 93106, (2) Chevron Energy and Technology Company, 100 Chevron Way, Richmond, CA 94802

Molecular structure-directing agents (SDAs) have been widely used to synthesize crystalline and nanoporous zeolites that exhibit a wide range of technologically important adsorption, ion-exchange, and/or catalytic reaction properties. In particular, organic species are typically used under alkaline conditions to prepare crystalline silicate, aluminosilicate, or other crystalline frameworks. The resulting structures depend on the molecular composition and architecture of the SDA species used and their interactions with the crystallizing inorganic framework. By varying the properties of the SDA (size, charge distribution, polarity, hydrophilicity, etc.), a wide variety of frameworks have been synthesized, although the specific molecular interactions that govern the formation of different structures are still not well understood. Here, we report results that establish important SDA-framework interactions of two closely related imidazole derivatives, N,N',2-trimethylimidazolium hydroxide and N,N'-diisopropylimidazolium hydroxide, that yield two very different zeolite structures (ITW and MTT respectively), when synthesized in the presence of similar synthesis solutions containing fluoride species. Even though the starting synthesis solutions for both zeolite structures are nearly identical in composition and hydrothermal treatment conditions, very different ITW or MTT zeolite structures are formed, indicating that framework selectivity is governed by the molecular interactions among the SDA molecules, silicate framework, and fluoride anions. Powerful one- and two-dimensional (2D) solid-state $^{29}Si\{^{19}F\}$, $^{29}Si\{^{13}C\}$, and $^{29}Si\{^{1}H\}$ HETeronuclear CORrelation (HETCOR) NMR measurements provide detailed and unambiguous information on the molecular interactions among the organic SDA and inorganic framework moieties of these complicated multicomponent systems. In addition, J-coupling-mediated double-quantum-filtered (DQF) $^{29}Si$-$^{29}Si$ NMR correlation experiments establish the connectivities among different $^{29}Si$ tetrahedral sites in the siliceous zeolite frameworks. In combination, these powerful 2D NMR methods provide new means for characterizing local interactions among structure-directing organic species and their different zeolite framework structures, which can consequently be elucidated in unprecedented detail. The results provide new insights on the molecular interactions that are ultimately responsible for imparting structural selectivity in zeolite syntheses.

## 137a Dissolution-Rate Enhancement of Fenofibrate by Adsorption Onto Silica Using Supercritical Carbon Dioxide

**Ganesh Sanganwar** and Ram B. Gupta. Chemical Engineering, Auburn University, Auburn, AL 36849-5127

Dissolution rate of a poorly-water-soluble lipid-regulating drug, fenofibrate, is increased by adsorbing the drug onto silica. The adsorption is achieved by first dissolving the drug in supercritical carbon dioxide and then depressurizing the solution on silica. Loadings of up to 27.5 wt% of fenofibrate onto silica are obtained. Since solvents are not used in the loading process, the fenofibrate/silica formulation is free of any residual solvent, and carbon dioxide freely leaves upon depressurization. The formulation was characterized using infrared spectroscopy, ultraviolet spectroscopy, x-ray diffraction, differential scanning calorimetry and scanning electron microscopy. Based on in-vitro dissolution study, a significant increase in the dissolution rate (~80% drug release in 20 minutes) of drug-silica formulation has been observed as compared to micronized fenofibrate (~20% drug release in 20 minutes). The high dissolution rate of drug-silica formulation can be attributed to increase in the surface area and decrease in the crystallinity of drug after adsorption onto silica. Two different formulations are compared: (A) amorphous fenofibrate/silica, and (B) slightly-crystalline fenofibrate/silica. The second formulation is found to be stable on storage.

Web Page: www.eng.auburn.edu/users/gupta/

# Mathematical modeling of delivery of chemotherapeutic drugs to human tissues: an analytical approach

Dwaipayan Mukherjee and Saikat Chakraborty
*Department of Chemical Engineering, Indian Institute of Technology –Kharagpur, Kharagpur-721302, INDIA*

## Introduction

Man's war against dangerous afflictions has led to the development of potent formulations and requires complicated dosage patterns. They require controlled and targeted delivery to maximize therapeutic efficacy and minimize side-effects. Chemotherapy has been recognized as a successful means of arresting cancer, which is the third biggest killer among humans. Cancerous tumors are characterized by their rapid spread across the diseased tissue and successively deteriorating conditions. However, it necessitates repeated dosage at specific times, and conventional delivery methods result in unnecessary delay in reaching the targeted location and expose other organs to the effect of the drugs. The drug delivery is thus desired to be precisely targeted and controlled.

This has resulted in previous attempts at designing novel, bio-implantable, polymer-coated delivery devices that can release the drug at a specified rate. We design and model a novel device that allows electrically regulated release of an ionic drug through an electroactive polymer membrane into a tumor-affected tissue. Doped electro active polymers like polypyrrole show considerable ability to conduct ions and even behave as a switch by alternating its oxidized and reduced states on reversal of electric potential.

Our drug delivery device consists of a cylindrical capsule comprising of a drug reservoir containing the ionic drug solution and a set of batteries for creating the desired electric field. The entire capsule is surrounded by a layer of selectively doped electroactive polymer. The device has been implanted at the center of the tumor such that the tumor has progressed symmetrically from all sides. The anionic drug diffuses out under the action of the electric field, permeates through the electroactive membrane as it reacts with the charged polymer groups and is released into the tissue, where it reacts with the biological cells.

## 1. Modeling

The transport of the anionic drug into the affected tissue occurs due to three factors: the concentration gradient, the electric field created by the batteries and the reaction of the anionic drug with the polymer groups inside the doped-polymer membrane.

*1.1 The potential distribution:* The potential field distribution is not an independent entity but is a function of the distribution of the ions in the medium as also the strength of the fixed charges within the membrane. This distribution is obtained by application of the Poisson-Boltzmann equation, which is given by:

$$\nabla^2\psi = -\frac{1}{\varepsilon}\left[\rho^* - 2qn_o \sinh\left(q\psi/_{kT}\right)\right], \tag{1.1}$$

where, $\psi$ is the potential distribution, $\rho^*$ is the fixed charge density, $\rho$ of the surrounding membrane normalized over the entire space of the reservoir, i.e. $\rho^* = \rho\left(\frac{V_m}{V_s}\right)$, $V_m$ and $V_s$ being the volumes of the membrane and solution respectively, $n_o$ is the no. density of ions in the drug solution and $q$ is the basic unit of electric charge. Eq.(1.1) is solved to get the potential distribution.

*1.2 Transport in the reservoir:* The transport in the drug reservoir in the presence of the electric field is given by the Nernst-Planck equation which may be expressed as:

$$\frac{\partial c_D}{\partial t} + \nabla\left[-D_s\left\{(\nabla c_D) + \frac{zF}{RT}c_D(\nabla\psi)\right\}\right] = 0, \qquad (1.2)$$

Some assumptions are made for the solution of this equation analytically. They are (i) the drug reservoir has infinite amount of drug, i.e., there is no depletion of drug within the reservoir, (ii) there are no convection effects in the system and (iii) the membrane is thin compared to the diameter of the reservoir. Based on these assumptions, the solution of eqn. (1.2) is given by

$$c = c_o^s\left(t - \frac{R-r}{D_s k_1}\right)\exp\left[\frac{k_2}{k_1}(R-r)\right], \qquad (1.3)$$

*1.2.2 Electric double layer:* The **Nernst-Planck-Poisson (NPP)** system of model equations was used to model the system during the positive scan. For the negative scan, the sign of the potential will be reversed and there will also be the formation of the electric double layer. We use the modified-Stern model to quantify the effect of the electric double layer on the applied electric potential and the **Nernst-Planck-Poisson-Modified Stern (NPPMS)** equation is

$$\frac{1}{D_s}\cdot\frac{\partial \hat{c}}{\partial t} = -k_1'\frac{\partial \hat{c}}{\partial r'} + k_2'(\hat{c} + c_b), \qquad (1.4)$$

solving which we get:

$$c = c_o^s\left(t + \frac{R-r}{D_s k_1'}\right)\exp\left[\frac{k_2'}{k_1'}(R-r)\right], \qquad (1.5)$$

*1.3 Transport in the polymeric membrane:* For modeling the transport in the membrane, we do not need to find out a separate potential distribution. Since the membrane (~ $10^{-3}$m) is very thin compared to the reservoir, we can assume the potential to follow a linear profile, with the same slope as that near the membrane.

*1.3.1 Positive scan:* The drug transport within the membrane is modeled with the same Nernst-Planck (NP) equation, with an extra reaction term being included due to the interaction between the drug and polymer group

$$\frac{\partial c_D}{\partial t} = \hat{D}_m \left( \frac{\partial^2 c_D}{\partial r^2} + \frac{1}{r}\frac{\partial c_D}{\partial r} \right) + D_m \frac{zFs}{RT} . \frac{\partial c_D}{\partial r} - k_f c_p c_D \quad , \tag{1.6}$$

where, $\hat{D}_m = D_m - RTk$

Taking $r' = \frac{r - R}{h}$ we have the solution of Eq.(1.8) as:

$$c_D = c_o \left( t - \frac{r'}{\mu} \right) \exp\left( - \frac{k_f P_o}{\mu} r' \right), \tag{1.7}$$

*1.3.2 Negative scan:* The same form of the NP equation is used here for the drug D but with the sign of the potential reversed.

$$\frac{\partial c_D}{\partial t} = \hat{D}_m \left( \frac{\partial^2 c_D}{\partial r^2} + \frac{1}{r}\frac{\partial c_D}{\partial r} \right) - D_m \frac{zFs}{RT} . \frac{\partial c_D}{\partial r} + k_r c_c \quad , \tag{1.8}$$

where, $\hat{D}_m = D_m - RTk$

For the complex C, we can write:

$$\frac{\partial c_c}{\partial t} = D_c \left( \frac{\partial^2 c_D}{\partial r^2} + \frac{1}{r}\frac{\partial c_D}{\partial r} \right) - k_r c_c , \tag{1.9}$$

Putting, $r' = \frac{r - R}{h}$ and $\mu' = \frac{D_m F s'}{RTh}$ we have the solution of Eq.(1.10) as:

$$c_D = c_o \left( t + {r'}/{\mu'} \right) + \frac{1}{k_3} \left[ k_r \left( t_p - {r'}/{\mu'} \right) - \frac{k_r}{k_3} \right] \exp\left( -k_r t - \frac{k_f P_o}{\mu'} r' \right) , \tag{1.10}$$

where $k_3 = k_r - k_f P_o$

*1.4 Transport and drug reaction in normal tissue:* The tissue is considered separated into two compartments – the normal tissue and the tumor-affected tissue. The dimensions of the tissue being too large compared to that of the drug reservoir, the system can be approximately described by Cartesian coordinates.

$$\frac{\partial c_D}{\partial t} = D . \frac{\partial^2 c_D}{\partial x^2} + R , \tag{1.11}$$

The reaction of the drug is modeled by Michaelis-Menten kinetics as $R = \frac{v_{\max} . c_D}{c_D + K_m}$ which for low values of concentration can be approximated by a 1st order equation, given by:

$$\frac{\partial c_D}{\partial t} = D.\frac{\partial^2 c_D}{\partial x^2} - kc_D . \tag{1.12}$$

*1.4.1 The negative scan:* The governing equation along with the boundary conditions is:

$$\frac{\partial c_D}{\partial t} = D.\frac{\partial^2 c_D}{\partial x^2} - kc_D \qquad \begin{array}{l} c(x,0)=0 \\ c(0,t)=a-b\left(e^{-\alpha t}-1\right) \\ c(1,t)=a_1+b_1e^{-\beta_1 t} \end{array} \tag{1.13}$$

The solution using suitable boundary conditions is given by

$$c_D(x,t) = \left(a+b-be^{-\alpha t}\right)(1-x)+\left(a_1+b_1e^{-\beta t}\right)x$$
$$+\sum_{n=1}^{\infty}\left[\frac{2Dn\pi\sin(n\pi x)}{P_n}\left\{(-1)^n(a_1+b_1)-a\right\}e^{-P_n t}\right], \tag{1.14}$$

where, $P_n = Dn^2\pi^2 + k$ , $a_1 = a+b$ , $\alpha = \beta = k$ .

*1.4.2 The positive scan:* The modeling for the positive scan will be carried out in much the same way as in the negative scan, but we use a Neumann boundary condition at the membrane-tissue interface and a Dirichlet condition at the other boundary. The spatial concentration profile at the end of the negative scan is used as the initial condition for the positive scan. The equation is

$$\frac{\partial c}{\partial t} = D.\frac{\partial^2 c}{\partial x^2} - kc , \tag{1.15}$$

$$c(x,0) = f(x) = g+hx$$

$$\frac{\partial c}{\partial x}(0,t) = 0 \tag{1.16}$$

$$c(1,t) = a_1 + b_1e^{-\beta t}$$

The solution is

$$c_D(x,t) = b_1e^{-kt} + 2\sum_{n=1}^{\infty}\left[b_1(-1)^n\frac{\cos(n-1/2)\pi x}{(n-1/2)\pi}e^{-M_n t}\right]$$
$$-2\sum_{n=1}^{\infty}\left[\frac{h\cos(n-1/2)\pi x}{(n-1/2)^2\pi^2}e^{-M_n t}\right]-2\sum_{n=1}^{\infty}\left[(-1)^n\frac{(g+h)\cos(n-1/2)\pi x}{(n-1/2)\pi}e^{-M_n t}\right], \tag{1.17}$$

where, $M_n = D(n-1/2)^2\pi^2 + k$ .

*1.5 Transport and drug reaction in tumor-affected tissue:* The drug diffusivity, *D* should have a lower value in the tumor region.

*1.5.1 The negative scan*

The corresponding solution for the tumor-affected region for the negative scan is:

$$c(x,t) = \left(a_1 + b_1 e^{-kt}\right)(1-x) + \left(a_1 + b_2 e^{-kt}\right)x$$
$$+\sum_{n=1}^{\infty}\left[\frac{2Dn\pi \sin(n\pi x)}{P_n}\left\{a_1 + b_1 - (-1)^n (a_1 + b_2)\right\} e^{-P_n t}\right], \qquad (1.18)$$

*1.5.2 The positive scan*

The corresponding solution for the tumor-affected region for the positive scan is:

$$c_D(x,t) = b_1 e^{-kt}(1-x) + b_2 e^{-kt} x + 2\sum_{n=1}^{\infty}\left[\frac{\sin(n\pi x)}{n\pi}\left\{p - (-1)^n (p+q)\right\} e^{-P_n t}\right]$$
$$+2\sum_{n=1}^{\infty}\left[\frac{\sin(n\pi x)}{n\pi}\left\{b_2 (-1)^n - b_1\right\} e^{-P_n t}\right]. \qquad (1.19)$$

## 2. Results and discussions

The results are shown for the complete 4 compartment model having the drug reservoir, the membrane, the normal tissue and the tumor-affected tissue. Diffusivity of the drug in the four zones – the reservoir, the membrane, the normal tissue and the tumor-affected tissue are taken as $1.6 \times 10^{-6}$ m²/sec, $1 \times 10^{-9}$ m²/sec, $1.2 \times 10^{-6}$ m²/sec and $1.2 \times 10^{-8}$ m²/sec respectively. The voltage scan times for both positive and negative scans is taken as 5 seconds. Drug concentration in the reservoir solution is 0.1M and the dopant concentration in the membrane is 10M. The membrane chosen is polypyrrole. The drug and biological data conform to the diffusion of the anti-tumor drug doxorubicin in rat-liver.

*2.1 Drug concentration:* Figures 2.1 and 2.2 show the spatial plots of drug concentration at various times for the 4 compartments. The fast disappearance of drug in the membrane in the positive scan is due to the fast ionic reaction which converts the drug into the neutral complex. The reverse reaction is also fast and hence a constant profile is obtained in the same region during the negative scan, as the fast evolving anions are unable to diffuse out.

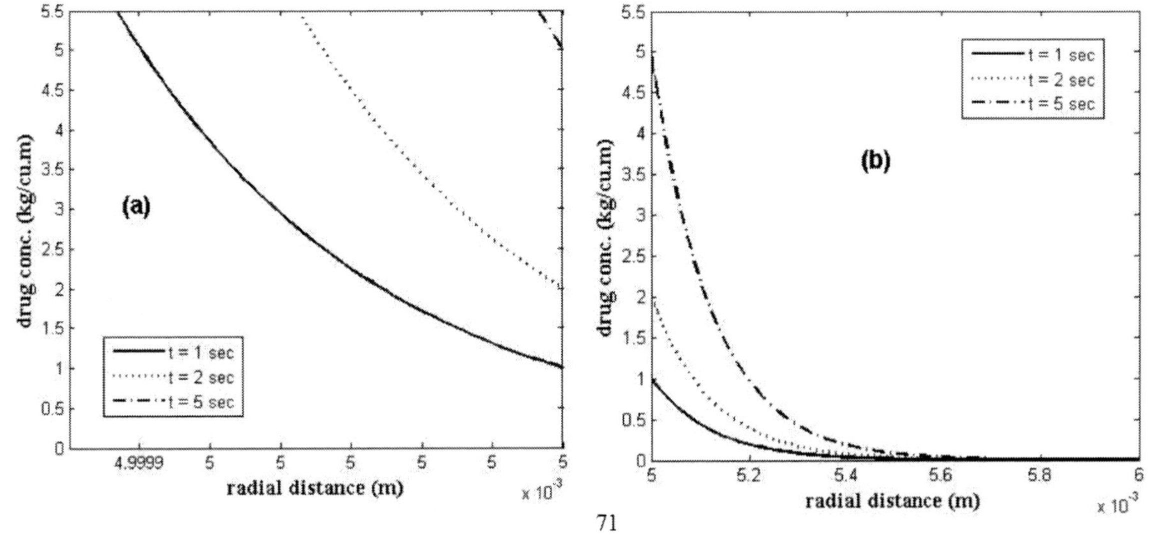

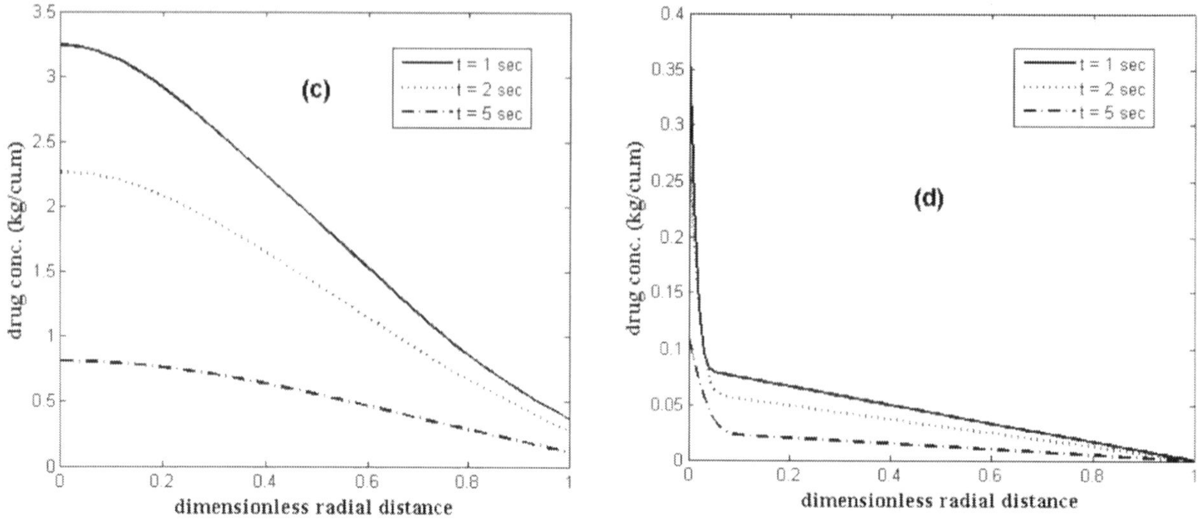


**Figure-2.1: Spatial drug concentration plots during the positive scan for (a) reservoir, (b) membrane, (c) normal tissue and (d) tumor-affected tissue**

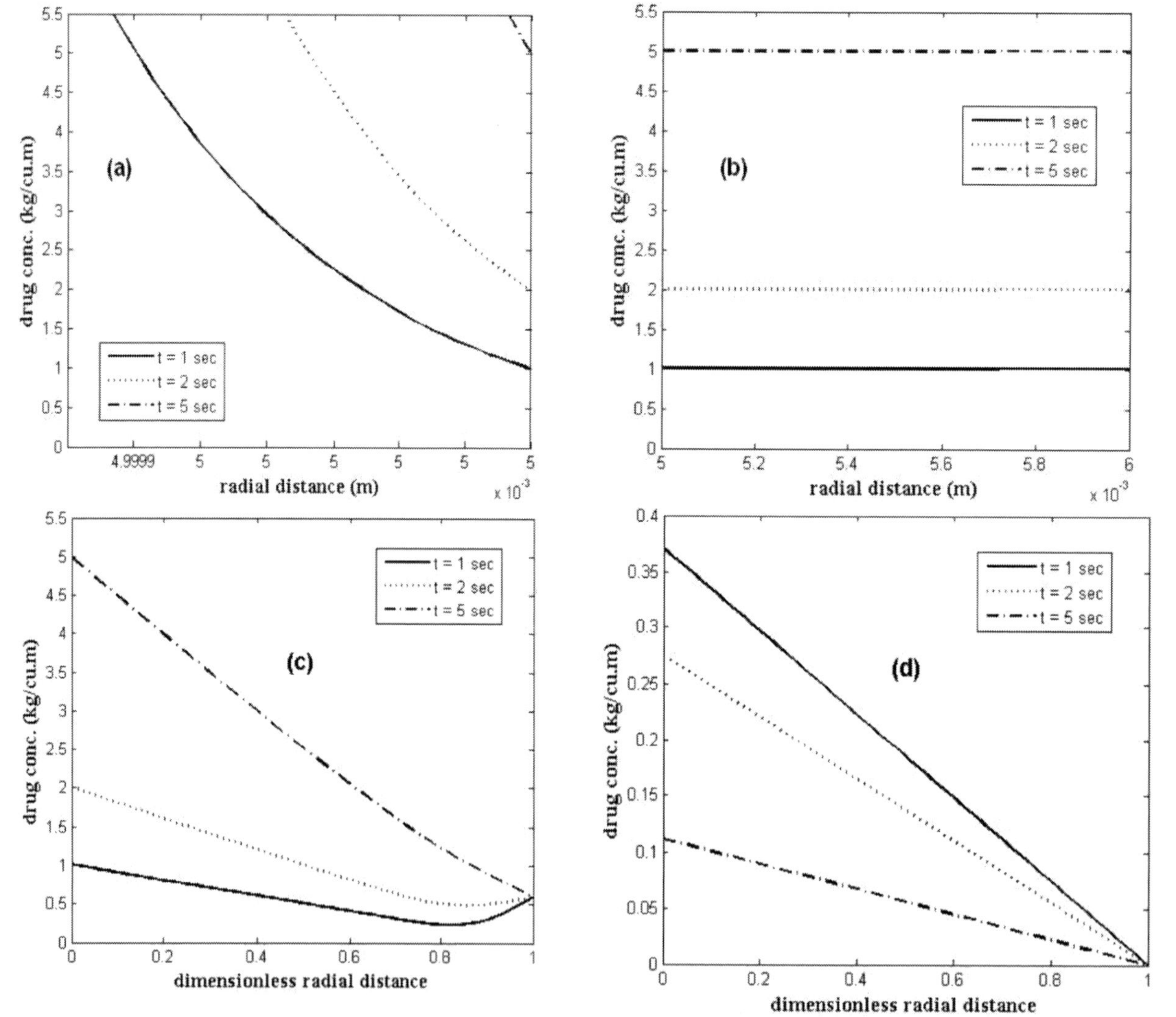


**Fig 2.2: Spatial drug concentration plots during the negative scan for (a) reservoir, (b) membrane, (c) normal tissue and (d) tumor-affected tissue**

## 3. Conclusion

Our modeling and simulation quantifies the drug concentration levels under normal conditions (characterized by the normal values of the parameters) and also for different parametric values. This can serve as an effective tool for predicting the drug level for a particular patient and help an oncologist make his decision regarding the optimal parametric values for a particular patient and his condition.

**Voltage:** Increase in voltage causes an increase in drug delivery rate and is the only option for drugs having a very high depletion rate. It is also an option for rapid drug supply in emergency situations.

**Scan times:** The study shows that the drug is stored in the membrane during positive scan which is released during the negative scan. Increase in positive scan times, with short negative scan times can supply the drug at high rate in short bursts, which may be required considering the physiological condition of the patient.

**Membrane thickness:** Reduction of membrane thickness increases the rate of drug delivery but also reduces the amount of drug that can be stored in the membrane during the positive scan. The thickness should be judiciously chosen keeping in mind whether the patient needs higher rate of drug supply or more amount of drug slowly over a longer period.

**Therapeutic concentration:** It is found that the solutions of the drug transport models predict a relation between various parameters of the boundary conditions. The therapeutic concentration – which in this model is a boundary condition – is also a design parameter that needs to be decided upon by the oncologist. So we have a relation between the therapeutic concentration required for the particular physiological condition and the voltage to be applied for the cure.

## References:


[1] Blacher S., Jost M., Melen-Lamalle L., Lund L.R., Romer J., Foidart J.M., Noel A., Quanification of in-vivo tumor invasion and vascularization by computerized image analysis, *Microvascular Research* 75 (2008) 169-178.

[2] Bolto B.A and Weiss D.E, Electronic Conduction in Polymers. II. The Electrochemical Reduction of Polypyrrole at Controlled Potential, Australian *Journal of Chemistry* 16 (1963) 1076-1089.

[3] Gustafson D. A., Pritsos C. A., Kinetics and mechanism of Mitomycin C bioactivation by Xanthene Dehydrogenase under aerobic and hypoxic conditions, *Cancer Research* 53 (1993) 5470-5474.

[4] Helfferich F., Ion Exchange; McGraw Hill, NY, 1962.

[5] Lim J., Whitcomb J., Boyd J., Varghese J., Transient finite element analysis of electric double layer using Nernst-Planck-Poisson equations with a modified Stern layer, *Journal of Colloid and Interface Science* 305 (2007) 159-174.

[6] Ricka J., Tanaka T., Swelling of ionic gels: Quantitative performance of the Donnan Theory, *Macromolecules* 17 (1984) 2916-2921.

[7] Skotheim T.A., Elsenbaumer R. L., Reynolds J. R., Handbook of Conducting Polymers, CRC Press, 1998.

[8] Zhang J., Ye Z., Lou Y., Metabolism of chlorambucil by rat-liver microsomal glutathione S-tranferase, *Chemico-Biological Interactions* 149 (2004) 61-67.

# 137c Modeling Skin Permeation with Quantitative Structure-Property Relationships

**B. J. Neely**, S. Golla, E. a. Whitebay, S. V. Madihally, R. L. Robinson Jr., and K. a. M. Gasem. School of Chemical Engineering, Oklahoma State University, 423 Engineering North, Stillwater, OK 74078

The largest organ of the body is the skin, which serves as a protective layer against the presiding environment; however, skin can be breached by chemicals, which may have either a beneficial or deleterious impact on the body. Transdermal drug delivery (TDD) has been used in a beneficial manner to provide a therapeutic avenue for systemic treatment of a variety of maladies.

The basic premise of skin permeation is the penetration of a chemical through the stratum corneum (SC) dermal layer and absorption into systemic circulation. In general, the chemical has two potential routes of entry: (a) through hair follicles and sweat ducts, and (b) across the SC. Hair follicles and sweat ducts occupy only a small fraction of the total skin surface area and are believed to be insignificant factors in TDD. To permeate through the SC, the chemical must first partition into the skin and then diffuse through the hydrophilic and/or lipophilic environment of the SC to the dermis. The ability of a chemical to permeate through the skin is quantified by the value of the permeation coefficient, Kp, which is related to Fick's law of diffusion.

The ability to predict accurately the dermal permeation of a chemical is important for the development of new therapeutic formulations in transdermal drug delivery and assessment for potential risk of environmental chemicals. Most early modeling efforts utilized traditional and linear quantitative structure-property relationship (QSPR) models; however, since most thermo-physical properties have non-linear relationships with chemical structure, traditional linear algorithms result often in inferior QSPR models. New models accounting for all the structural features of importance to skin permeation, which avoid the pitfall of over-reliance on linear models, are required.

Our models for skin permeation, as expressed by the permeation coefficient, improve on other literature models in several aspects, including (a) the use of a larger dataset consisting of diverse chemical structures, (b) selection of a dataset consisting of both literature and statistically determined descriptors (c) use of multiple QSPR software to provide descriptors assuring both model superiority and stability, (d) use of nonlinear transformations to obtain the most suitable set of descriptors, and (e) use of robust non-linear neural networks with multiple randomizations and initializations to ensure network stability. Employing these state-of-the-art nonlinear algorithms, we have modeled accurately skin permeation with an absolute average deviation, root mean square error, and correlation coefficient of 8.2, 0.38, and 0.93, respectively. These results support our transdermal drug delivery research in the identification of candidate chemical penetration enhancers.

# 137d Modeling Local Pharmacokinetics in Drug-Eluting Stents: The Coupling Effects of Luminal Flow, Transmural Flow and Drug Binding

**Jahid Ferdous** and Chuh Khiun Chong. School of Chemical & Biomedical Engineering, Nanyang Technological University, 62 Nanyang Drive, Singapore, 637459, Singapore

Drug-eluting stent (DES) is a common approach to mitigate restenosis, a major drawback of stent-based treatment in coronary disease. However, the long-term effect of this mode of treatment is unknown. A detailed study of the drug delivery system is expected to give a better insight about what is required to improve the design of DES to achieve its long-term therapeutic effect. Physiological transport forces, drug physiochemical properties and drug interaction with the vascular tissue are the main factors which govern the local pharmacokinetics of DES. We hypothesized that these factors should be coupled in modeling drug delivery. In this study, a 2-dimensional model which drug transfer was coupled with both luminal and transmural flow and drug binding was used to investigate the drug release, deposition and distribution in the arterial wall. Two model drugs, paclitaxel (hydrophobic) and heparin (hydrophilic), were used to examine how the nature of the drug affects the pharmacokinetics. Moreover, drug interaction with the binding sites was treated as a reversible binding process which is a function of free and bound drug concentration in the arterial wall. Our results show that, for hydrophobic drug, the effect of coupling drug transfer with both luminal and transmural flow and drug binding is significant. On the other hand, for hydrophilic drug, transmural flow shows negligible effect on drug delivery. Incorporation of drug binding in the model increases markedly the retention time of drug in the arterial wall. It is therefore important to consider the coupling factors in modeling the local pharmacokinetics of DES. This study provides further insights which have not been offered in reported models based on either luminal or transmural flow, with or without drug binding.

## 137e Assessment of the Time Required for Controlled-Release Devices to Reach Steady State

**Laurent Simon**, Otto H. York Department of Chemical Engineering, New Jersey Institute of Technology, University Heights, Newark, NJ 07102

Assessing whether a desired release rate will be achieved in a certain period of time is a challenge that transdermal patch manufacturers must address. An engineering-based design methodology is suggested in this contribution to respond to patient need for a more accurate and timely release of active pharmaceutical ingredient(s) to the target site. New design equations, based on principles of transport phenomena, are derived to facilitate the rapid development of controlled-release devices. Concepts, such as an effective time constant, traditionally used to analyze dynamical systems, are invoked to estimate the time it takes to reach a steady-state flux in passive and physically-enhanced delivery systems. Three case studies were investigated to illustrate the methodology: benzocaine transport through ethylene-vinyl acetate membranes, iontophoretic delivery of amitriptyline HCl across human skin and heat-enhanced transport of desoxycorticosterone through silicone membranes. The predictions agree with drug-release profiles.

# 137f a Study of the Release of Doxorubicin in Porous Alumina

**Bradley R. Gordon**[1], Charles E. Luckett[2], Douglas S. English[2], Sheryl H. Ehrman[1], and Daniel D. Lim[2]. (1) Department of Chemical and Biomolecular Engineering, University of Maryland, College Park, MD 20742, (2) Department of Chemistry and Biochemistry, University of Maryland, College Park, MD 20742

A model is presented for the diffusion and release of Doxorubicin (Dox) and Sulforhodamine 101 (SR101) in porous alumina. The diffusion of Dox and SR101 were studied using fluorescence recovery after photobleaching (FRAP) at pH's 5.0 and 7.4 over the temperature range 273K to 298K. Data from these experiments were fitted to first-order kinetics to predict the effects of pH and temperature on the adsorption of Dox and SR101. It was observed that the diffusion of Dox is dependent upon the pH whereas the release is not. It was further found that the diffusion of SR101 is independent of pH and that the release is temperature dependent. These observations suggest that the pH induced charge interactions between the diffusing species and the alumina surface play a significant role in the uptake and release from porous alumina.

## 162a Controlling and Manipulating Supported Phospholipid Monolayers as Soft Resist Layers for Fabricating Chemically Micropatterned Surfaces

**Chang-Ying Xue** and Kun-Lin Yang. Department of Chemical and Biomolecular Engineering, National University of Singapore, 4 Engineering Dr 4, Singapore, 117576, Singapore

Chemically micropatterned surfaces have broad applications in many fields. In this presentation, we report a new method for preparing chemically micropatterned surfaces by controlling and manipulating supported phospholipid monolayers as soft resist layers with molecular-level precision. First, we introduce self-assembled supported phospholipid monolayers on solid surfaces and use microcontact lift-up processes to create micropatterned phospholipid monolayers (with micrometer resolution) on the surface. Next, the micropatterned phospholipid monolayers can function as "soft" resist layers to protect underlying solid substrates and create either positive or negative chemically micropatterned surfaces during subsequent treatments. Unlike traditional "hard" resist layers which can only be removed by using harsh chemical treatments, this novel soft resist layer only comprises a single layer of compact phospholipid; therefore, it can be easily removed by water rinsing after the preparation of micropatterns. This method is also versatile. It can be applied to prepare protein microarray or silver patterns on solid substrates.

## 162b Scalable Meniscus-Directed Fabrication of Assemblies from Microspheres and Nanoparticles on Hydrophobic and Superhydrophobic Surfaces

Vinayak Rastogi[1], Daniel M Kuncicky[1], Lindsey B Jerrim[1], and **Orlin D. Velev**[2]. (1) Chemical and Biomolecular Engineering, North Carolina State University, College of Engineering 1, Box 7905, 911 Partners Way, Raleigh, NC 27695, (2) Department of Chemical and Biomolecular Engineering, North Carolina State University, College of Engineering 1, Box 7905, 911 Partners Way, Raleigh, NC 27695

The controlled drying of liquid suspension droplets deposited on surfaces or confined in thin menisci is one of the simplest and most efficient ways to assemble particles and shape them in two and three dimensions. The organization of the particles is a result of a complex combination of convective transport and capillary forces at the liquid/air interface. We will demonstrate how the process can be engineered for controlled assembly of self-contained supraparticles in drying sessile suspension droplets on surfaces with intermediate or very high hydrophobicity. The dynamic shape of the liquid surface guiding the particle assembly changes as a function of the contact angle on the substrate, liquid interfacial tension, droplet volume, particle size and volume fraction. Depending on the combination of these parameters, the droplet drying process can result in the formation of particle crystals in the form of disks, hemispheres, spheres, "doughnuts", or ellipsoids. The supraparticle assemblies formed can also be made anisotropic, layered, magnetic or biologically active and in these roles they may find applications in various technologies. The method is very simple, inexpensive and scalable.

## 162c Centerline Placement and Alignment of Carbon Nanotubes In Cylindrical Droplets of Nanometer Diameter

**Richa Sharma**, Chemical Engineering, MIT, 66-565, 25 Ames St., Cambridge, MA 02139 and Michael S. Strano, 66-566 Department of Chemical Engineering, Massachusetts Institute of Technology, 77 Massachusetts Avenue, Cambridge, MA 02139.

The hydrodynamic flow patterns inside high aspect ratio evaporating cylindrical droplets have utility for parallel alignment, positioning and placement of nanoparticles. We successfully create ~200 nm diameter cylindrical droplets for the first time using alternating hydrophobic and hydrophilic SAM layers made by a micro-contacting printing technique. We find that as the diameter of the droplet is reduced below 1000 nm (950-175 nm), a reversal of the internal flow is evident with highly accurate parallel alignment and placement of single walled carbon nanotubes along the centerline with 95 % precision. In contrast deposition is along the edges for droplets larger than 3 um in diameter. This flow reversal is a direct consequence of the evaporation mode switching for the narrow droplets. A model is developed to describe the transition from radially outward to inward flow. This technique should be useful for precisely aligning and placing individual or bundled anisotropic nanoparticles (i.e. carbon nanotubes) with arbitrary surface chemistries.

References:

1. Sharma, R.; Lee, C. Y.; Choi, J. H.; Chen, K.; Strano, M. S. Nano Letters 2007, 7,2693-2700, "Nanometer Positioning, Parallel Alignment, and Placement of Single Anisotropic Nanoparticles Using Hydrodynamic Forces in Cylindrical Droplets".

2. Sharma, R.; Strano, M. S.;"Centerline Placement and Alignment of Anisotropic Nanotubes in High Aspect Ratio Cylindrical Droplets of Nanometer Diameter" (submitted).

## 162d Large-Scale Fabrication of Sub-100 Nm Periodic Nanostructures by Spin-Coating

Chih-Hung Sun, Wei-Lun Min, and **Peng Jiang**. Chemical Engineering, University of Florida, Gainesville, FL 32611-6005

Large-scale fabrication of periodic nanostructures with resolution beyond optical diffraction limitation is of great technological importance in developing next-generation electronic, optical, magnetic, and bio-analytical devices. Bottom-up self-assembly and templating nanofabrication provide a much simpler, faster, and inexpensive alternative to top-down nanolithography (e.g., electron-beam lithography, focused ion beam, and nanoimprint lithography) in creating periodic nanofeatures. Unfortunately, this bottom-up approach suffers from low throughput and incompatibility with mature microfabrication, limiting the mass-production and on-chip integration of practical devices. Here we report a spin-coating technique that combines the simplicity and cost benefits of bottom-up self-assembly with the scalability and compatibility of standard top-down nanofabrication. Spin-coating enables wafer-scale production of colloidal crystals consisting of sub-100 nm particles. A large variety of functional periodic nanostructures ranging from self-cleaning antireflection coatings and magnetic nanodots to reproducible surface-enhanced Raman scattering (SERS) substrates and plasmonic nanohole arrays have also been templated from the shear-aligned nanoparticle arrays. The optical and SERS properties of the templated nanosystems have been characterized by both experimental measurement and theoretical simulation.

Web Page: www.che.ufl.edu/jiang

## 162e Fabrication of Polymeric Nanochannel/microwell Arrays by DNA Combing and Imprinting

**Jingjiao Guan**, NSF Center for Affordable Nanoengineering of Polymeric Biomedical Devices, The Ohio State University, 1381 Kinnear Road, Suite 100, Columbus, OH 43212 and Ly James Lee, Chemical and Biomolecular Engineering, The Ohio State University, Columbus, OH 43210.

Nanoscale channels are of great interest for various applications and a functional nanochannel-based device also requires microscale components such as microchannels and microwells for sample loading. We report a novel non-cleanroom process to generate centimeter-sized and highly-ordered arrays of nanochannels/microwells by DNA combing and imprinting (DCI). In this method, DNA nanowires are formed first by molecular combing performed on an array of micropillars and are coated with metals by sputtering coating to produce metallic nanowires with controllable diameter. The metallic nanowires are then converted to polymeric nanochannels by imprinting using the nanowires as sacrificial templates. At the same time, the micropillars are converted to polymeric microwells by imprinting. Compared to other micro/nanofabrication methods, DCI is simple, inexpensive, and capable of generating large and well-defined nanochannel/microwell arrays with potential applications such as biomedical sensing and separation.

# 162f Attachment of Single Bacterium Cell, DNA and Single Molecules on Chemically Modified Graphene Sheets: Avenue to Build Smart Electrochemical Circuitry

Nihar Mohanty and **Vikas Berry**. Chemical Engineering, Kansas State University, 1011 Durland Hall, Kansas State University, Manhattan, KS 66506

Recently there has been a great interest in graphene nanostructures owing mainly to its unique electrical properties where conductivity exhibits low Johnston noise, low charge scattering and ballistic transport. Study of graphene's chemical modification, its structure and its integration with other biological/ chemical systems will open avenues to further broaden its impact to include areas like biophysics, solar cells, chemical circuitry etc. We are studying electronic and structural properties of graphene-biohybrids and chemically modified graphene (CMG) nanostructures. Specifically, we have built graphene-bacteria ensembles, DNA-tethered graphene-structures, polyelectrolyte adsorbed CMG nanostructures. We are interested in the effect of chemical modification of graphene on its structure, such as formation of wrinkles on surface and creation of folds etc. Further we are also investigating the effect of inherent wrinkles and folds on the modification of CMGs. The electrical properties of all the CMG-sheets are distinctly different from pristine graphene. We are studying the electrical properties of (a) CMG's hybrids with single bacterial cell attachment, (b) CMG with DNA (single strand and double strand) tethered and hybridized on surface and (c) CMG with polyelectrolyte-layer assembled on surface. These hybrids function as: (a) single bacterium devices, (b) DNA hybridization sensor and (c) charge polarity sensitive chemical-detector, respectively. We are also characterizing the system for (1) stress application on single bacterial cell on CMG, (2) single molecular attachment on CMG and (3) selective 'wrapping' of CMG-sheets on bacterium.

# Dispersion of Nanoclay in Linear Low-Density Polyethylene (LLDPE) and Maleated Linear Low-Density Polyethylene (LLDPE-g-MA) in Supercritical Carbon Dioxide

*Matt Factor and Sunggyu Lee*

*Department of Chemical & Biological Engineering*
*Missouri University of Science & Technology*
*Rolla, MO 65409-1230*

## Abstract

Addition of nanoclay to polymer is often desirable due to the ability of the nanoclay to improve the thermo-mechanical and barrier properties as well as flame retardancy of the polymer. The degree of property enhancement by the nanoclay is largely dependent on how well it is dispersed throughout the polymeric system. However, complete or nearly complete dispersion is difficult to achieve. Supercritical carbon dioxide is one technique that has the potential to achieve a high level of dispersion. Organo-nanoclay Cloisite 93A dispersion was attempted in linear low-density polyethylene (LLDPE) and linear low-density polyethylene grafted with maleic anhydride (LLDPE-g-MA) via supercritical carbon dioxide at various processing conditions. According to x-ray diffraction (XRD) analysis, both LLDPE and LLDPE-g-MA pellets experience nanoclay dispersion into the pellet, but these polymers have different run conditions that result in a better nanocomposite. This is most likely due to the grafted maleic anhydride that increases the compatibility between LLDPE and Cloisite 93A. Besides being a means of effective dispersion, it is conceivable that supercritical carbon dioxide may be used to exfoliate nanoclays which can further enhance the resultant nanocomposite properties.

## Introduction

The need is increasing for polymers to have their properties tailored for specified applications. One way to tailor the properties of a polymer would be to add fillers to it. A useful filler to enhance a polymer's mechanical, thermo-mechanical and barrier properties is nanoclay[1-5]. Nanoclays are environmentally friendly, in ample supply, and inexpensive[4,6]. Nanoclays have a high interfacial area when intercalated or exfoliated and thus the ability to greatly enhance a polymer's properties at low loadings[1,2,6]. The difficulty with nanoclays is completely exfoliating and uniformly dispersing them throughout the polymer matrix in order to achieve the maximum level of enhancement[1,7]. One method to increase the possibility of intercalation and exfoliation of nanoclays is to modify them to be compatible with the desired polymer matrix or processing medium. This article will focus on using modified nanoclays that are compatible with the processing medium.

Supercritical fluids pose a unique processing environment because they have a liquid-like density and a gas-like diffusivity, making them tunable fluids[1,8]. Carbon dioxide is an advantageous supercritical fluid and processing medium because it has a low toxicity, is highly abundant, inexpensive, non-combustible, and has easily achievable critical conditions ($T_C$ =

31.1°C, $P_C$ = 7.4MPa)[1,8,9]. Supercritical carbon dioxide ($scCO_2$) has been successfully used for a wide variety of applications such as extraction, purification, precision cleaning[9], depolymerization, polymer grafting[10] and producing polymer membranes[8]. In addition to these applications, supercritical carbon dioxide has also been used to infuse different materials into polymers[11] and to exfoliate nanoclays[1]. Some polymers tend to swell in supercritical carbon dioxide[11], which allows particles of materials that are soluble in this medium to more easily infuse into the swollen polymeric matrix. One such example of this is the infusion of sodium nitrite into a biodegradable polymer[11]. And when nanoclays are in a supercritical carbon dioxide medium, the carbon dioxide molecules can infuse between the intercalated clay platelets and break them apart into an exfoliated structure as they expand during depressurization[1]. Other techniques have been employed to aid in the delamination of clays within a polymer such as melt compounding[3,7], extrusion[2,7], solution casting, and *in-situ* polymerization[1,2,8].

This article is concerned with the infusion of modified organic nanoclays into linear low-density polyethylene (LLDPE) and linear low-density polyethylene grafted with maleic anhydride (LLDPE-g-MA). The graft copolymer of LLDPE-g-MA was chosen for this study because maleic anhydride modifications are reported to aid in the dispersion of clay in a polymeric matrix[3,7]. Organo-nanoclay Cloisite 93A was chosen for this experiment because it has a high affinity for supercritical carbon dioxide as compared to other nanoclays, allowing easier transport from this medium into the polymer.

Clays have been successfully exfoliated and powders have been infused into polymers using $scCO_2$. The purpose of this study is to determine if clay can be infused into LLDPE or LLDPE-g-MA using $scCO_2$. Later, it will be tested if the two processes of exfoliating the clay and then infusing it into LLDPE or LLDPE-g-MA can be combined into one step. Once infusion is determined to have occurred, exfoliation and uniform dispersion of the nanoclay will be attempted.

## Experimental

### *Materials*

Cloisite 93A was obtained from Southern Clay Products, Inc. A quaternary ammonium salt (methyl, dehydrogenated tallow ammonium) modifies natural montmorillonite to create Cloisite 93A. The modifier concentration is 90meq/100g of clay. Ninety percent of the particles by volume are less than 13μm, 50% are less than 6μm, and 10% are less than 2μm. And the density of the clay is 1.88g/cc. All clay is dried at a temperature around 80°C in an oven for at least 24 hours before processing.

The polymers used in this experiment were linear low-density polyethylene (LLDPE, Dowlex 2517) and linear low-density polyethylene grafted with maleic anhydride (LLDPE-g-MA, Polybond® 3109). The LLDPE is in pellet form and has a melting point of 124°C, a specific gravity of 0.917 and a melt flow index of 25g/10min. The LLDPE-g-MA is in the also in the form of pellets with a density of 0.926g/cc and a melt flow index of 30g/10min. The grafting level of maleic anhydride is 1wt% in LLDPE-g-MA and the melting point is 123°C.

***Processing Procedure***

The polymer samples and Cloisite 93A are placed in a stainless steel thimble and mechanically mixed until the polymer is thoroughly coated with the clay. Ten grams of the polymer and 2.5g of Cloisite 93A were used, giving a respective weight ratio of 4 to 1. The thimble is then put inside a stainless steel reactor with an impeller that is turned using a MagneDrive®. Cooling water is sent through the head of the reactor to help control the temperature. The impeller is started during heating to help keep the environment a uniform temperature. The impeller is kept at around 300rev/min. Once the desired temperature and pressure are attained, the run time is started. All runs have a run time of 3 hours and are carried out in a batch mode. After the run is complete, the reactor is depressurized. The processing conditions are displayed in Table 1.

**Table 1.** Processing Conditions in $scCO_2$ for Polymer-Clay Nanocomposites

| Run | T (°C) | P (MPa) | Wt. Ratio (Clay:Polymer) |
|---|---|---|---|
| 1 | 60.0 | 10.3 | 0.25 : 1 |
| 2 | 98.9 | 10.3 | 0.25 : 1 |
| 3 | 60.0 | 17.2 | 0.25 : 1 |
| 4 | 98.9 | 17.2 | 0.25 : 1 |
| 5* | 98.9 | 17.2 | 0 : 1 |

* LLDPE-g-MA only

***Characterization***

The infused polymer pellets are formed into coins that are 1mm thick so they can be analyzed by x-ray diffraction (XRD). The coins are made by heating them on a metal plate at 195°C until they are completely melted. Then a heated metal plate is placed on top of this lower plate with spacers in between them. This top plate is held in place until the polymer is hardened and the coin can be removed. All XRD analysis was done with a Philips X-Pert diffractometer with a Cu $K\alpha_1$ radiation source having a wavelength of 0.154056nm. The generator voltage was 45kV and the tube current was 40mA. The XRD data angle range was from 1° to 10° with an increment of $2\theta$.

Differential scanning calorimetry (DSC) analysis was conducted on LLDPE-g-MA nanocomposite samples that weighed between 5 and 10mg. The instrument is a TA DSC 2010 that performs its analysis on the sample in a nitrogen environment. The nanocomposites are heated from room temperature to 190°C at a rate of 10°C/min. This temperature is held for three minutes and then the samples are cooled at the same rate to about 35°C. The same heating/cooling processes are then repeated again and the data is collected. The first heating/cooling is to remove any thermal history within the sample. The second heating/cooling is to obtain the melting and crystallization temperatures. The pressure is at typical atmospheric conditions.

## Results and Discussion

Cloisite 93A was processed with LLDPE in supercritical carbon dioxide. Four different run conditions were investigated in order to see if infusion occurred at any of these conditions and which conditions were best for the possible infusion. XRD analysis was conducted on the polymer pellets to determine if infusion had occurred. Cloisite 93A creates a peak on the XRD plot at a Bragg angle around 3.4° which corresponds to a $d_{001}$-spacing of 2.59nm [3]. All runs display similar peaks around the same angle, indicating that infusion of Cloisite 93A occurred[1,2,3,12]. The sample LLDPE-3 has the lowest almost non-existent peak, which could indicate either the clay is well dispersed or almost no clay is in the polymer[1,2,3,6], most likely the latter. Runs 1, 2 and 4 have clear peaks from the Cloisite 93A with $d_{001}$-spacings respectively of 2.73, 2.81, and 2.73nm. The increase in $d_{001}$-spacings as compared to pure Cloisite 93A could be due to the $scCO_2$ and/or polymer diffusing in between the platelets and pushing them apart, making them closer to an exfoliated structure[1,2,3,6,7]. Sample LLDPE-2 with a low pressure and a high temperature had the largest increase in spacing between platelets. This sample experiences an increased platelet spacing possibly since the $CO_2$ molecules and polymer chains are more mobile in a low pressure environment, making it easier for them to work their way between the platelets. And the high temperature helps the infusion of clay by softening the polymer. According to Horsch et al.[1], larger intensities in XRD peaks can indicate an increase in tactoid size and parallel registry of the clay. The run with the highest conditions (LLDPE-4) had the largest peak intensity, possibly indicating that this environment could create larger clay tactoids within the polymer. The two low-pressure runs with similar peaks had a lower intensity, which might indicate that lower pressures create smaller tactoids and more exfoliated clay structures within the polymer. Figure 1 contains the XRD plot of LLDPE-Cloisite 93A nanocomposites.

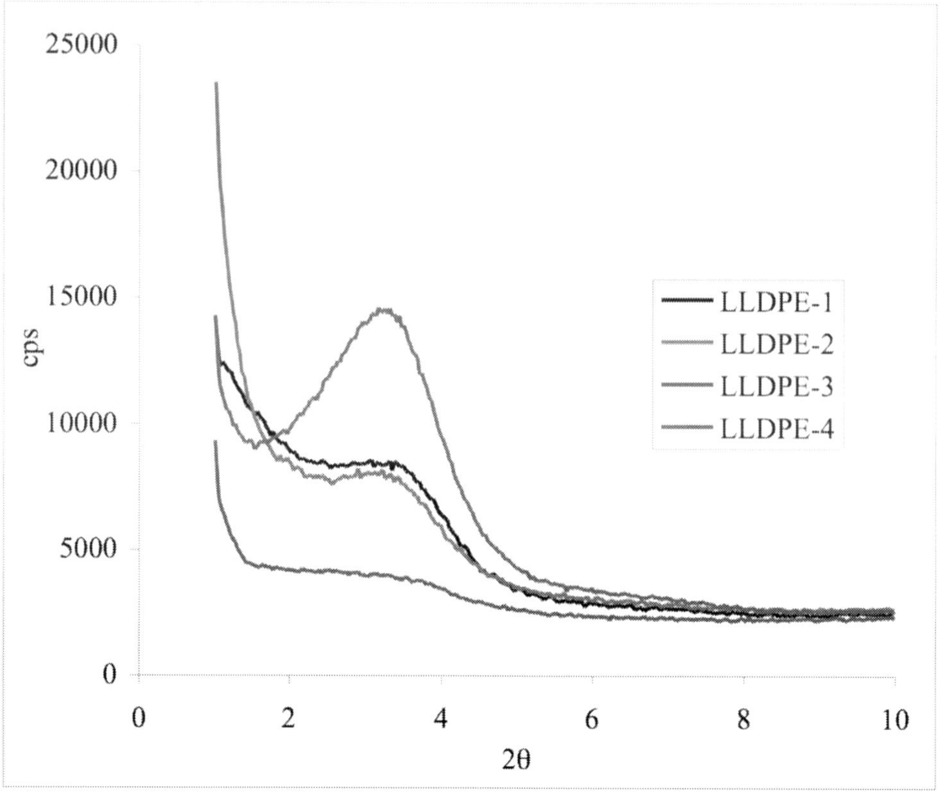


**Figure 1.** XRD Plot of LLDPE Processed with Cloisite 93A at Various Conditions (Pure Cloisite 93A Plot Left Off so Nanocomposite Peaks Could be Seen)

LLDPE-g-MA pellets were processed with Cloisite 93A in $scCO_2$ at various temperatures and pressures. The resultant polymer was then analyzed using DSC and XRD. Differential scanning calorimetry analysis revealed that the melting temperature of the LLDPE-g-MA after $scCO_2$ processing with Cloisite 93A was altered by a maximum of 0.46°C (LLDPE-g-MA-1) and that the crystallization temperature changed by a maximum of 0.80°C (LLDPE-g-MA-2). The changes seen from the addition of the nanoclay to the LLDPE-g-MA are not very significant, leading one to infer that the processing conditions of the nanocomposite should not change much from those for pure LLDPE-g-MA. Melting and crystallization temperatures for the nanocomposites are summarized in Table 2.

**Table 2.** Melting Point ($T_{mp}$) and Crystallization Point ($T_{cp}$) of LLDPE-g-MA-Cloisite 93A Nanocomposites Measured by DSC Analysis

| Sample | $T_{mp}$ (°C) | $T_{cp}$ (°C) |
|---|---|---|
| Pure LLDPE-g-MA | 122.58 | 108.67 |
| LLDPE-g-MA-1 | 122.12 | 108.83 |
| LLDPE-g-MA-2 | 122.56 | 109.47 |
| LLDPE-g-MA-3 | 122.47 | 108.93 |
| LLDPE-g-MA-4 | 122.66 | 109.08 |
| LLDPE-g-MA-5 | 122.60 | 108.85 |

From the XRD analysis, it is apparent that 93A was successfully infused into the LLDPE-g-MA for run conditions 1, 2 and 4 since the 93A peak is present[1,2,3,12] (Figure 2). Runs 3 and 5 exhibit slight peaks in the XRD plot. But at this time it is unknown what the peak for run 5 is from, especially since run 5 was processed without 93A. The peak could be from the breakdown of maleic anhydride or some residual clay in the system could have found its way into the LLDPE-g-MA in run 5. Concerning the peak found in LDPEMA-3, a possible scenario could be that the clay is almost completely dispersed, giving little to no peak in the plot[1,2,3,6]. The $d_{001}$-spacings for the clay peaks found in runs 1, 2, and 4 are respectively, +0.24nm, +0.15nm, and -0.10nm in relation to that of 2.44nm for 93A[3]. These differences in distances may not be very large, but may still be significant, as a trend seems to exist. The $d_{001}$-spacings indicate that in runs 1 and 2, where the pressure was low, the spacing between clay platelets increased due to processing. In run 1, the lowest temperature and pressure environment, the largest increase in spacing between clay platelets occurred. Run 2 might not have realized such a large increase in spacing because the high temperature could have degraded the clay modifier[1], working against the $scCO_2$ and polymer getting between the platelets to expand the gap. And at a higher pressure seen in run 4, the spacing between clay platelets decreased. This decrease in spacing might in fact be due more to the high processing temperature also used in this run that could have caused the clay to lose some of its modifier[1]. Also, the higher pressure environment could have made the polymer chains less mobile so fewer were able to intercalate with the nanoclay particles. The increase in distance between platelets is favored over a decrease because it can indicate that polymer has found its way between the platelets, bringing the clay closer to an exfoliated structure[1,2,3,6,7].

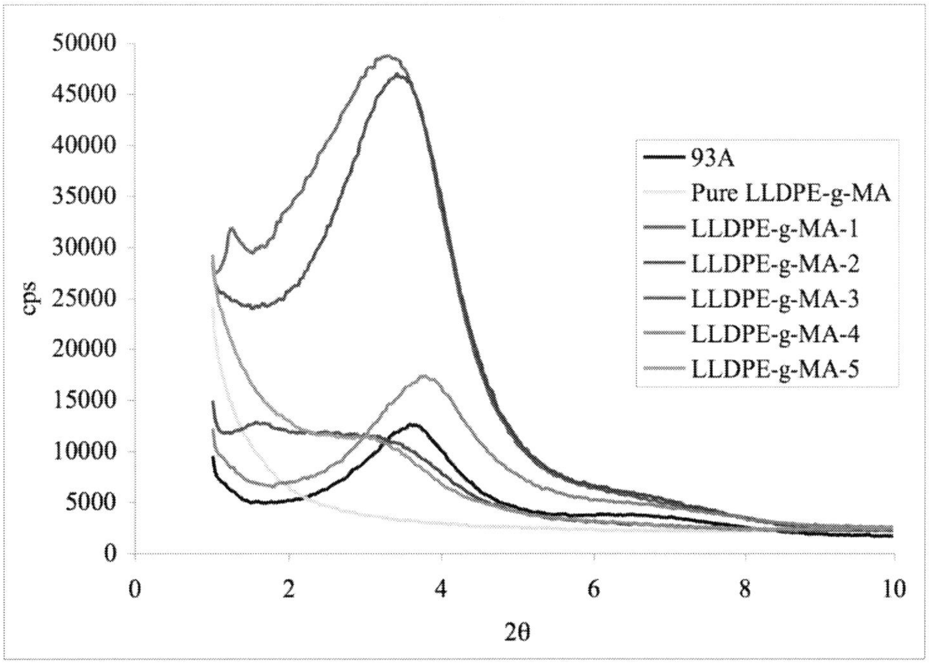


**Figure 2.** XRD Plot of LLDPE-g-MA Processed with Cloisite 93A at Various Conditions

## Conclusion

Indicated by XRD analysis, Cloisite 93A was successfully infused into both LLDPE and LLDPE-g-MA using supercritical carbon dioxide. Different peak relations are seen between LLDPE and LLDPE-g-MA because the grafted maleic anhydride alters the transport of the clay particles. The high temperature and low pressure run for LLDPE had the largest increase in spacing between clay platelets, indicating that these conditions for LLDPE are better when trying to achieve an intercalated or exfoliated clay structure. For LLDPE-g-MA nanocomposites, the melting and crystallization temperatures remained near to those for pure LLDPE-g-MA, revealing that the processing conditions of the nanocomposites should be similar as those for pure LLDPE-g-MA. LLDPE-g-MA experienced its best infusion conditions on the low temperature and pressure run, since the clay had the largest separation between platelets, bringing the process closer to clay intercalation or exfoliation. It is proposed that the maleic anhydride in LLDPE-g-MA allows the clay to more easily transport into the polymer at a lower temperature than that for LLDPE, which is why the largest separation between clay platelets was seen in a low temperature run for LLDPE-g-MA as opposed to a high temperature run for LLDPE. And a lower temperature would degrade less of the clay modifier, giving LLDPE-g-MA an advantage over LLDPE when using modified clays. In the future, FT-IR analysis will be used to determine the percentage of clay infused and TEM analysis will be used to help determine the degree of exfoliation of clay.

***References***


1. Horsch, S.; Serhatkulu, G.; Gulari, E.; Kannan, R. M. Supercritical $CO_2$ Dispersion of Nano-Clays and Clay/Polymer Nanocomposites. *Polymer* **2006**, 47, 7485-7496.
2. Morgan, A. B.; Gilman, J. W. Characterization of Polymer-Layered Silicate (Clay) Nanocomposites by Transmission Electron Microscopy and X-Ray Diffraction: A Comparative Study. *Journal of Applied Polymer Science* **2003**, 87, 1329-1338.
3. Wang, K. H.; Choi, M. H.; Koo, C. M.; Choi, Y. S.; Chung, I. J. Synthesis and Characterization of Maleated Polyethylene/Clay Nanocomposites. *Polymer* **2001**, 42, 9819-9826.
4. Ma, X.; Yu, J.; Wang, N. Production of Thermoplastic Starch/MMT-Sorbitol Nanocomposites by Dual-Melt Extrusion Processing. *Macromolecular Materials and Engineering* **2007**, 292, 723-728.
5. Luyt, A. S.; Geethamma, V. G. Effect of Oxidized Paraffin Wax on the Thermal and Mechanical Properties of Linear Low-Density Polyethylene-Layered Silicate Nanocomposites. *Polymer Testing* **2007**, 26, 461-470.
6. Chen, B. Polymer-Clay Nanocomposites: An Overview with Emphasis on Interaction Mechanisms. *British Ceramic Transactions* **2004**, 103, 6, 241-249.
7. Ryu, S. H.; Chang, Y. W. Factors Affecting the Dispersion of Montmorillonite in LLDPE Nanocomposite. *Polymer Bulletin* **2005**, 55, 385-392.
8. Xu, Q.; Pang, M.; Li, J.; Jiang, Y. Application of Supercritical Carbon Dioxide in the Preparation of Biodegradable Polylactide Membranes. *Journal of Applied Polymer Science* **2004**, 94, 2158-2163.
9. Shieh, Y. T.; Su, J. H.; Manivannan, G.; Lee, P. H. C.; Sawan, S. P.; Spall, W. D. Interaction of Supercritical Carbon Dioxide with Polymers. I. Crystalline Polymers. *Journal of Applied Polymer Science* **1996**, 59,695-705.
10. Clark, K. Synthesis of Maleated Poly(Vinylidene Fluoride) in Supercritical Carbon Dioxide Medium. Ph. D. Dissertation, University of Missouri – Columbia, Columbia, MO, 2004.
11. Leavitt, L. The Development of VCI-Infused Biodegradable Polymeric Resin Using Supercritical Carbon Dioxide. Personal Communication, University of Missouri – Columbia, Columbia, MO, 2006.
12. Park, C. I.; Park, O. O.; Lim, J. G.; Kim, H. J. The Fabrication of Syndiotactic Polystyrene/Organophilic Clay Nanocomposites and Their Properties. *Polymer* **2001**, 42, 7465-7475.

# 164b Biodistribution of Dendrimer-Based Nanodevices for Brain Delivery of Therapeutics

**Rangaramanujam M. Kannan**[1], Hui Dai[2], Sujatha Kannan[2], and Roberto Romero[3]. (1) Chemical Engineering, NICHD PRB, Wayne State University, 5050 Anthony Wayne Drive, Detroit, MI 48202, (2) Pediatrics, NICHD PRB, Wayne State University, 5050 Anthony Wayne Drive, Detroit, MI 48202, (3) National Institute of Child Health and Human Development, National Institutes of Health, Bethesda, MD 20892

Aim:

Central nervous system (CNS) infections are formidable diseases with high rates of morbidity and mortality. Since the majority of antimicrobial agents discovered so far do not cross the blood-brain barrier (BBB), the treatment of CNS infections is a major challenge issue. Delivering therapeutic agents in a targeted manner is highly desirable, but tough to achieve. The advent of dendritic polymers offers a new length scale (~3-10 nm) in drug delivery, with significant versatility to incorporate multiple active molecules such as targeting moieties and imaging agents.

In this study, we examine the feasibility of achieving brain-targeted drug delivery by dendrimer (G4 hydroxyl terminated), which were labeled with fluoroisothiocyanate (FITC) through subdural microinjection, intravenous, or intranasal infusions in a well-established rabbit model of cerebral palsy..

Methods:

Pregnant New Zealand White rabbits underwent laparotomy and were injected with 20 Escherichia coli lipopolysaccharide along the length of the uterus on day 28 of gestation. The pups were born spontaneously at term (31 days). The Biodistrbution of dendrimer-FITC or free-FITC in pups brain was compared with that endotoxic group(LPS 20ug/kg) or control group(PBS). 5ug Dendrimer G4-OH-FITC in 10 microl of PBS were infused into CSF through subdural microinjection (10ul/min). Intravenous injection and intranasal infusions were performed using 500ug Dendrimer G4-OH-FITC in 100ul PBS. The same amount Free-FITC in Dendrimer G4-OH-FITC was performed by subdural microinjection, intravenous and intranasal infusions, as a control. At 6h or 24h, two pups from each group were sacrificed and brain was fixed, frozen section, and the brain section of pups were immunostained with tomato-lectin and GFAP, and observed by fluorescent microscope and confocal microscope.

Results:

Upon subdural microinjection in the endotoxin group, the dendrimer-FITC was found to be localized in actived microglial & astrocytes cells, around the lateral ventricle, corpus callosum, hippocampus and internal capsule. In the control group, which normally don't have activated microglia or astrocytes in these regions, did not show appreciable uptake. For comparable doses, the dendrimer-FITC intensity in the endotoxin group was significantly higher than those in the control group. Upon intravenous injection, the dendrimer-FITC were found to be localized in internal ventricle, hippocampus and commissura fornicis region. Dendrimer-FITC is taken up specifically by microglial cells on intravenous injection only in endotoxin group. As a control free-FITC uptake was studied. Free-FITC uptake appears to be non-specific, showing measurable fluorescence only on subdural microinjection. No Free-FITC uptake followed by intravenous injection in both endotoxic group and control group. For intranasal infusions, no Dendrimer-FITC and Free-FITC uptake was seen in both endotoxin group and control group.

Conclusion:

Our findings indicate that dendrimer nanodevices can be used to specifically target activated microglial cells and astrocytes in neuroinflammatory disorders. It suggests that brain-targeted drug delivery can be achieved by infusion of dendrimer nanodevices through subdural microinjection and intravenous infusion.

## 164c Nanoscale Reversible Loading and Unloading In Layer-by-Layer "Exponential" Polymer Films

**Sudhanshu Srivastava**[1], Vincent Ball[2], Paul Podsiadlo[1], Peter Ho[1], and Nicholas Kotov[1]. (1) Chemical Engineering, University of Michigan, 2300 hayward street, Ann Arbor, MI 48109, (2) Institut National de la Santé et de la Recherche Médicale

Layer-by-layer (LBL) films have been shown in most of the studies for loading of active molecules only during the preparation by using the species of interest as active constituents in the film build up. The significance of finding a unique approach to reversibly load-and-unload nanoparticles (NPs) in an already prepared LBL films is four-fold. (a) It would enable fast and universal preparation of NP-based coatings with a variety of functionalities. (b) NPs mobility inside the polymeric matrix would allow new methods for self-assembly processes. (c) Dynamic exchange process is essential in the advancement of fine separation tools for NPs. (d) Last but not the least, to create important opportunities for biomedical applications using organic/inorganic nanocolloids, proteins, DNA, RNA, etc. in controlled-release devices. In this study we demonstrate that exponentially-growing LBL films have the ability to load/ unload semiconductor NPs or other nanoscale species in a controlled fashion. To demonstrate this we used exponentially growing LBL films made from poly(diallyldimethylammonium chloride) (PDDA) and polyacrylic acid (PAA). As prepared (PDDA/PAA)n films, were exposed to a CdTe suspension (~ 4 nm). Release of the incorporated CdTe and reproducibility was demonstrated by immersing the CdTe-loaded films in pure water at pH = 9. The variations of the structure of the LBL layers (capping/topping layer), pH of water and the charge on NPs can be used to control the incorporation and release of the NPs making possible sensing, protective, biological, and optoelectronic functionalities. The similar procedure was also demonstrated for the loading of SWNT and nanorods.

## 164d Glassy Worms: From Rheology to Solvent Tuned Flexibility and Nano-Molded Rods

Kandaswamy Vijayan[1], Andre E. Brown, **Karthikan Rajagopal**[2], and Dennis E. Discher[2]. (1) Physics and Astronomy Graduate Group, University of Pennsylvania, Philadelphia, PA 19104, (2) Chemical and Biomolecular Engineering, University of Pennsylvania, Philadelphia, PA 19104

Amphiphilic diblock copolymers are robust building blocks for constructing functional supramolecular structures such as worm-like and spherical micelles and bilayers (polymersomes). One of the advantages of using polymeric blocks is that their chemical nature can be used to control attributes such as the flexibility of the structures formed. Herein, we report on the flexibility of worm-like micelles prepared from polyethyleneoxide-b-polystyrene (PEO-b-PS) diblock copolymers in which polystyrene forms the hydrophobic core and polyethyleneoxide forms the hydrophilic shell. In its melt state polystyrene is a glassy polymer (tg = 100oC) and we find that this glassiness is preserved even when confined to the nanoscale dimension of the worm micelle core. Estimation of the apparent persistence length of worm micelles formed from two PEO-b-PS diblock copolymers that differ in their molecular weights indicates that it is far less flexible than previously characterized worm micelles with fluid cores. The glassiness of the core is also evident by lack of fluorescence recovery after photo bleaching. Interestingly, the rigidity of the worm micelle is interspersed with flexible 'hinges' due to defects, the periodicity of which is dependent on the molecular weight of polystyrene block. In addition, the hinge flexibility exhibits temperature dependence up to 80oC which is indicative of increased thermal motion around the hinges rather than any obvious melting of polystyrene. Addition of block specific solvents such as xylene or chloroform homogeneously fluidizes the hydrophobic core and imparts flexibility to the worm-like micelle along its contour. This allows for 'molding' the worm micelles in flow and thus enabling the creation of high aspect ratio nanoscale rod-like structures with tunable flexibility.

## 164e Molecular Dynamics In Amorphous Polymers and Self-Assembling Molecular Glasses

Daniel B. Knorr Jr., **Jason P. Killgore**, and Rene M. Overney. Chemical Engineering, University of Washington, Box 351750, Seattle, WA 98195

A major challenge in molecular engineering is to understand and control inter- and intra-molecular dynamics. Often, such systems are designed by a trial and error process due to the lack of quantitative information relating to molecular interactions occurring in the desired device configuration (e.g., thin films). Here, we provide a framework to obtain such information based on nanoscale experimental investigations of molecular mobility in two systems of interest: (1) a model polymer system, monodisperse, atactic polystyrene and (2) a series of molecular glasses that form self-assembled 'physically-linked' polymers by aromatic-pefluorophenyl dispersion interactions.

We show that, for the model system, critical relaxation temperatures and apparent Arrhenius-type activation energies can be deduced via a variety of scanning probe-based techniques including shear modulation force microscopy (SM-FM) and intrinsic friction analysis (IFA). Furthermore, we show that, with regard to IFA data, direct separation of cooperative (entropic) contributions to the apparent Arrhenius activation energy from non-cooperative (enthalpic) contributions is possible by employing a combination of absolute rate theory and the Erying model. As such, the degree of cooperativity for various molecular relaxations in the system is readily apparent.

The second system, a series of three self-assembling molecular glasses with phenyl-perfluorophenyl, naphthyl-perfluorophenyl and anthryl-perfluorophenyl moieties, was also experimentally investigated on the nanoscale using SM-FM and IFA. In these systems, we were able to determine specific temperatures regimes of varying cooperativity. Apparent Arrhenius activation energies and the degree of cooperativity were determined for these regimes, providing fundamental insight into the underlying cooperative molecular relaxation phenomena.

## 164f Reorientational Dynamics of Pegylated Dendrimers

**Sanja Ristic** and Jovan Mijovic. Othmer-Jacobs Department of Chemical and Biological Engineering, Polytechnic University, 6 Metrotech Center, Brooklyn, NY 11201

Dendrimers are an exciting and promising novel class of macromolecular architecture due to their well defined, highly branched, compartmentalized structure in the nanometer size range. The globular morphology of these polymers enables them to encapsulate important molecules such as therapeutic drugs within the interstitial space of their branches for subsequent targeted delivery. Attaching biocompatible PEG on the dendrimer surface increase their water solubility, drug loading capacity and the rate of drug release. To understand and assess these capabilities and to usher the way for the optimization of highly specific tasks in complex environments, such as *in vivo* targeting applications, it is essential that we acquire a comprehensive knowledge of the dynamics of PEGylated dendrimers on the molecular level.

In this work, an investigation was carried out on the dynamics of systems composed of generation 3 of poly(amidoamine) - PAMAM dendrimers in blends with poly(ethelene glycol) – PEG with molecular weight of 2000g/mol, as well as the dynamics of dendrimers with linear PEG chains covalently attached to the surface groups. The results were generated over a broad range of temperature and frequency by Dielectric Relaxation Spectroscopy (DRS) and Dynamic Mechanical Spectroscopy (DMS). The principal objective is to elucidate how dynamics vary as a function of temperature, concentration of dendrimers in blends and the number of covalently attached PEG chains in the PEGylated dendrimers.

The initial DRS results of PEG-dendrimer blends reveal a shift of the $\beta$ process in dendrimers, assigned to the local fluctuations of branch ends that include amino groups, to the lower frequency with increasing concentration of PEG. In PEGylated dendrimers, no effect of concentration on the average relaxation times was observed, but the dielectric relaxation strength was found to increase with increasing number of covalently attached PEG molecules. A complete quantitative picture of the dynamics of PEG-dendrimer blends and PEGylated dendrimers will be presented.

## 166a Nano-Sized Bioactive Glass In a Biodegradable Polymer: How Advantageous Is Nano-Size?

**Dirk Mohn**[1], Tobias J. Brunner[1], Superb K. Misra[2], Aldo R. Boccaccini[2], and Wendelin J. Stark[3]. (1) Department of Chemistry and Applied Biosciences, ETH Zurich, Wolfgang-Pauli-Strasse 10, Zurich, 8093, Switzerland, (2) Department of Materials, Imperial College London, London, SW7 2BP, United Kingdom, (3) Chemistry and Applied Biosciences, ETH Zurich, Wolfgang-Pauli-Str. 10, ETH Hönggerberg, Zurich, 8093, Switzerland

Composites of biodegradable polymers and bioactive ceramics in the form of biocompatible scaffolds are widely used in the field of tissue engineering with emphasis towards hard tissue regeneration [1, 2]. Bioactive glasses can add bioactive properties to a polymer matrix. In order to investigate the effect of particle size of the fillers we have introduced micron- or nano-sized bioactive glass (NBG) particles [3] in different concentrations within a poly(3-hydroxybutyrate) (P(3HB)) matrix. Mechanical and thermal properties, topography, protein adsorption, *in vitro* degradation behaviour, and cell proliferation [4] were compared.

The addition of NBG particles induced a nano-structured topography (Figure) on the surface of the composites not visible on micron-sized particles containing composites. This surface effect observed for NBG composites considerably increased the protein adsorption and had a reinforcement effect on the composite (Figure). Immersion in SBF revealed a high level of *in vitro* bioactivity for P(3HB)/NBG composites. Proliferation of MG-63 osteoblast-like cells on the various composites demonstrated a good cytocompatibility of all composite materials.

This study revealed that such nanoparticles are a most interesting bioactive filler material for biodegradable polymers in order to prepare advanced composites for tissue engineering.

Figure: Scanning electron microscopy images of a planar section of P(3HB) with 30 wt% nano-sized bioactive glass particles (left). Modulus comparison for various concentrations of micron- and nano-sized bioactive glass particles in P(3HB) composites. $^{**}p < 0.01$

References

[1] K. Rezwan, Q.Z. Chen, J.J. Blaker and A.R. Boccaccini, Biomaterials, 2006, 27, 3413-31.

[2] S. Loher, V. Reboul, T.J. Brunner, M. Simonet, C. Dora, P. Neuenschwander and W.J. Stark, Nanotechnology, 2006, 17, 2054-61.

[3] T.J. Brunner, R.N. Grass and W.J. Stark, Chem. Commun., 2006, 13, 1384-6.

[4] S.K. Misra, D. Mohn, T.J. Brunner, W.J. Stark, S.E. Philip, I. Roy, V. Salih, J.C. Knowles and A.R. Boccaccini, Biomaterials, 2008, 29, 1750-61.

10 µm

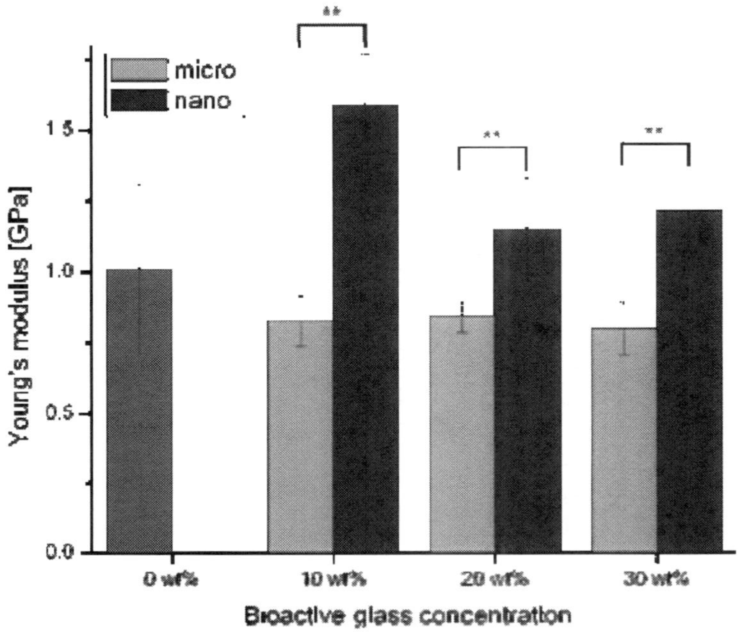
micro
nano
**
**
**
Young's modulus [GPa]
0.0
0.5
1.0
1.5
0 wt%
10 wt%
20 wt%
30 wt%
Bioactive glass concentration

## 166b Nanorods Modulate Cell Adhesion and Survival

Jiyeon Lee[1], Byung Hwan Chu[1], Fan Ren[1], Benjamin Keselowsky[2], and **Tanmay Lele**[1]. (1) Chemical Engineering, University of Florida, Museum Road Bldg 723, Gainesville, FL 32608, (2) Biomedical Engineering, University of Florida, Museum Road Bldg 723, Gainesville, FL 32608

The ability to control the behavior of cells interacting with materials is desirable for the success of implanted devices such as biosensors or drug delivery devices. In this study, we investigated the use of ZnO nanorods for modulating the adhesion and viability of NIH 3T3 fibroblasts, umbilical vein endothelial cells and capillary endothelial cells. Cells adhered far less to ZnO nanorods than the ZnO flat substrate. The few cells that adhered on the nanorods were not viable and rounded. Cells were unable to assemble focal adhesions and stress fibers on nanorods. Time-lapse imaging revealed that cells that initially adhered to nanorods were unable to spread. This suggests that it is the lack of initial spreading, rather than the long-term exposure to ZnO that causes cell death. We conclude that ZnO nanorods are potentially useful as an adhesion-resistant biomaterial capable of reducing viability in anchorage-dependent cells.

# Assessment of Cell-Material Interactions on 3D Nanostructured Titania-Polymer Surfaces Toward the Improvement of Osseointegration of Orthopedic and Dental Implants

*Jason Howard, Shraddha Patel, Jeffery Fontenot, Stephen McMullen, Austin Thompson, Jillann Walker, David K. Mills, and Scott A. Gold*

*Institute for Micromanufacturing, Louisiana Tech University, 911 Hergot Ave., Ruston, LA, 71272-0043*

## Introduction

As the life expectancy of the general population continues to rise, so has the number of people experiencing varying types of physical decline. The number of individuals experiencing degeneration or loss of musculoskeletal function is increasing among the older adults in society. This increase has facilitated a need for a means of managing such a loss of functioning. One area of current focus is in the incorporation of bone regenerative properties into implantable biomaterials.

Titanium is widely used for orthopedic and dental implants for its good biocompatibility, corrosion resistance, and excellent mechanical properties. Titanium, however, is bioinert, which means that upon implantation into the body it cannot directly bond to the bone. The process of implant integration with the bone following implantation involves the encapsulation of the titanium implant by a fibrous tissue that isolates the implant from the surrounding bone [1]. One way to overcome this problem has been through various types of surface modifications of the titanium to promote osseointegration (or attachment to the bone) of the implant. It is well known that the adhesion of cells to the surface of a biomaterial is vitally important for *in vivo* integration [2].

The current struggle lies in the ability of titanium to attach to the bone *in situ* and the quality of said attachment at its insertion. The current average functional lifetime of bone implants range from 10 to 15 years [3]. In the US over 3 million musculoskeletal procedures are conducted each year with the majority being fracture repair and total joint replacements [4]. While many of the procedures result in success, implant loosening and failure still remain significant problems. One of the established causes of these implant failures is the result of implant degradation and osteolysis, which is bone tissue death [5]. As the bone tissue becomes less viable at the biomaterial interface, the implant becomes less fixed to the bone, eventually requiring surgery to correct. Recent studies have reported that 25% of hip replacement surgeries were corrections to previously failed implants [6], while up to 29% of patients that received hip implants died in the year post-surgery [7]. In order to overcome this problem, bone implant materials must be designed to activate rapid bone growth to accommodate any implant loosening or degradation to ensure a secure attachment to the bone. To achieve this goal, the surface of the implant material needs to be able to attract osteoblasts and provide a viable environment for osteogenesis of the cells. In order to overcome this problem, bone implant materials must be designed to activate rapid bone growth to accommodate any implant loosening or degradation to ensure a secure attachment to the bone. To achieve this goal, the surface of the implant material needs to be able to attract osteoblasts and provide a viable environment for osteogenesis of the cells. Thus, there is a growing interest in modifying the surface at the biomaterial interface to allow the growth of healthy osteoblasts and to promote their long-term survival, such that the revision surgeries can be circumvented.

In this current work, we utilize porous titania, formed by electrochemical anodization of titanium as a template for the fabrication of polymer nanostructures on a surface. Osteoblasts are then grown on this modified surface and their viability compared to that on a flat and unmodified porous titanium surfaces using MTT and PicoGreen assays. This surface modification approach shows much promise for improving osseointegration of titanium implants.

## Methods

### *Anodization*

Anodization is a widely used surface modification technique for many metals. By varying the current density and the anodization time, different pore diameters and depths can be fabricated [8]. In this study, this surface modification process was performed using a hydrofluoric acid (HF) solution to produce nanoporous titanium (shown in Figure 1), which then served as the template for the polymer wetting resulting in a polymeric nanostructured surface for cell seeding experiments. All anodization experiments were performed using an electrolyte consisting of 1 wt.% HF + 50 vol% methanol. The current density was varied from 10 to 30 mA cm$^{-2}$ for an anodization time of 30 minutes.

### *Template Wetting*

Template wetting nanofabrication process utilizes wetting phenomena to create uniform coatings of low surface energy materials on high surface energy porous surfaces. This technique has been successfully used to produce polymer nanotubes (Figure 1) and nanowires for several polymers, including PMMA, and PVDF [9]. The polymer nanotubes were fabricated using the template wetting process on the anodized titanium templates by wetting with the polymer poly(vinylidenefluoride) (PVDF) solution. PVDF is biocompatible, and thus has been used widely as a surface modification agent of biomaterials. Following evaporation of the solvent, reactive ion etching (RIE) was used to remove any residual polymer solvent. The titania was chemically etched to exposed the polymer nanotubes.

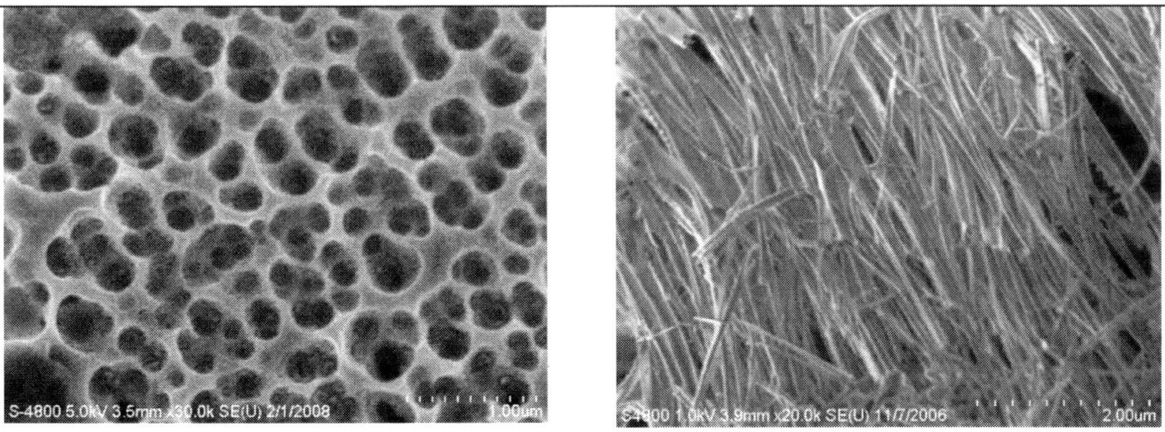


**Figure 1**: SEM images of porous titania formed by anodizaiton (left) and polymer nanotubes (right) created via template wetting.

### *Cell Characterization*

*MTT assay (cell viability)*: The MTT assay is used as a test of cell viability through the quantification of adherent cells. This assay utilizes the formation of purple formazan crystals from the mitochondrial dehydrogenases of viable cells cleaving the tetrazolium ring. This purple solution is then measured at 570 nm using a spectrophotometer. The amount of formazan formed can be correlated to the resulting viability or cytotoxicity of the cells on the surface.

*PicoGreen assay (DNA quantification)*: The PicoGreen assay provides another measure of cell viability. This assay uses the fluorochrome PicoGreen which selectively binds to dsDNA. Bound PicoGreen is then measured at 260 nm using a spectrophotometer. The amount of bound PicoGreen can be correlated to the resulting cell viability.

## Results and Discussion

ImagePro Plus software was used to analyze the SEM images of the anodized titanium samples to determine inner pore diameter. The trend between the inner pore diameter and current density observed in Figure 2 is similar to the trend found in the literature [10], however the inner pore diameters achieved in this study were not as large as the reference diameters for the given current densities. These results confirm that a consistent pore size can be achieved.

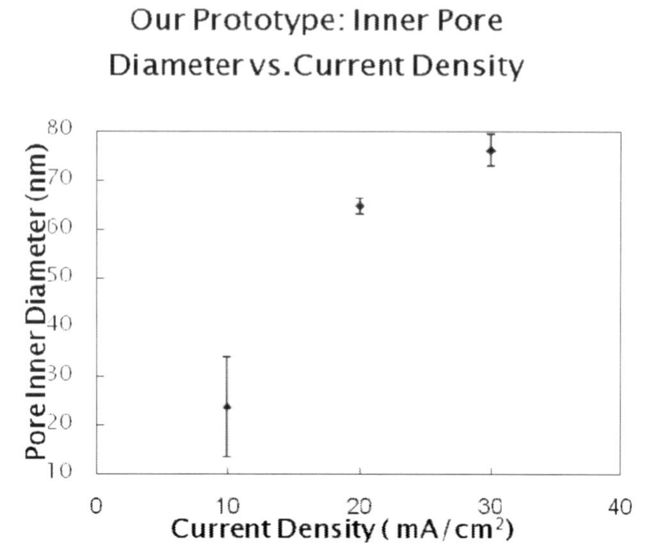


**Figure 2:** Pore inner diameter vs. current density

Following anodization, the titanium samples were then coated with a PVDF polymer solution. Energy-dispersive x-ray spectroscopy (EDS) confirmed the presence of polymer on the surface of the wetted titanium samples.

The results of the MTT assay (Figure 3) show that the polymer-coated modified titanium demonstrated better osteoblast viability than the smooth titanium and the micro rough industry standard titanium. Figure 3 shows the 3 day MTT assay results with a 95% confidence interval. All values in Figure 3 have been normalized using a baseline control for each day. These initial results are promising for the future of the nanoporous structures that will be created with the process.

**Conclusion**

Titanium oxide nanoporous films were grown on titanium surfaces via anodization and wetted with a PVDF polymer coating on the surface of the modified titanium. Cell proliferation using MTT assays showed that the best proliferation was on modified polymer-coated titanium compared with the industry standard roughened titanium and smooth titanium.

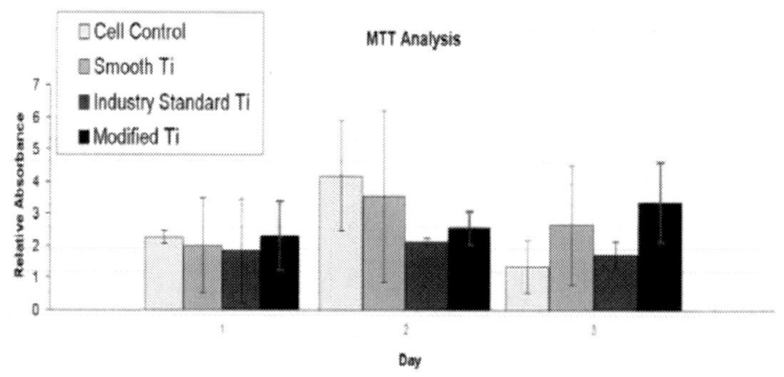


**Figure 3:** MTT analysis of osteoblast viability on titanium

**References**

1. Das, K., Bose, S. and Bandyopadhyay, A. (2007), "Surface modifications and cell-materials interactions with anodized Ti," *Acta Biomaterialia*, 3, pp. 573-585.

2. Monsees, T., Barth, K., Tippelt, S., Heidel, K., Gobunov, A., Pompe, W. and Funk, R. (2005), "Effects of different titanium alloys and nanosized surface patterning on adhesion, differentiation, and orientation of osteoblasts-like cells," *Cells Tissues Organs*, 180, pp. 81-95.

3. Orosz, G., Hannan, E., Magaziner, J., Koval, K., Gilbert, M., Aufses, A., Straus, E., Vespe, E. and Siu, A. (2002), "Hip fracture in the older patient: reasons for delay in hospitalization and timing of surgical repair," *J. Am. Geriatr. Soc.*, 50, pp. 1336-1340.

4. Praemer, A., Furner, S. and Rice, D. (1992), "Musculoskeletal conditions in the United States," *Amer. Acad. Of Orthopaedic Surgeons*.

5. Sinha, R. (2000), "The effects of osteolysis and aseptic loosening," *Medscape Orthopaedics & Sports Medicine eJournal*, October 20.

6. Webster, T. (2003), "Nanophase ceramics as improved bone tissue engineering materials," *Am. Ceram. Soc. Bull.*, 82, pp. 1-8.

7. Orosz, G., Hannan, E., Magaziner, J., Koval, K., Gilbert, M., Aufses, A., Straus, E., Vespe, E. and Siu, A. (2002), "Hip fracture in the older patient: reasons for delay in hospitalization and timing of surgical repair," *J. Am. Geriatr. Soc.*, 50, pp. 1336-1340.

8. Yao, C. and Webster, T. (2006), "Anodization: a promising nano-modification technique of titanium implants for orthopedic applications," *J. Nanosci. Nanotechnol.*, 6, pp. 2682-2692.

9. Steinhart, M., Wehrspohn, R., Gosele, U. and Wendorff, J. (2004), "Nanotubes by template wetting: a modular assembly system," *Angew. Chem. Int. Ed.*, 43, pp. 1334-1344.

10. Kaneco, S., Chen, Y., Westerhoff, P. and Crittenden, J. (2007), "Fabrication of uniform size titanium oxide nanotubes: Impact of current density and solution conditions," *Scripta Materialia*, 56, pp. 373-376.

## 166d Optimization of Preosteoblast Proliferation Rate on Whey Protein Gels for Bone Tissue Regeneration

**Mia Dvora** and James E. Henry. Cain Department of Chemical Engineering, Louisiana State University, 110 Chemical Engineering, South Stadium Rd, Baton Rouge, LA 70803

Bone tissue engineering relies on the ability of osteoprogenitor cells to proliferate rapidly and populate an implantable scaffold prior to differentiation and mineralization. The quality of regenerated tissue achieved is in part governed by the number of cells available for tissue formation. It is therefore our goal to determine the effects of the material variables of a scaffold on growth kinetics, and to use that information to tune the material for rapid cell proliferation.

In the work presented, a material based in a sol-gel made of whey protein isolate (WPI) is designed for use as a scaffold for bone regeneration. We have previously demonstrated that these biopolymer gels support preosteoblast proliferation – an attribute vital to scaffold success. In order to optimize the composition of a WPI gel composite for this application, the growth curves of preosteoblast cells are constructed by culturing MC3T3-E1 cells on WPI gels of varying protein, polysaccharide, and salt concentrations. The curves are used to extract kinetic parameters, which are in-turn compared to one another to elucidate the influence of each component on the behavior of the preosteoblasts – leading to the material that supports the highest cell proliferation rate.

For this optimization study, MC3T3-E1 cells are cultured on WPI discs for different lengths of incubation. At each time point, the cells are harvested and their DNA is purified. The DNA is quantified using fluorescence, and used as a direct measure of cell number. Scanning electron micrographs validate the bulk kinetic data; images demonstrate that the proliferating cells exhibit the desired phenotype. The optimal gel composition and structure inferred from this study are a key in creating a material that will support the growth of new bone tissue in a clinical setting.

# Enhanced Osteoblast Adhesion, Proliferation and Differentiation on Nanocrystalline Diamond Coatings

Lei Yang [1,a], Thomas J. Webster [1,2,b] and Brian W. Sheldon [1,c]
[1] Division of Engineering, Brown University, Providence, RI 02912, USA
[2] Department of Orthopedics, Brown University, Providence, RI 02912, USA
Email: [a] Lei_Yang@brown.edu; [b] Thomas_Webster@brown.edu; [c] Brian_Sheldon@brown.edu

Abstract:
Osteoblast (OB, bone forming cell) functions on diamond are critical for considering the use of anti-abrasive diamond coatings on orthopedic implants. In this study, OB functions including adhesion (up to 4 hrs), proliferation (up to 5 days) and differentiation (up to 28 days) on various diamond coatings were investigated. Two kinds of diamond coatings (nano- and submicron-crystalline diamond) were fabricated through microwave plasma enhanced chemical vapor deposition and characterized by scanning electron microscopy (SEM) and atomic force microscopy (AFM). Surface roughness and hydrophobicity of the diamond coatings increased dramatically as grain size grew. These coatings were tested for OB adhesion and proliferation by counting adherent cells after incubation of 4hrs up to 5 days. OB differentiation on diamond coatings after incubation from 1 to 3 weeks was investigated by measuring total intracellular protein synthesis, alkaline phosphatase activity, and calcium deposition. Results demonstrated enhanced OB functions on nanocrystalline diamond coatings compared to submicron grain size coatings. The long-term (up to 3 weeks) functions of OB on nanocrystalline diamond coatings were promoted compared to submicron diamond, polished silicon and glass coverslips. In summary, these results provided insights into the application of diamond coatings in orthopedics, which can potentially improve wear problems of current implants and prolong their lifetime.

Keywords: Osteoblast; adhesion; proliferation; differentiation; nanocrystalline diamond

## 1. Introduction

An increasing demand for total hip and knee replacements requires improvements in durability and biocompatibility properties from today's orthopedic implants. In order to prolong the 10-15 year lifetime of current implants and reduce the need for revision surgeries, promoting bone cell functions at the bone-implant interface and fostering resistance to bio-chemical (erosion) and mechanical (especially wear) reactions are of great importance. Along these lines, it has been reported that diamond is one of the most robust materials for orthopedic applications to date, however, bone cell functions on diamond remain unclear.

Recent researchers have revealed several promising approaches to improve bone cell interactions on implants by introducing nanoscale features [1]. This rationale has been also applied to diamond coatings on orthopedic implants. Specifically, nanocrystalline diamond (NCD) coatings have been investigated for their promise in orthopedic applications and recent studies have demonstrated their strong potential. For example, Amaral *et al.* [2] fabricated NCD on a $Si_3N_4$ substrate and demonstrated improved osteoblast (OB, bone forming cells) proliferation and synthesis of OB differentiation markers, like alkaline phosphatase and extracellular matrix calcium mineralization, compared to polystyrene tissue culture plates. In addition, diamond with combined micron and nano features promoted osteoblast proliferation the most compared to polystyrene tissue culture plates after 3 days [3].

However, a thorough study concerning both short and long-term functions of OB (specifically, adhesion, proliferation and differentiation) has not been completed on diamond of various micron to nanometer surface features. The present work revealed a systematic investigation on OB functions from 4 h up to 3 weeks on both NCD with grain sizes <100 nm and submicron diamond (SMD) with grain sizes 100-600 nm. Interactions between diamond surface properties and OB functions were also related here.

2. Materials and Methods

Diamond coatings on polished silicon were fabricated through microwave plasma enhanced chemical-vapor-deposition (MPCVD) using $Ar$-$H_2$-$CH_4$ plasma. The concentration of $CH_4$ was kept at 1 vol.% and $H_2$ varied from 5 vol.% to 20 vol.% for NCD and SMD, respectively. The silicon substrates were seeded with nano diamond powders with average sizes around 50 nm before growth. The growth pressure and temperature were 140 torr and 800 °C, respectively, with a growth time of 2 h.

The surface topography of NCD and SMD was examined by scanning electron microscopy (SEM, LEO 1530VP, Zeiss, field emission gun at 4 kV) and atomic force microscopy (AFM, AutoProbe, Park Scientific Instrument, silicon nitride tips, contact mode at a scanning speed of 1 Hz). The root mean square (RMS) roughness of the coatings was calculated from AFM data based on $10 \times 10 \mu m$ scans. Contact angles on diamond surfaces were measured through the sessile drop shape method.

Human osteoblasts (OB; CRL-11372, ATCC, population number between 10 and 15) were used for adhesion, proliferation and differentiation tests. For the adhesion assays, the cells were seeded at a density of 3500 cells/cm$^2$ on NCD, SMD, polished silicon and microscope glass coverslips (CG) and were cultured in Dulbecco's Modified Eagle's Medium (DMEM) supplemented with 10% fetal bovine serum (FBS) and 1% penicillin/streptomycin (P/S). After incubating in a $CO_2$ humidified incubator (5% $CO_2$, 37°C) for 4 h, the cells were washed three times with phosphate buffering solution (PBS) and the adherent cells were fixed with 10% Formalin (Fisher Scientific, Co.) and stained by DAPI (Sigma Aldrich Co.). The stained cells were counted under a fluorescence microscope (FM, Axiovert 200M, Zeiss) to calculate OB adhesion density (cell per unit area of the substrate). For OB proliferation tests, the procedures were similar, except that the seeding density of OB was 3000 cells/cm$^2$ and the cells were cultured up for 1, 3 and 5 days.

For OB differentiation tests, OB were seeded at a density of $10^5$ cells/cm$^2$ and were cultured in DMEM supplemented with 10% FBS, 1% P/S, 50 $\mu$g/ml L-ascorbate (Sigma) and 10 mM $\beta$-glycerophosphate (Sigma). The cells were incubated at 5% $CO_2$ and at 37 °C for 7, 14, and 21 days. Medium was replaced every second day. At the end of each prescribed time period, supernatant medium was removed and the remaining OB were lysed using distilled water and three freeze-thaw cycles. This protocol only removes intracellular as well as cell membrane-bound proteins and does not remove proteins contained in the extracellular matrix [4]. Total protein content in the cell lysates was determined spectrophotometrically using a commercially available kit (Pierce Chemical Co.) following the manufacturer's instructions. For the measurement of intracellular alkaline phosphate (ALP) activity, cell lysates were analyzed by a commercial ALP activity detection kit (Upstate Cell Signal Solutions, USA) and details can be found elsewhere [5]. The ALP activity was normalized by total intracellular protein synthesis and substrate surface area. For the calcium deposits, OBs were removed from substrates (specifically, diamond coatings as well as silicon and CG controls) after 21 days of incubation and the substrates were treated with 0.6 M HCl overnight at room temperature. The amount of calcium present in the acidic supernatant was quantified using a commercially available kit (Pointe Scientific Co., MA) and following manufacturer's instructions [5]. The amount of calcium deposited was normalized over the substrate surface areas.

## 3. Results and Discussion

### Surface properties of the diamond coatings

NCD coatings fabricated with 5% $H_2$ revealed spherical grains with sizes between 30~100 nm and nanoscale roughness with RMS roughness values of 19.8nm (Fig.1). In contrast, SMD coatings fabricated with 20% $H_2$ had a wide range of grain sizes from 100nm to 600 nm and consequently a much higher RMS roughness value (59.5 nm) than NCD (Fig.1). A previous study demonstrated that diamond grain sizes increase dramatically as the $H_2$ concentration increases during the MPCVD process [6]. This trend agrees with the topographical results mentioned above.

Contact angle measurements revealed a small difference between NCD (88.7±6.8°) and SMD (99.5±5.8°). NCD had a smaller contact angle although both of them were very hydrophobic surfaces for biological applications.

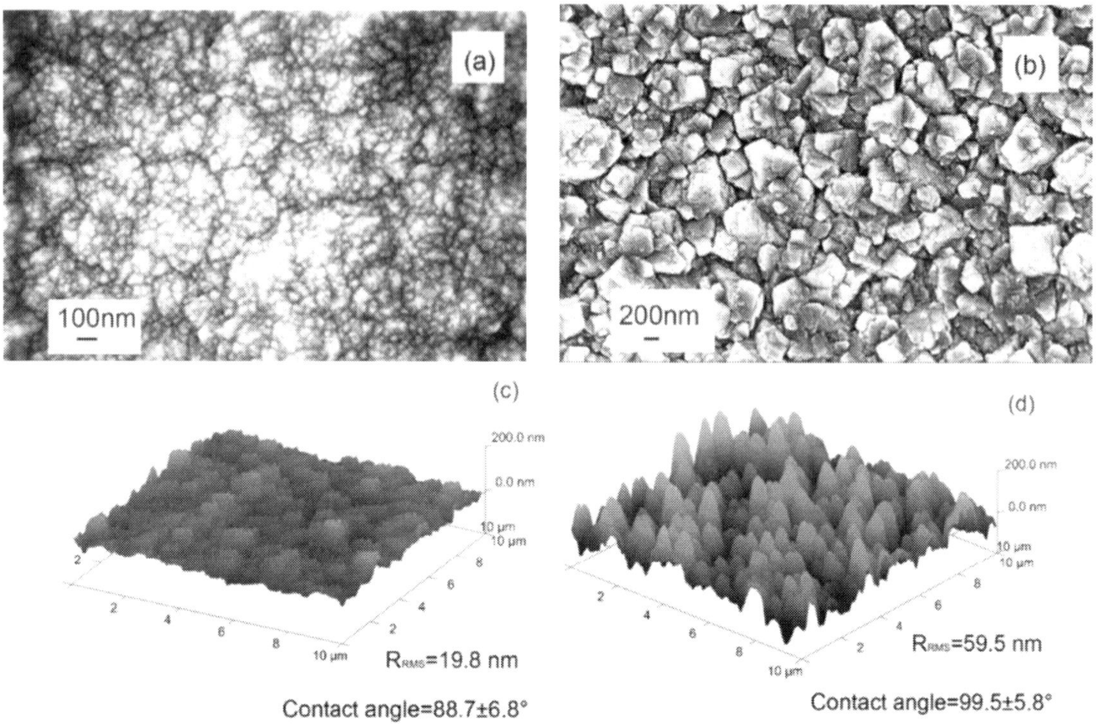


Fig.1 SEM and AFM images of NCD (a and c) and SMD (b and d).

OB adhesion and proliferation on the diamond coatings

The results of OB adhesion and proliferation on NCD and SMD are shown in Fig. 2(a) and (b), respectively. NCD promoted OB adhesion more than SMD and the average adherent OB density on NCD was almost twice as that on SMD. This could be attributed to the lower contact angle on NCD compared to SMD, indicating a more hydrophilic surface than SMD and resulting in enhanced protein adsorption favorable for cell adhesion [7]. However, it was noticed that OB adhesion on both diamond coatings were much poorer than on silicon and CG controls, which could be explained by the more hydrophilic properties of silicon and CG (contact angles ~30°). For all the time periods up to 5 days, OB proliferated more on NCD compared to SMD. After 5 days, OB proliferation on both diamond coatings was higher than on silicon controls, although it was still much lower than on CG. These results indicated that NCD promoted OB proliferation compared to SMD and silicon controls.

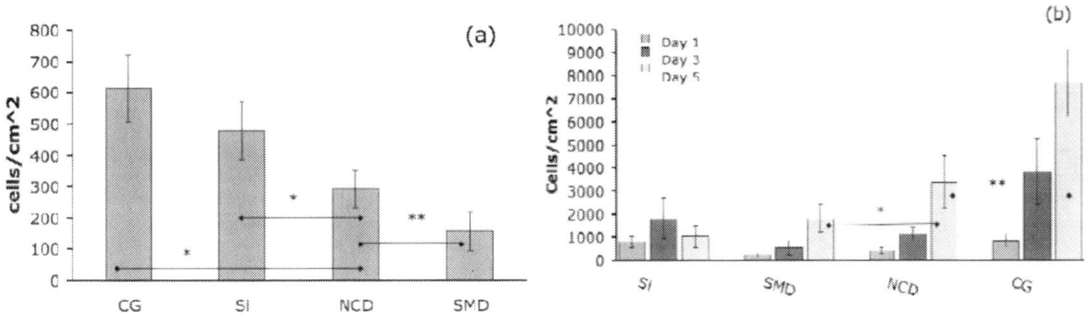


Fig. 2 OB adhesion (a) and proliferation (b) on NCD and SMD coatings. Data=mean value ± standard deviation; n=9; *p<0.01, **p<0.05

OB differentiation on the diamond coatings

The results of total intracellular protein synthesis on diamond coatings are shown in Fig. 3. After 21 days, OB on NCD synthesized more intracellular protein than on silicon and CG, and slightly more than on SMD although the difference was not significant. The results of ALP activity on diamond coatings are revealed in Fig. 4. OB on NCD showed significantly enhanced ALP activity than on SMD and CG, indicating possibly more OB differentiation on NCD than on SMD and CG. This result was further confirmed by the enhanced calcium deposition of OB on NCD, as shown in Fig. 5. All the OB differentiation makers demonstrated that NCD enhanced OB differentiation more than SMD for almost all of the time periods. NCD significantly promoted the OB functions (ALP activity and calcium deposition, which are important indicators of OB differentiation) after 21 days compared to CG and silicon control.

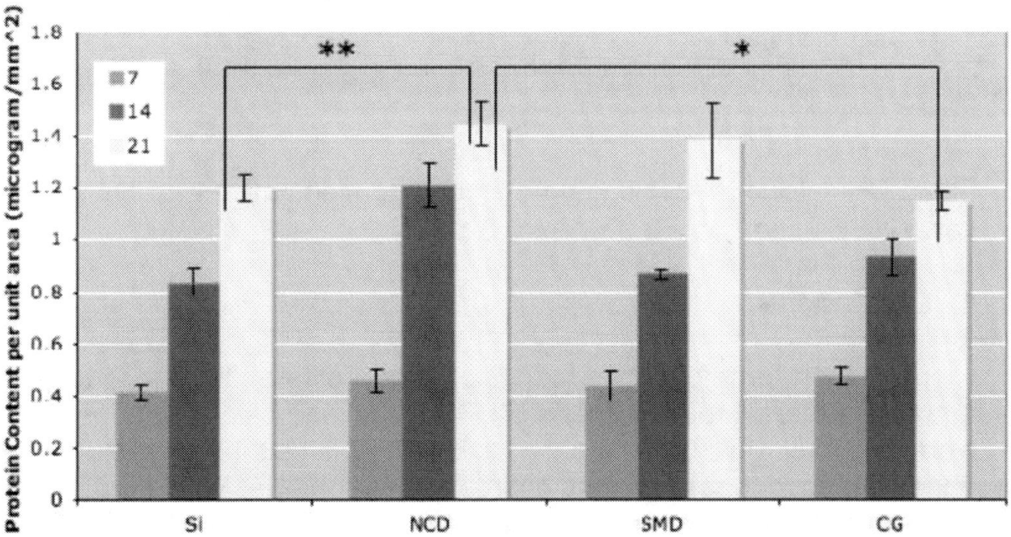


Fig. 3 Total intracellular protein synthesis by OBs on diamond coatings. Data=mean value ± SEM; n=9; *p<0.05, **p<0.07

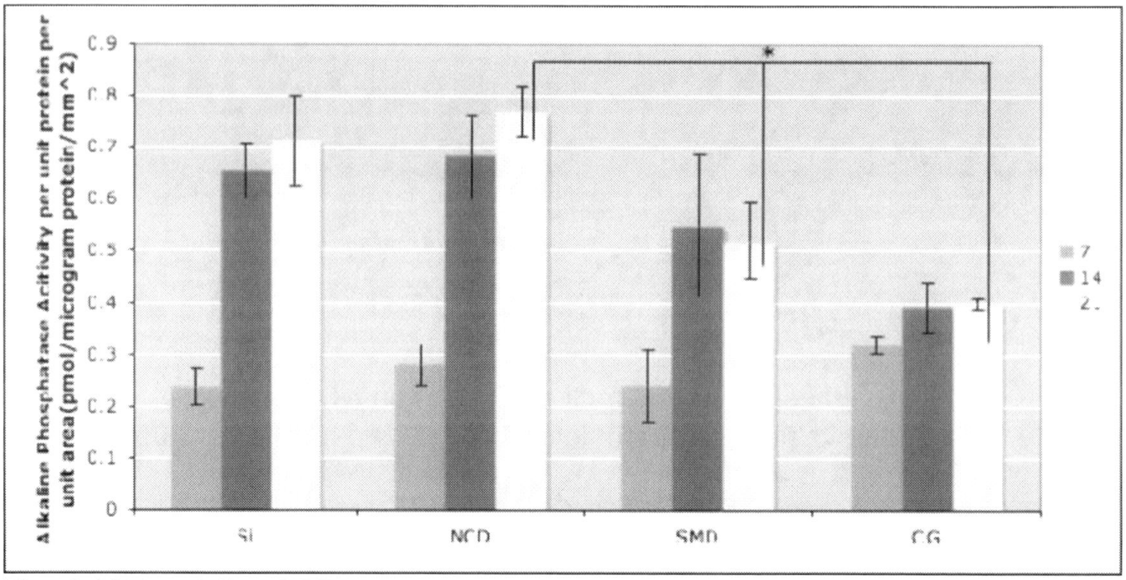


Fig. 4 ALP activity of OBs on diamond coatings. Data=mean value ± SEM; n=9; *p<0.05.

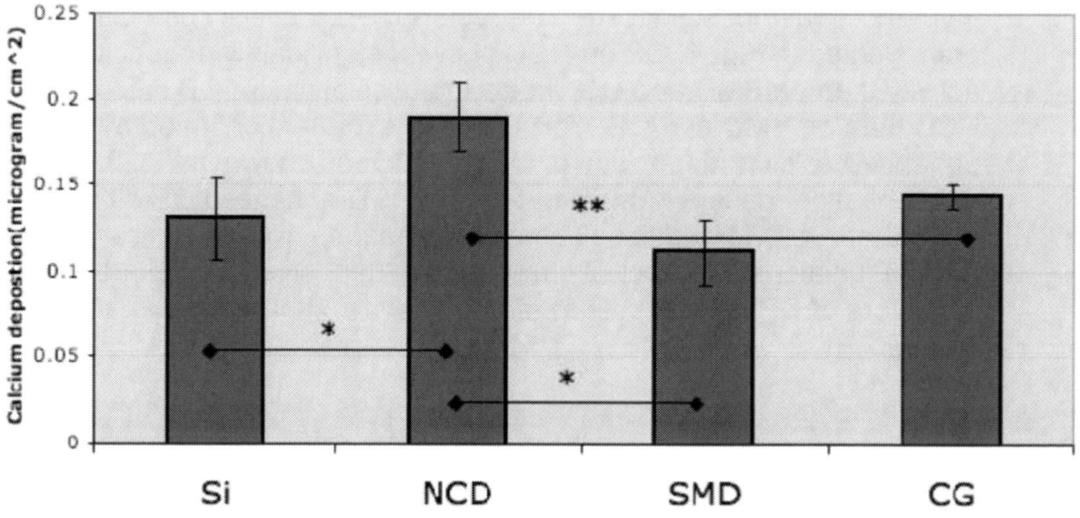


Fig. 5 OB calcium deposition on diamond coatings after 21 days. Data=mean value ± SEM; n=9; *p<0.03, **p<0.09.

## 4. Conclusions

Both short-term functions (adhesion and proliferation) and long-term functions (total protein synthesis, ALP activity, and calcium deposition) of OBs were promoted on NCD compared to SMD. This could be correlated to the nanoscale roughness and less hydrophobic properties of the NCD coatings. NCD coatings also promoted OB differentiation compared to silicon and CG after long time periods (2~3 weeks), although OB adhersion and proliferation were less on NCD. These results provided important information for applying of diamond coatings on orthopedic implants.

References

[1] Balasundaram G. Nanomaterials for Better Orthopedics. In: Webster TJ, editor. Nanotechnology for the Regeneration of Hard and Soft Tissues. Singapore: World Scientific; 2007, p. 53-78.

[2] Amaral M, Dias AG, Gomes PS, Lopes MA, Silva RF, Santos JD, Fernandes MH. Nanocrystalline diamond: In vitro biocompatibility assessment by MG63 and human bone marrow cells cultures. J Biomed Mater Res A 2007 Dec 17 [Epub ahead of print].

[3] Bacakova L, Grausova L, Vacik J, Fraczek A, Blazewicz S, Kromka A. Improved adhesion and growth of human osteoblast-like MG 63 cells on biomaterials modified with carbon nanoparticles. Diamond & Related Materials 2007; 16: 2133–2140.

[4] Webster TJ, Hellenmeyer EL, Price RL. Increased osteoblast functions on theta+delta nanofiber alumina. Biomaterials 2005; 26: 953–960.

[5] Liu H, Slamovich EB, Webster TJ. Increased osteoblast functions among nanophase titania/poly(lactide-co-glycolide) composites of the highest nanometer surface roughness. J Biomed Mater Res 78A; 2006: 798–807.

[6] Yang L, Webster TJ, Sheldon BW. Topographcial Effects of Micro/Nano Crystallized Diamond for Controlling Osteoblast Adhesion. Extended abstract for AICHE Annual Meeting, 2007.

[7] Wilson CJ, Clegg RE, Leavesley DI and Pearcy MJ. Mediation of Biomaterial–Cell Interactions by Adsorbed Proteins: A Review. Tissue Eng 2005; 11:1-18.

# Self-assembled, Nanometer-rough Cartilage Sealants for Orthopaedic Applications

*Yupeng Chen[1, 2], Rajesh A. Pareta[2], Hicham Fenniri[4] and Thomas J. Webster[2, 3]*

*[1]Chemistry Department, [2]Division of Engineering, [3]Department of Orthopaedics, Brown University, Providence, 02912, RI, USA.*
*[4]National Institute for Nanotechnology and Department of Chemistry, National Research Council and University of Alberta, Edmonton, AB T6G ZV4, CANADA.*

## Introduction

Articular cartilage defects are widespread, because of intensive sports activities, genetic disorders (like rheumatoid arthritis) and osteoarthritis. Due to cartilage loss, patients usually have significant pain and are limited in joint motion. Today, more than 20 million Americans suffer from arthritis [1]. It is the leading cause of disability in people over the age of 55 in the United States [2]. Except for artificial joint replacements (usually including metals) with their associated short functional life-span, there is no appropriate long-lasting treatment for cartilage damage. Currently, many researchers have focused on promoting cartilage tissue growth to repair the joints of patients instead of relying on complete joint replacement surgeries [3] [4].

Autologous chondrocyte implantation/transplantation (ACI/ACT) is a cellular based therapy already in clinical use [5]. For ACI/ACT, chondrocytes are seeded onto synthetic [6~8] or naturally derived [8~9] matrix scaffolds (implant materials) that would be sutured into the joint. However, current histological and clinical data of ACI/ACT are conflicting and highly controversial. One of the complications associated with these methods is the lack of cellular and consequently tissue binding between engineered cartilage and natural tissue [10].

Because nanophase materials (or materials with one dimension less than 100nm in at least one direction) are able to mimic the dimensions of constituent components of natural cartilage, bone and the transitional zone of cartilage (for example, due to the presence of collagens, glycosaminoglycans (GAGs) and hydroxyapatite), they have great potential to be successful alternative implant materials to what is being proposed today.

One type of nanomaterials, helical rosette nanotubes (HRNs, Figure 1), are novel biomimetic self-assembled supramolecular structures, whose basic building blocks are guanine (G) and cytosine (C) DNA base-pairs which can solidify into a viscous gel [11~13]. The G^C heteroaromatic bicyclic base possesses the Waston-Crick donor-donor-acceptor of guanine and the acceptor-acceptor-donor of cytosine. G^C undergoes a hierarchical self-assembly process under physiological conditions to form a six-membered supermacrocycle by the formation of 18 hydrogen bonds. Because of electrostatic forces, base stacking interactions and hydrophobic effects, the rosettes form a stable stack with an inner channel 11Å in diameter. An amino acid side chain (lysine) was chosen to impose chirality and surface chemistry on the HRNs. Importantly, the lysine side chain (with an amine group and a carboxyl group) provided the possibility to functionalize a variety of drugs onto HRNs (for example, to functionalize peptide growth factors through peptide bonds). In addition, some small bioactive molecules could be trapped in the inner channels of HRNs by hydrogen bonds; this is another choice for drug delivery by HRNs. Hopefully, two drug carrier sites for HRNs could show different release rates to fit specific requirements of drugs. Thus, because HRNs can: 1) increase viscosity when heated to low temperatures (40~60°C); 2) be easily functionalized with a high-density of peptides and drugs; 3) mimic the nanostructure of natural cartilage, bone and the transitional zone between them; 4) create a surface environment which improves protein (like collagen and fibronectin) absorption as well as enhance cell adhesion and subsequent functions; 5) be injected into cartilage defect sites and in situ repair damaged cartilage; and 6) biodegrade with the recovery of natural tissue, HRNs are expected to serve as a novel, multifunctional implant material to regenerate cartilage/bone.

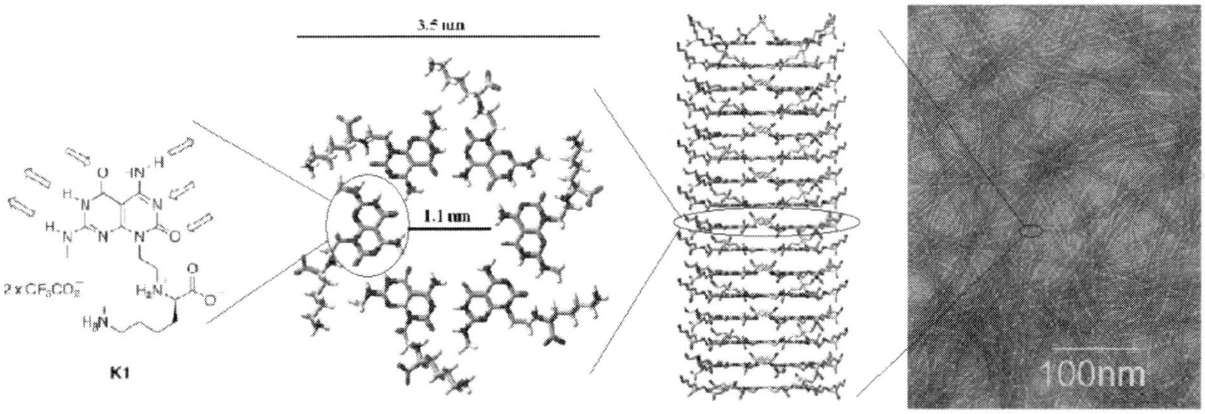


Figure 1. HRNs (with a lysine side chain) undergo spontaneous self-assembly under physiological conditions.

To further mimic the natural cartilage/bone interface and to improve the mechanical properties of engineered cartilage/bone, select biomaterials can be added to HRNs, like alginate ($(C_6H_8O_6)_n$). Alginate, just like GAGs (glycosaminoglycans, one of the major components in cartilage), belongs to the polysaccharide family. It is considered a biocompatible material for cells as it generally does not interfere with cellular functions. Many studies have involved alginate as the material to encapsulate mesenchymal stem cells and promote their osteogenic/chondrogenic differentiation [14, 15]. In this study, the mechanical properties of select hydrogel/HRN composites were tested. In addition, electrospinning was used to generate three-dimensional, implantable, composite HRN fibers containing fibroblast-like type-B synoviocytes (SFB cells). Importantly, results showed that HRNs enhanced hydrogel adhesive strength to fractured collagen and created a scaffold with nanometer-rough surface structures that promoted SFB cells adhesion and viability. In this manner, this study provided an alternative cartilage regenerative material produced by nanotechnology techniques that can be injected as a liquid, solidify at body temperatures under short periods of time, have suitable mechanical properties to cartilage, and promote SFB cell viability and adhesion.

## Materials

SFB cells were isolated from synovial membranes, which were digested with 0.1% trypsin and 0.4% collagenase II from knee joints of 4-months old female pigs. DMEM/F-12 Ham media supplemented with 10% fetal bovine serum (FBS; Hyclone) and 1% penicillin/streptomycin (P/S; Hyclone) was used to culture the cells under standard cell culture conditions, that is, a sterile, 37 °C, humidified, 5% $CO_2$ / 95% air environment.

Collagen casings #320 purchased from Nippi (Japan) were cut into 10.0 ± 0.5 cm * 2.0 ± 0.05 cm * 0.050 ±0.002 mm pieces. 70% ethanol was used to clean the collagen casing surface. PEG (molecular weight 20,000) was purchased from Sigma.

## Methods

*Mechanical Testing*

PEG was dissolved in double distilled water to a concentration of 0.2mg/mL. HRNs were added to the PEG solution to reach different concentrations. 50µL of a PEG solution was added to collagen casing pieces in a 2cm*2cm adhesive area. Two pieces of collagen casings were adhered together by the adhesive area, then heated at 60°C for 10mins. After 24 hours, until two pieces of collagen casings adhered together tightly, a tensile force was applied on the collagen casing. The tensile strength increased gradually until the adhesive area separated. The maximum tensile load was recorded by the computer.

*Contact Angle Measurements*

PEG with/without HRNs were placed on the collagen casings and contact angles measured using a static contact angle meter. An auto pipette was employed with the meter to ensure that the volume of the solution droplet was the same (20 µL) for each collagen casing.

*Electrospinning Encapsulations*

For electrospinning experiments, three different hydrogel solutions were prepared separately in DMEM medium: 10% polyvinyl acetate (PVA), 5% PVA and 3% alginate, and 5% PVA, 3% alginate and 0.01 mg/mL of HRNs solution. Each solution was placed into a 10 ml syringe. The positive lead from a high voltage supply was connected via an alligator clip to the external surface of the needle. A rectangular aluminum foil was used as a static collector and connected to the ground. The tip-to-collector distance was 15 cm. The flow rate of the solution was 2 ml/h and was dispensed automatically. The voltage was kept at 20 kV. TEM and SEM were used to analyze the nano/sub-micro fibers of the composite.

*Cell Adhesion and Viability*

For cell viability and adhesion experiments, SFB cells were isolated from the knees of 4-month old female pigs. SFB cells were mixed with the hydrogel solution at a concentration of 500,000 cells/mL. The hydrogel solution containing cells was electrospun onto glass slides (2 cm diameter). The glass slides with cells were then cultured in DMEM medium with 10% FBS in a sterile, 37 °C, humidified, 5% $CO_2$ / 95% air environment for 1 day. Then, cells were stained and counted under a fluorescence microscope to exam cell viability and adhesion.

*Microscopy Studies*

For TEM analysis, a drop of HRN-K1 with PEG was placed on the TEM grids. Some samples were dried at room temperature directly while others were adhered on collagen casing surfaces and dried in 60 °C for 10 mins. Then, by applying the tensile force, TEM grids were pulled apart from the collagen casing. After negatively staining, all samples were imaged by TEM (Philips, EM410 PW6008) at 10,000×~235,000× operating at 100 kV acceleration voltages.

*Statistical Analysis*

Statistics were performed using a one-tail T-Test for n=3~9 where $p<0.05$ was considered statistically significant.

## Results

*Mechanical Testing and TEM Analysis*

Results from this study showed that as the HRN concentration increased, the adhesive strength of PEG on collagen casings also increased, as shown in Figure 2. Transmission electron microscopy (TEM) pictures showed that in the PEG solution, 0.1mg/mL HRNs aggregated into longer ribbons (Figure 3A), which may have increased the mechanical strength of the HRNs. In Figure 3B, after heating at 60°C for 10 mins, HRNs assembled into long ribbons and some separation points were found after fracture during the tensile tests. These results demonstrated that HRNs strongly adhered onto collagen casings, thus, increased the adhesive strength of PEG between two pieces of collagen casings.

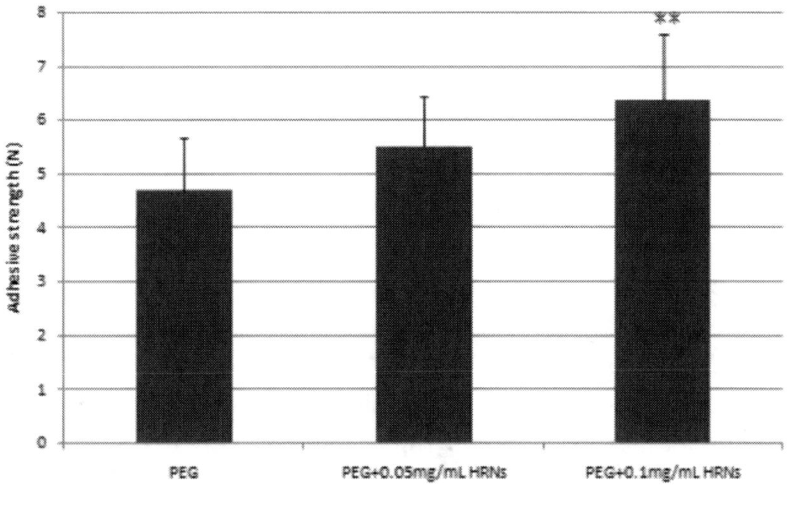

Figure 2. Increased adhesive strength of PEG in the presence of HRNs. Data are mean ± SEM (N=1, n=5). **$p<0.1$ compared to PEG (no HRNs).

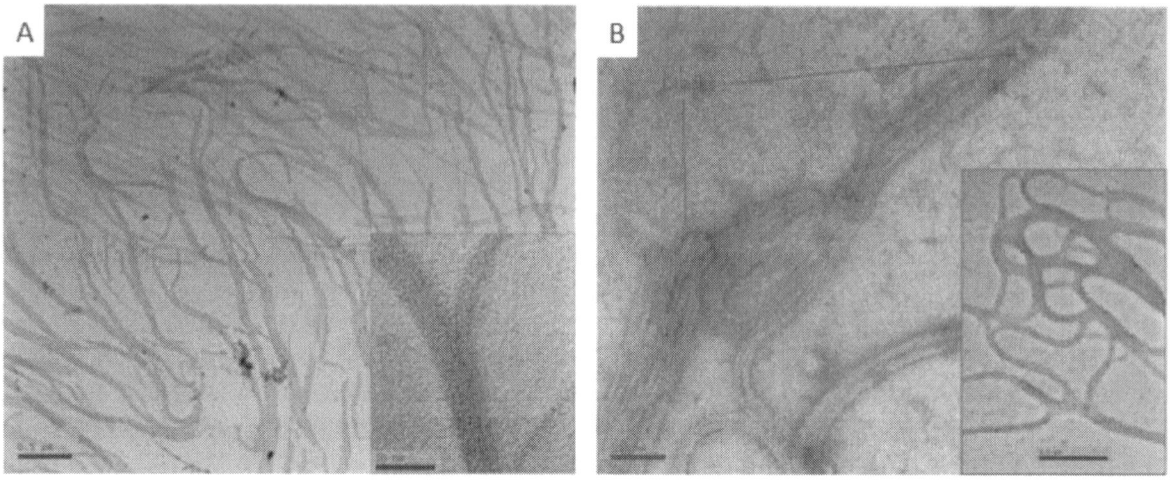


Figure 3. TEM pictures of the unheated HRN/PEG solution on a TEM grid without tension (A) and the heated HRN/PEG solution with tension (B).

*Contact Angle Studies*

For contact angle studies, as shown in Figure 4, with increasing HRN concentrations, the contact angle of the hydrogel solution on collagen casings decreased, which provided evidence that HRNs promoted the surface contact of PEG to increase adhesive strength.

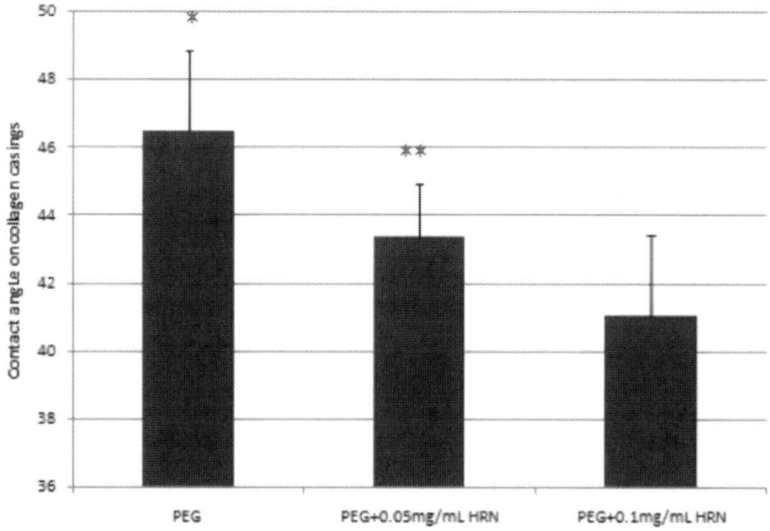


Figure 4. Decreased contact angles of PEG solution in the presence of HRNs on collagen casings. Data are mean ± SEM (N=1, n=6). *$p<0.05$ compared to controls (PEG only), **$p<0.05$ compared to controls (PEG only).

*Electrospinning Encapsulation*

TEM analysis showed the PVA/alginate hydrogel was electrospun into sub-micro fibers with diameters from 200~700 nm. Moreover, on these fibers, HRNs were clearly observed when add to created nanorough scaffolds as marked by yellow arrows, as shown in Figure 5A. In Figure 5B, no HRNs are present in the hydrogels, thus, the nano-structures on the hydrogel fibers can not be observed.

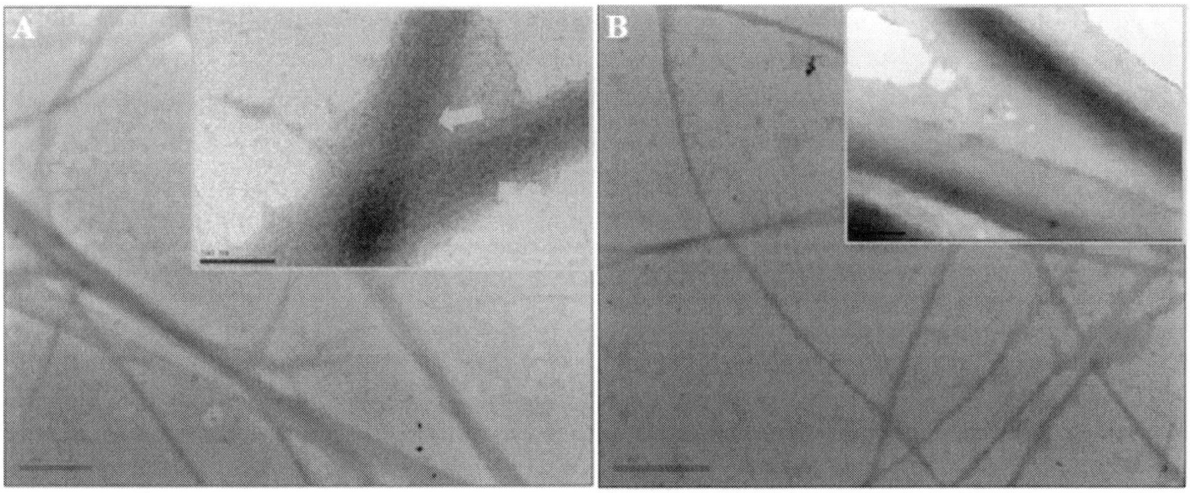


Figure 5. Electrospinning hydrogel fibers with HRNs (A) and without HRNs (B).

*Cell Viability and Adhesion Studies*

After 1 day of cell culture, cell numbers for the PVA/alginate/HRN samples were the highest compared to any other scaffold created in this study. In addition, cell viability (live cell numbers/total cell numbers) was similar on all of the scaffolds, as shown in Figure 6.

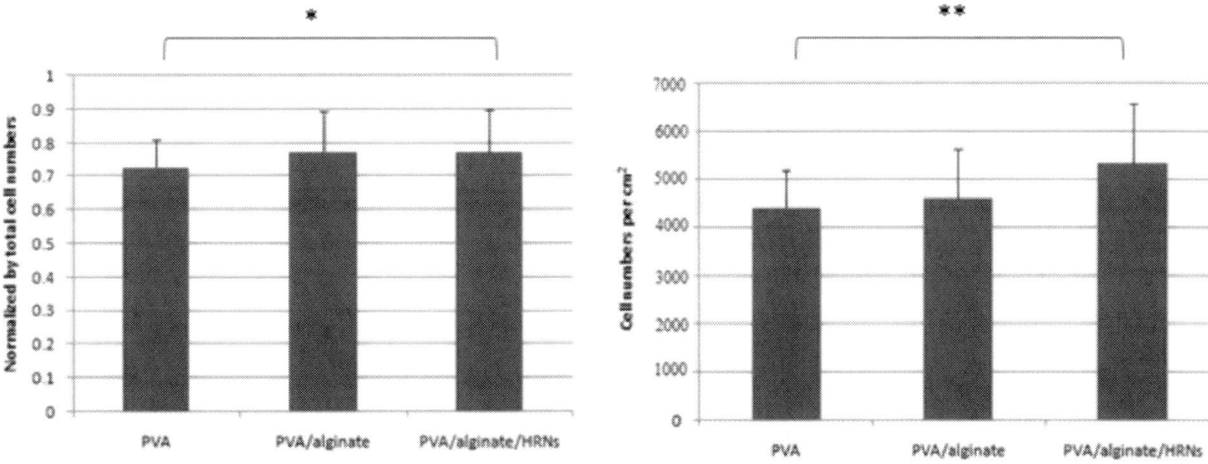


Figure 6. SFB cell viability (left) and adhesion (right) after 1 day culture. Data are mean ± SEM. *p>0.1 compared to PVA alone, *p<0.1 compared to PVA alone. HRNs increased cell numbers.

## Discussion and Conclusions

Results of this study demonstrated that HRNs (novel, nanostructured self-assembled materials) increased the adhesion strength to collagen and can be directly electrospun with synovial cells into a 3D fiber mesh at the site of surgery to increase cell adhesion. HRNs may accomplish this by increasing surface energy of hydrogel composites (as demonstrated by decreased contact angles) and provide for nanostructured surface features known to promote cell functions. HRNs appear to provide a supportive matrix for cartilage precursor cells and that they can be used to deliver drugs or growth factors useful in cartilage tissue engineering. Therefore, electrospun HRNs might be a useful scaffold for engineering a cartilage bio-composite for grafting into cartilage defects.

### Acknowledgements

This study was supported by NIH (1R21AG027521) and the Office of Research and Development, RR&D Service, Department of Veterans Affairs. We would like to thank Dr. Roy Aaron and Dr. Deborah Ciombor for providing SFB cells.

### References

1. Tuan RS, Boland G, Tuli R., Adult mesenchymal stem cells and cell-based tissue engineering. Arthritis Research and Therapy, 2002, 5, 32-45.
2. PR Newswire, Milestone Scientific Signs Agreement With Carticept Medical to Collaborate on CompuFlo(TM) Injection System for Treatment of Arthritic Joints, 2007.
3. Langer R, Vacanti JP. Tissue engineering. Science 1993; 260: 920–926.
4. Temenoff JS, Mikos AG. Review: tissue engineering for regeneration of articular cartilage. Biomaterials 2000; 21: 431–440.
5. Brittberg M, Lindahl A, Nilsson A, Ohlsson C, Isaksson O, Peterson L. Treatment of deep cartilage defects in the knee with autologous chondrocyte transplantation. N Engl J Med 1994; 331: 889–895.
6. Puelacher WC, Mooney D, Langer R, Upton J, Vacanti JP, Vacanti CA. Design of nasoseptal cartilage replacements synthesized from biodegradable polymers and chondrocytes. Biomaterials 1994; 15:774–778.
7. Freed LE, Marquis JC, Nohria A, Emmanual J, Mikos AG, Langer R. Neocartilage formation *in vitro* and *in vivo* using cells cultured on synthetic biodegradable polymers. J Biomed Mater Res 1993; 27:11–23.
8. Grande DA, Halberstadt C, Naughton G, Schwartz R, Manji R. Evaluation of matrix scaffolds for tissue engineering of articular cartilage grafts. J Biomed Mater Res 1997; 34: 211–220.
9. Nehrer S, Breinan HA, Ramappa A, Shortkroff S, Young G, Minas T, Sledge CB, Yannas IV, Spector M. Canine chondrocytes seeded in type I and type II collagen implants investigated in vitro. J Biomed Mater Res 1997; 38: 95–104.
10. Brun P, Abatangelo G, Radice M, Zacchi V, Guidolin D, Daga Gordini D, Cortivo R. Chondrocyte aggregation and reorganization into three-dimensional scaffolds. J Biomed Mater Res 1999; 46: 337–346.
11. Chun AL, Moralez JG, Fenniri H, Webster TJ. Helical rosette nanotubes: a biomimetic coating for orthopedics? Biomaterials 2005; 26: 7304-7309.

12. Chun AL, Moralez JG, Fenniri H, Webster TJ. Helical rosette nanotubes: A more effective. orthopaedic implant material. Nanotechnology 2004; 15: s234-s239.
13. Fenniri H, mathivanan P, Vidale KL, Sherman DM, Hallenga K, wood KV and Stowell JG. Helical rosette nanotubes: design, self-assembly and characterization. J. Am. Chem. Soc. 2001, 123, 3854-3855.
14. Eslaminejad MB, Mirzadeh H, Mohamadi Y, Nickmahzar A. Bone differentiation of marrow-derived mesenchymal stem cells using beta-tricalcium phosphate-alginate-gelatin hybrid scaffolds. J Tissue Eng Regen Med. 2007 Nov-Dec;1(6):417-424.
15. Kurth T, Hedbom E, Shintani N, Sugimoto M, Chen FH, Haspl M, Martinovic S, Hunziker EB. Chondrogenic potential of human synovial mesenchymal stem cells in alginate. Osteoarthritis Cartilage. 2007 Oct;15(10):1178-1189.

## 188a Control of Nanoscale Flow in Thin Wetting Films by Electric Field

**Jairus Kleinert**, Sejong Kim, and Orlin D. Velev. Chemical and Biomolecular Engineering, North Carolina State University, Campus Box 7905, Raleigh, NC 27695-7905

We report a novel method for manipulating flow in wetting films of nanoscale thickness using electric fields. Aqueous films of tens of nanometers thickness were formed on hydrophilic mica surfaces under saturation humidity conditions. A linear relationship was established between the fluorescence intensity of dye in a film and film thickness. On this basis, we estimated the films formed to be less than 100 nm thick. Applying an electric field engendered tangential flow in the film by DC electroosmosis. The direction and speed of the film flow, visualized with fluorescent markers, were readily controlled by the external DC electric field. The film flow rate was characterized as a function of ionic strength, pH of the fluid, and external electric field. The resulting flow characteristics were consistent with the theory of electroosmosis. To manipulate the flows in a controlled geometry, we selectively hydrophobized the substrate by microcontact printing to make two-dimensional "virtual wall" channels. Flow velocity profiles of these confined nanoflows appear to be complicated and different from the profiles seen in conventional microfluidics and are currently being analyzed. This nanofluidics methodology will provide a basis for single molecule transport and manipulation as well as an improved nanoscale and molecular level understanding of the interactions of fluids with surfaces.

## 188b Electric Field-Assisted Convective Assembly of Large-Domain Colloidal Crystals

**Jairus Kleinert**, Sejong Kim, and Orlin D. Velev. Chemical and Biomolecular Engineering, North Carolina State University, Campus Box 7905, Raleigh, NC 27695-7905

Two-dimensional colloidal crystals are difficult to assemble rapidly and without defects. Convective assembly is an easily implemented and controlled technique for depositing colloidal crystals by evaporation of a microparticle suspension from a withdrawing meniscus, but this method is slow and does not produce large crystal domains. Alternatively, dielectrophoresis has been used to produce large single-orientation crystals rapidly but offers no simple way to immobilize the crystals. We report a new method of convective assembly combined with an AC electric field to increase the long-range order and deposition rate of latex microsphere crystals. With the electric field, electrowetting on dielectric allowed control of the shape of the meniscus and flow on the microscale. Film spreading at the meniscus caused by electrowetting increased the evaporation rate and allowed for more rapid crystal deposition. Another effect of the electric field was dipolar chaining and organization of particles into large single-orientation crystal domains. We have achieved up to a five-fold increase in the rate of crystal deposition and an order of magnitude increase in crystal domain size compared to convective assembly without an electric field. This work provides a basis for establishing an improved large-scale colloidal crystal assembly process.

## Design of an efficient Quartz Crystal Nanobalance: a Finite Element Study

Reetu Singh and Venkat R. Bhethanabotla*
Sensors Research Laboratory, Department of Chemical and Biomedical Engineering, University of South Florida, Tampa, Florida, 33620, USA.

### Abstract

The mass sensitivity (> 1 micrograms) of current commercial analytical instruments, such as thermal gravimetric analyzers, severely limits their utility for measurement of valuable but poorly soluble materials such as synthetic proteins or DNA fragments. A quartz crystal microbalance (QCM), based on a transverse shear mode piezoelectric crystal operating at high frequencies, is gaining popularity in chemical and bio-sensing applications due to higher mass sensitivities as compared to the traditional analyzers and lesser sensitivity to vibrations. However, these devices suffer from non-uniformity of sensitivity distribution along the sensor surface thereby limiting their use for the determination of mass. Overcoming this limitation would lead to the development of a robust sensor with improved mass sensitivities and reduced sensitivity to vibrations, as compared to the currently available microbalances. The sensitivity profile can be influenced by a number of factors the electrode design and surface properties of the crystal. In the current work, we develop a finite element (FE) model of the QCM to investigate the mass sensitivity and its radial distribution on the sensor surface for various electrode designs. Such a model will aid in the development of versatile nano-balances with a uniform sensitivity distribution.

* Author for correspondence at email: venkat@eng.usf.edu

Introduction

A quartz crystal nanobalance (QCN), based on a transverse shear mode piezoelectric crystal operating at high frequencies, is gaining popularity in chemical and bio-sensing [1] applications due to higher mass sensitivities as compared to the traditional analyzers and lesser sensitivity to vibrations. One potential use of a QCM is as a 'droplet gravimeter' which can measure the mass of non volatile residues (NVR) in a droplet of solvent [2]. The transverse shear mode QCM device can provide nanogram level sensitivity thereby enabling detection limits of parts per million. However, these devices suffer from non-uniformity of sensitivity distribution over the sensor surface thereby limiting their use for the determination of mass. Overcoming this limitation would lead to the development of a robust 'nanobalance' with improved mass sensitivity and increased frequency stability, as compared to the currently available microbalances. This can potentially revolutionize the mass measurement systems, in applications such as droplet gravimetry. The sensitivity profile of a QCN can be influenced by a number of factors such as the electrode design, size, and surface properties of the crystal. It has been established the mass sensitivity decays with distance from the center of the electrode. Controlling this sensitivity variation to achieve a constant sensitivity across the sensor surface, so that the mass sensitivity is independent of the placement of mass, will result in an efficient nanobalance.

Mathematical modeling provides an efficient and cost-effective way to investigate the performance of QCN devices before an actual device is fabricated [3, 4]. The finite element technique provides the ability to accurately model complex structures, and to optimize design parameters. In the current work, we develop a finite element (FE) model of the QCN to investigate the mass sensitivity and its radial distribution on the sensor surface for various electrode designs. Such a model will aid the development of a versatile nanobalance with a uniform sensitivity distribution. The same model can be extended to study fluid-device interaction by utilizing a full bidirectional coupling with the fluid domain, when the device is used in liquid-sensing applications.

## 1. Computational details:

An AT-cut quartz crystal, cut at 35.35 degrees to the z-axis, is a well known substrate for use as quartz crystal microbalance (QCM) due to its high sensitivity [5]. The QCN device (operating frequency 9 MHz) was also based on an AT-cut quartz crystal, cut at 35.35 degrees to the z-axis, with gold electrodes on the top and bottom surfaces. The device was 8.0 mm diameter and 0.185 mm thick with 5.0 mm diameter gold electrodes having a thickness of $0.15*10^{-3}$ mm on either side of the QCN. The density of gold and QCN were 18500 kg/m$^3$ and 2649 kg/m$^3$, respectively. These dimensions and properties are summarized in Table 1. The Young's modulus and Poisson ratio for gold were $92.05*10^3$ MPa and

0.42, respectively. The anisotropic elastic modulus matrix $C^s$ and and piezoelectric constant matrix e were taken from Lu *et al* [4].

$$C^s(MPa) = \begin{bmatrix} 86.74*10^3 & 267.15*10^3 & -8.25*10^3 & 0 & -3.66*10^3 & 0 \\ & 102.83*10^3 & -7.42*10^3 & 0 & 9.92*10^3 & 0 \\ & & 129.77*10^3 & 0 & 5.70*10^3 & 0 \\ sym & & & 68.81*10^3 & 0 & 2.53*10^3 \\ & & & & 38.61*10^3 & 0 \\ & & & & & 29.01*10^3 \end{bmatrix}$$

$$e(C/m^2) = \begin{bmatrix} 0.171 & 0 & 0 \\ -0.0187 & 0 & 0 \\ -0.152 & 0 & 0 \\ 0 & -0.0761 & 0.067 \\ 0.067 & 0 & 0 \\ 0 & 0.067 & -0.095 \end{bmatrix}$$

The model was meshed using 24, 8131 three dimensional 20 node coupled field solid elements, in Ansys® [6]; nevertheless a very small piezoelectric coupling was applied for the gold electrode. The meshed model is depicted in Fig. 1. The applied voltage was 1 V and a damping coefficient of 2% was used. A harmonic analysis was carried out to study the frequency response in the range of 8.9 to 10.1 MHz, employing a total of 200 steps/intervals. The frequency response was then utilized to obtain the displacement profiles on the device surface, which in turn are used to compute the sensitivity distribution on the surface [7].

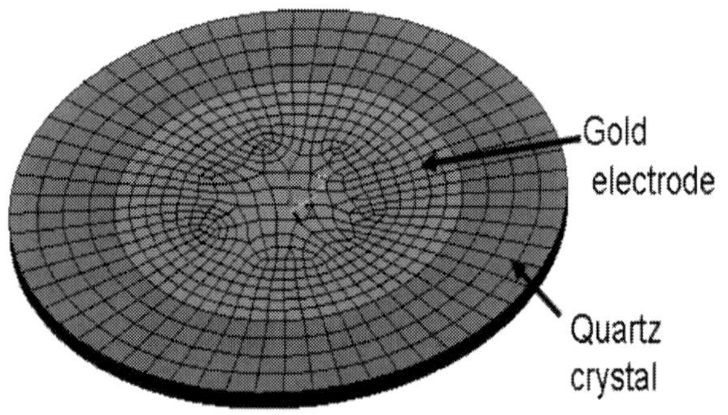


**Figure 1:** Schematic diagram showing the finite element model of the quartz crystal nanobalance with the gold electrodes

**Table 1:** Dimensions and properties of the QCN device components

| | Piezoelectric crystal | Gold electrode |
|---|---|---|
| Diameter (mm) | 8 | 5 |
| Thickness (mm) | 0.185 | $0.15*10^{-3}$ |
| Density (kg/ $m^3$) | 2649 | 18500 |

## 2. Results and Discussion

The distributions of the radial and total displacement of the device surface are shown in Figs. 2 (a,b) which also indicate the first thickness shearing mode of the device. Figure 3 reflects the shearing motion of the device polarized in the X-direction, as also reported in the literature [8, 9]. Our results indicate that the displacement peaks at the center of the electrode and decays rapidly on moving away from the center (Fig. 4), which is in qualitative agreement with the experimental results reported in the literature [10]. In addition, the distribution of the displacement is elliptical rather than circular and is largely bounded by the electrode surface. The peak static displacements are of the order of a picometer (0.923 pm) and are mainly in the x direction. The displacements in the y and z directions are almost 2 orders of magnitude smaller than in the x-direction thereby suggesting the primarily X directional nature of the shearing waves.

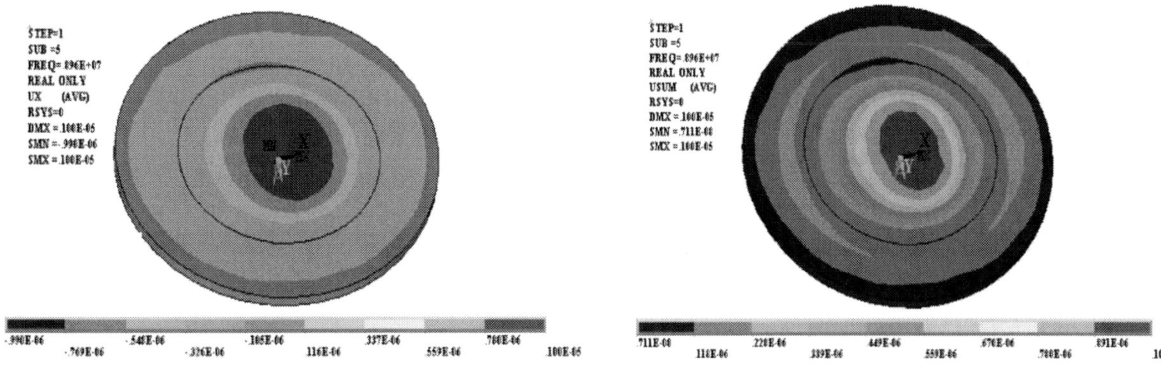

**(a)** **(b)**

**Figure 2:** Contour plot showing the (a) radial displacement. (b) Total displacement of the device surface. The displacements shown above are in micrometers.

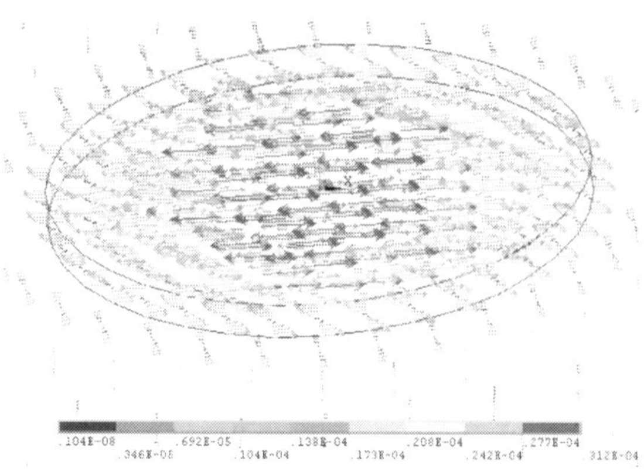


**Figure 3**: Vector plot showing the shearing motion of the device

Considering the influence of mechanical resonance, the dynamic displacements can be obtained from the static displacements by using the Quality factor (Q) [11]

$$U_{dyn}=U_{stat}*Q \tag{1}$$

For a 9 MHz crystal, Q in air is reported to be 84195 [12]. This gives the value of peak displacement to be

$$\begin{aligned} U_{dyn} &= 0.923*10^{-6}\ (\mu m)*84195 \\ &= 77.7\ nm \end{aligned} \tag{2}$$

Thus, peak displacement for the current 9 MHz device, at a peak driving voltage of 1 V is 77.5 nm, which is in agreement with the experimentally observed displacements of a QCM obtained using scanning tunneling microscopy under vacuum [13]. The sensitivity distribution for the device surface was computed using the equation

$$S = \frac{|u(r)|^2}{2\pi \int_0^{\infty} r|u(r)|^2\,dr} C_f \tag{3}$$

$C_f$ is Sauerbrey's sensitivity constant, u(r) is the displacement on the device surface and r is the distance from the center. Cf = 1.78e+011 Hz/kg

The device sensitivity distribution is shown in Fig. 4a The sensitivity shows qualitative agreement with the sensitivity distribution obtained using analytical models [7] as evident from the distributions normalized using the corresponding maximum values depicted in Fig. 4b. Josse et al have analyzed n-m type electrode designs (n and m denote the unequal diameters of the top and bottom electrodes) using the analytical model [7]. The future work is focused on the investigation of the n-m type electrodes using the FE model developed in the current work, which can be thoroughly compared with the results from analytical models presented by Josse et al [7]. The developed FE models can then be reliably extended to study ring designs in order to explore the possibility of obtaining a flat sensitivity distribution based on known analytical results in addition to the investigation of more complex electrode configurations which cannot be addressed by the analytical models.

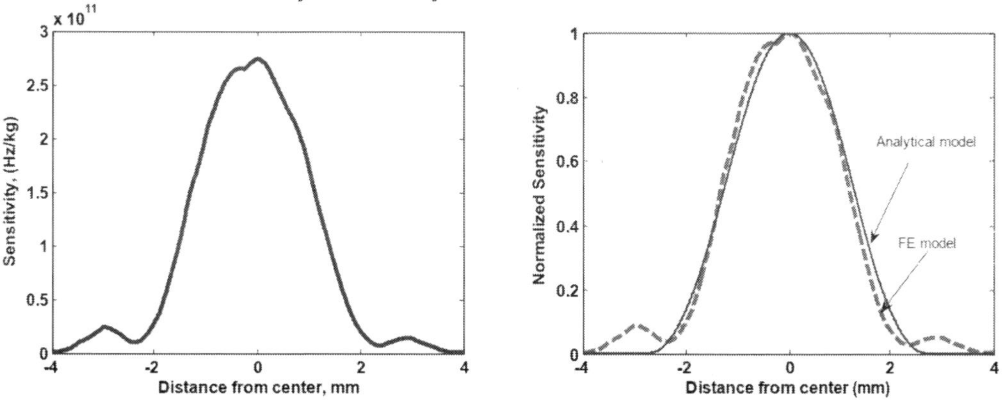


(a) (b)

**Figure 4: (a)** Sensitivity distribution on the device surface (b) Normalized Sensitivity distribution from the current FE model vs. analytical model of Josse et al. [7]

## 3. Conclusions

In this work, a three dimensional finite element model for a quartz crystal nanobalance device was developed to investigate the displacement and sensitivity distributions on the surface. Our results indicate that the displacement distribution of the device surface peaks at the centre and decays rapidly on moving to the device periphery; most of the displacement is confined to the electrode region. In addition, the shearing motion of the device is polarized in the X-direction as also reported in the literature. The developed model predicts surface displacements in accordance with the experimental values reported in the literature. Further efforts are underway to investigate different electrode designs to identify an electrode configuration which can provide high and uniform sensitivity across the device surface.

## Acknowledgements

The authors gratefully acknowledge the computational resources provided by Academic computing and Engineering computing at the University of South Florida, Tampa, FL. This work was supported by NSF grants CHE-0722887 and ECCS-0801929, and IIP-07122360.

## References


1 J. W. Grate, Chemical Reviews **100**, 2627 (2000).

2 A. L. Smith, Journal of ASTM International **3**, 1 (2006).

3 D. H. Wu, Y. J. Tsai, and Y. T. Yen, Sensors and actuators B **92** 337 (2003).

4 F. Lu, H. P. Lee, P. Lu, et al., Sensors and Actuators A **119**, 90 (2005).

5 B. A. Auld, *Acoustic Fields and Waves in Solids* (John Wiley & Sons, New York,1973)

6 Ansys Inc, v11.0, Canonsburg, PA.

7 F. Josse and R. W. Cernosek, Analytical Chemistry **70**, 237 (1998).

8 D. S. Ballantine, S. J. Martin, A. J. Ricco, et al., *Acoustic Wave sensors: Theory, design, and Physico-chemical applications* (Academic Press, San Diego, 1997)

9 S. J. Martin, V. E. Granstaff, and G. C. Frye, Analytical chemistry **63**, 2272 (1991).

10 A. C. Hillier and M. D. Ward, Analytical Chemistry. **64**, 2539 (1992).

11 B. A. Martin and H. E. Hager, Journal of Applied Physics **65**, 2627 (1989).

12 L. Rodríguez-Pardo, J. Fariña Rodríguez, C. Gabrielli, et al., IEEE Sensors Journal. **5**, 1251 (2005).

13 B. Borovsky, B. L. Mason, and J. Krim, Journal of Applied Physics. **88**, 4017 (2000).

## 188d The Electrical Conductivity Properties of Polythiophene/TiO2 Nanocomposites Prepared In the Presence of Surfactants

**Ayse Gul Yavuz**[1], Aysegul Uygun[1], and Songul Sen[2]. (1) Chemistry, Suleyman Demirel University, Isparta, Turkey, (2) Chemistry, Mehmet Akif Ersoy University, Burdur, Turkey

Titanium dioxide is a widely studied material due to its unique optical, electrical, chemical properties, high photo catalytic activity and photoelectric conversion efficiency. It is well known for its photovoltaic and photo catalytic properties. Among conducting polymers, various derivatives of polythiophene (PT) have been investigated extensively because of their interesting semiconducting, electronic and optical properties, combined with processing advantages and good mechanical characteristics. Furthermore, there are some of the disadvantages associated with conducting polymer composite materials, such as: their conductivity is highly dependent on synthesis conditions, there is often an insulating layer formed on the conductor which effects on the degree of electronic conductivity and the composite may become mechanically or thermally unstable due to heavy loading of the conducting particles.

In this work, TiO2 nanoparticles were modified by small amount addition of polythiophene to improve the dispersibility of TiO2 nanoparticles and enhance the photocatalytic activity of composites. Also, effect of surfactant onto properties of PT/TiO2 nanocomposites was investigated using anionic, cationic and nonionic surfactants. The structural, crystallographic, morphological and thermal properties of the nanocomposites were investigated by using SEM, FTIR, XRD and DTA/TG techniques. The electronic conduction aspects of the produced composites were investigated depending on the temperature and the surfactant type by measuring DC electrical conductivity. The observed consequences of the conductivity were compared to the measured results of the SEM, XRD and DTA/TG. The type of electrical conductivity behavior and the reason/reasons of the different conduction performances of the fabricated composite materials were discussed.

## 188e A Mechano-Chemical Model of Growth Termination In Vertical Carbon Nanotube Forests

**Jae-Hee Han**[1], Charles Robert Welch[2], Charles P. Marsh[3], Thomas A. Carlson[3], and Michael S. Strano[4]. (1) Chemical Engineering, Massachusetts Institute of Technology, 77 Massachusetts Avenue, NE47-577, Cambridge, MA 02139, (2) Information Technology Laboratory, U.S. Army Engineer Research and Development Center, 3909 Halls Ferry Road, Vicksburg, MS 39180, (3) Construction Engineering Research Laboratory (CERL), U.S. Army Engineer Research and Development Center, 2902 Newmark Drive, Champaign, IL 61822, (4) 66-566 Department of Chemical Engineering, Massachusetts Institute of Technology, 77 Massachusetts Avenue, Cambridge, MA 02139

Understanding the mechanisms by which vertical array of carbon nanotubes (CNTs) terminate their growth may lead to the production of aligned materials of infinite length. Both single- and doubled-walled carbon nanotubes (SWNTs and DWNTs) films demonstrate characteristic yet unexplained deflections of the top surface near the edges and corners of the film. We show that this upturn in the surface can be explained by assuming a mechanical coupling between neighboring nanotubes.[1] A Monte Carlo simulation of film growth is able to qualitatively reproduce the shape by assuming that the coupling is limited by the enthalpy of the carbon forming reaction. The shape of the surface is approximately conic with hyperbolic cross sections that allow for the calculation of a threshold force ($F_{max}$ = 34 to 51 nN for SWNTs and 25 to 27 nN for DWNTs) and elastic constant ($k$ = 384 to 547 N/m for SWNTs and 157 to 167 N/m for DWNTs) from the images of experimentally synthesized films. Despite differences in nanotube type and precursor chemistry, the values appear consistent. The origin of the mechanical coupling is discussed.

Reference

[1] Han, J. H.; Graff, R. A.; Welch, B.; Marsh, C. P.; Franks, R.; Strano, M. S. *ACS Nano* **2008**, *2*, 53-60.

# 188f Electronically Homogeneous Semiconducting Single-Walled Carbon Nanotubes Solid Bundles

**Jae-Hee Han**[1], Ryuichiro Maruyama[2], Woo-Jae Kim[1], Chang Young Lee[1], Jong Hyun Choi[1], Daniel A. Heller[1], and Michael S. Strano[3]. (1) Chemical Engineering, Massachusetts Institute of Technology, 77 Massachusetts Avenue, NE47-577, Cambridge, MA 02139, (2) Solid State Ionics Lab., Advanced Material Laboratories, Sony Corporation, Atsugi 2TEC 6F, No.2, 4-16-1 Okata, Atsugi, Kanagawa, Japan, (3) 66-566 Department of Chemical Engineering, Massachusetts Institute of Technology, 77 Massachusetts Avenue, Cambridge, MA 02139

The ability to synthesize and manipulate uniform bundles of electronically pure single walled carbon nanotubes (SWNTs) remains central to the advancement of these systems in nano-electronic applications. The assembly of separated SWNT into bundles (parallel aggregates) that are electronically homogeneous and of controlled size is currently beyond engineering capabilities. With the advances of both density gradient[1,2] and dielectrophoresis methods[3], we have performed the first synthesis and measurements of electronically homogeneous solid-state bundles of SWNTs.[4] Several interesting properties are noted. We find that bundles of semi-conducting SWNT that are free of all metallic impurities emit photoluminescence (PL) in the near infrared. The emission is an important quality control metric, and allows for photophysical characterization of such systems. TEM image reveals a dielectrophoretically synthesized 36 nm bundle of (6,5)-enriched SWNTs bundle attached to the very end of a supporting chemically etched tungsten tip. A vertical conductance measurement characterizes the resistance with length and reveals non-ohmic, step changes in current at constant potential characteristic of ballistic conductance. A scattering length, $L$, of approximate 4 to 18 nm is calculated from the trace. We also find that such solid bundles show unprecedented significantly bright PL emission as imaged using an InGaAs camera at both 658 and 785 nm laser excitations, confirming a high degree of semiconductor purity. Controls with un-enriched samples show no emission. The $G/G_0$, where $G$ and $G_0$ are conductance and quantum conductance $2e^2/h \approx 1/13$ (k $\Omega$)$^{-1}$ respectively, ranges from 0.3 to 2.2 for a typical un-enriched SWNT bundle. High purity semi-conducting bundles exhibit significantly low ratios at about 0.002 to 0.004. Some interesting applications involving environmentally-related PL shift from these semiconducting solid bundles is discussed.

References

1. Arnold, M. S.; Green, A. A.; Hulvat, J. F.; Stupp, S. I.; Hersam M. C. *Nature Nanotech.* **2006**, *1*, 60-65.

2. Kim, W. -J.; Usrey M. L.; Strano, M. S. *Chem. Mater.* **2007**, *19*, 1571-1576.

3. Lee, C. Y.; Baik, S.; Zhang, J.; Masel, R. I.; Strano, M. S. *J. Phys. Chem. B* **2006**, *110*, 11055-11061.

4. Han, J. H.; Maruyama, R.;, Kim, W. -J.; Lee, C. Y.; Choi, J. H.; Heller, D. A.; Strano, M. S. **2008** (submitted).

## 188g Dynamics of Surfactant-Suspended Single Walled Carbon Nanotubes In a Centrifugal Field

**Nitish Nair**[1], Woo-Jae Kim[1], Richard D. Braatz[2], and Michael S. Strano[3]. (1) Department of Chemical Engineering, Massachusetts Institute of Technology, Building 66-580, 25 Ames St, Cambridge, MA 02141, (2) Department of Chemical and Biomolecular Engineering, University of Illinois at Urbana-Champaign, 600 South Mathews Avenue, Box C-3, 293 Roger Adams Laboratory, Urbana, IL 61801-3602, (3) 66-566 Department of Chemical Engineering, Massachusetts Institute of Technology, 77 Massachusetts Avenue, Cambridge, MA 02139

A hydrodynamic model is used to describe the motion of surfactant-suspended single walled carbon nanotubes in a density gradient, while being subjected to a centrifugal field. The number of surfactant molecules adsorbed on each nanotube determines its effective density, and hence, its position in the gradient after centrifugation has been completed. Analysis of the spatial concentration distributions of CoMoCAT nanotubes suspended with 2 w/v% sodium cholate yielded 2.21, 2.36 and 2.74 surfactant molecules adsorbed per nanometer along the length of the (6,5), (7,5) and (8,7) nanotubes, respectively. The estimates are commensurate with experimental values reported in the literature. Based on these results, a linear relation is proposed between SWNT diameter and the adsorbate population, which can be used to predict the fate of a species in the separation process. Since the density of the surfactant-nanotube assembly is highly sensitive to the number of adsorbed molecules, a perturbation would cause it to be enriched at a different location in the gradient. The level of sensitivity is also reflected in the 95% confidence levels that are reported in this work.

## 188h Annealing Studies on Functionalized Metallic Single-Walled Carbon Nanotube Thin Film

**Woo-Jae Kim**, Department of Chemical Engineering, Massachusetts Institute of Technology, 77 Massachusetts Avenue, building 66-580, Cambridge, MA 02139 and Michael S. Strano, 66-566 Department of Chemical Engineering, Massachusetts Institute of Technology, 77 Massachusetts Avenue, Cambridge, MA 02139.

We carried out annealing studies on transparent thin films made with functionalized metallic single-walled carbon nanotubes (SWNT), and also investigated the effect of functional groups on the conductivity of SWNT thin films. Metallic SWNT films were prepared by selective functionalization using 4-hydroxybenzene diazonium salts followed by density-induced centrifugation, as reported in our previous work. Sheet resistance of functionalized metallic SWNT film is extremely high (23,564 kÙ/sq at 75% transmittance); however it reduces down to 1.2kÙ/sq, i.e., 5% of the un-annealed value, when the film is annealed at above 400oC in N2 flow, because all the functional groups are completely removed as confirmed by Raman and FT-IR experiment. Furthermore, this value is lower by 20% than that of un-separated SWNT film, indicating that metallic SWNT film shows higher conductivity than un-separated SWNT film. However, when the metallic SWNT film is annealed at higher temperature (600oC), the transparency of the film is increased, resulting in the increase of sheet resistance, due to the strong bundling of the SWNT, and the sheet resistance further increases due to the isolation of metallic SWNT from the network. We also demonstrated the effect of surfactant impurities on the conductivity of SWNT films.

## 188i Modeling Sustained Chain Reactions In Carbon Nanotubes

**Nitish Nair**, Department of Chemical Engineering, Massachusetts Institute of Technology, Building 66-580, 25 Ames St, Cambridge, MA 02141 and Michael S. Strano, 66-566 Department of Chemical Engineering, Massachusetts Institute of Technology, 77 Massachusetts Avenue, Cambridge, MA 02139.

We have developed a computational technique for tracking chain reactions along functionalised one-dimensional lattices, specifically single walled carbon nanotubes (SWNT). A hybrid Molecular Dynamics/Monte Carlo algorithm has been used for a coarse-grained model of a SWNT. The parametric space comprises the activation energies of the functional groups, and the fraction of enthalpy released during the reaction which is converted into translational kinetic energy. With the help of these two handles, we have estimated the stabilities of the proposed nanostructures and explored boundaries within which a reaction that has been initiated at one end of the nanotube will be sustained along its length. Finally, the effect of the density of loading of the functional groups has been studied with respect to the aforementioned details.

## 188j A Quantitative Description of Absolute Endocytosis and Exocytosis Rates of Nanoparticles

**Hong Jin**, Chemical Engineering, Massachusettes Institute of Technology, Building 66-580, 25 Ames St., Cambridge, MA 02139, Daniel A. Heller, Department of Chemistry, University of Illinois at Urbana-Champaign, 104 RAL, Box 93-5, MC-712, 600 S. Mathews Ave., Urbana, IL 61801, Richa Sharma, Chemical Engineering, MIT, 66-565, 25 Ames St., Cambridge, MA 02139, and Michael S. Strano, 66-566 Department of Chemical Engineering, Massachusetts Institute of Technology, 77 Massachusetts Avenue, Cambridge, MA 02139.

The endocytosis of nanoparticles is important to drug delivery, bioimaging and toxicology. However, there is no quantitative model linking internalization rate with nanoparticle geometry. Current theory suggests that the rate is bracketed by elastic energy release as the cell membrane wraps the nanoparticle and an entropic penalty as additional cell surface receptors are recruited to continue the process. While successful in predicting a maximum uptake at a given radius, the approach requires this value be a threshold below which endocytosis is impossible and this is in direct contradiction to experiment. To address this, we develop a deterministic kinetic model and assert that nanoparticles smaller than the predicted cutoff can cluster on the cell surface after diffusion, thereby gaining the free energy necessary to surmount the entropic barrier for uptake. We focus on $d(GT)_{15}$ wrapped SWNT[1], and use its intrinsic band gap fluorescence to track interactions with 3T3 cells using a 2D InGaAs imaging array coupled to an inverted microscope with a perfusion stage[2].

The model accurately describes uptake rates from single particle tracking for single walled carbon nanotubes(SWNT)[2] of lengths from 130±18 to 660±40 nm and literature data for Au nanoparticles of diameters from 14 to 100 nm. Remarkably, both have a maximum rate near 25 nm when scaled using an interaction radius for membrane diffusion. The endocytosis rate constant of SWNT ($10^{-3}$ $min^{-1}$) is nearly 1000 times that of Au nanoparticles ($10^{-6}$ $min^{-1}$). The recycling (exocytosis) rate constants are similar in magnitude ($10^{-4}$ to $10^{-3}$ $min^{-1}$) for poly(D,L-lactide-co-glycolide), SWNT and Au nanoparticles and across cell lines and generally decrease with increasing size. The ability to understand and predict the cellular uptake of nanoparticles quantitatively should find utility in designing nano-systems with controlled toxicity, efficacy and functionality.

(1) Jin, H.; Jeng, E. S.; Heller, D. A.; Jena, P. V.; Kirmse, R.; Langowski, J.; Strano, M. S. Macromolecules 2007, 40, 6731-6739.

(2) Jin, H.; Heller, D. A.; Strano, M. S. Nano Lett. 2008, ASAP.

# 188k Synthesis and Surface Chemistry of Oligonucleotide-Templated Nanocrystals

**Jong Hyun Choi**[1], Kok Hao Chen[2], Jae-Hee Han[1], Amanda Chaffee[2], Alice Chang[1], and Michael S. Strano[3]. (1) Chemical Engineering, Massachusetts Institute of Technology, 77 Massachusetts Avenue, 66-580, Cambridge, MA 02139, (2) Chemical and Biomolecular Engineering, University of Illinois, Roger Adams Laboratory, 600 S. Mathews Avenue, Urbana, IL 61801, (3) 66-566 Department of Chemical Engineering, Massachusetts Institute of Technology, 77 Massachusetts Avenue, Cambridge, MA 02139

Semiconductor and magnetic nanoparticles hold unique optical and magnetic properties, and great promises for bio-imaging and therapeutic applications. Here, we present a novel scheme of synthesizing fluorescent PbS and magnetic Fe3O4 nanoparticles (NPs) using DNA and RNA oligonucleotides with various sequences. The bonding chemistry between oligonucleotides and NPs is studied using FTIR, suggesting that the different chemical moieties of nucleobases passivate the NP surface. Strong coordination of oligonucleotides provides the chemical and colloidal stabilities, leading to high particle yields. The PbS NPs have a variable fluorescence quantum yield depending on the sequence used, and a narrower size distribution than Fe3O4 NPs. Numerical simulations confirm that the reactivity of oligonucleotides is responsible for surface capping and growth of these two types of NPs. The experimental and numerical results show that the oligonucleotide sequence plays a crucial role in creating NPs, and this capping chemistry is applicable for various types of nanocrystals. This new synthesis scheme should be valuable in interfacing inorganic NPs with biomolecules for a variety of biological applications.

## 188l Nanotherapeutic Design Using Predictions of Nanoparticle-Protein Interactions

**Esther S. Jeng**, Chemical Engineering, Massachusetts Institute of Technology, Building 66-580, 77 Massachusetts Ave., Cambridge, MA 02139 and Michael S. Strano, 66-566 Department of Chemical Engineering, Massachusetts Institute of Technology, 77 Massachusetts Avenue, Cambridge, MA 02139.

While many biological processes can be controlled by small molecule inhibitors in the binding site of a specific protein, control of most biological processes requires disruption of protein-protein interactions. Nanoparticles provide a scaffold that allows for the design of specific functional groups on a surface for binding to a larger protein-protein interfacial site. Nanoparticles can also be further functionalized with reporter molecules for tracking and sensor purposes. Although many different nanoparticle materials have been functionalized and subsequent binding with proteins has been measured experimentally, these binding events have not been modeled to date. We present a model to predict binding energies of functionalized nanoparticles with biomolecules. A standalone nanoparticle functionalizer and builder were created to allow for binding calculations between target proteins and moieties of various charges, sizes, and configurations. This model can inform about potential candidates for binding systems or reject structures to direct subsequent synthesis of the nanoparticles.

# 188m Thermal Anisotropy In Carbon Nanotubes Measured by Transient Thermotransmittance and Anisotropic Effects on Reaction Rates In Composites

**Joel T. Abrahamson**[1], Nitish Nair[1], and Michael S. Strano[2]. (1) Department of Chemical Engineering, Massachusetts Institute of Technology, Building 66-153, 25 Ames St, Cambridge, MA 02139, (2) 66-566 Department of Chemical Engineering, Massachusetts Institute of Technology, 77 Massachusetts Avenue, Cambridge, MA 02139

Transient thermotransmittance (TTT) offers a way to measure thermal conductivity and interfacial resistance in nanomaterials. TTT measurements on aligned arrays of carbon nanotubes help us to study their anisotropic heat transport properties. We can also examine the effects of covalent functionalization. Nor is TTT limited to materials made only of nanotubes. An attractive target for study is a novel nanocomposite where single-walled carbon nanotubes (SWNT) are surrounded by highly energetic metal nanoparticles or organic molecules. Our simulations of the energy balances for a SWNT coated with reactive Zr metal and oxide indicate that coupling to thermal transport in the nanotube accelerates reaction front velocity along the length of the Zr annulus from 530 to 5100 mm/s. This offers a proof-of-concept for one-dimensional anisotropic energetic materials, which could find new applications in inorganic synthesis and novel propellants. Nanotube conductivity and the relative sizes of the annulus and the nanotube limit the reaction velocity enhancement, rather than interfacial resistance. This is due to the large temperature differences between nanotube and annulus at the reaction front.

[1] J. T. Abrahamson, N. Nair, M. S. Strano. Nanotechnology. 19, (2008), 195701

[2] P. M. Norris et al. Rev. Sci. Instr. 74, (2003), 400

## 188n Centerline Placement and Alignment of Single Carbon Nanotubes In Cylindrical Droplets of Nanometer Diameter

**Richa Sharma**, Chemical Engineering, MIT, 66-565, 25 Ames St., Cambridge, MA 02139 and Michael S. Strano, 66-566 Department of Chemical Engineering, Massachusetts Institute of Technology, 77 Massachusetts Avenue, Cambridge, MA 02139.

The hydrodynamic flow patterns inside high aspect ratio evaporating cylindrical droplets have utility for parallel alignment, positioning and placement of nanoparticles. We successfully create ~200 nm diameter cylindrical droplets for the first time using alternating hydrophobic and hydrophilic SAM layers made by a micro-contacting printing technique. We find that as the diameter of the droplet is reduced below 1000 nm (950-175 nm), a reversal of the internal flow is evident with highly accurate parallel alignment and placement of single walled carbon nanotubes along the centerline with 95 % precision. In contrast deposition is along the edges for droplets larger than 3 um in diameter. This flow reversal is a direct consequence of the evaporation mode switching for the narrow droplets. A model is developed to describe the transition from radially outward to inward flow. This technique should be useful for precisely aligning and placing individual or bundled anisotropic nanoparticles (i.e. carbon nanotubes) with arbitrary surface chemistries.

## 188o Complexity In Nanocrystal Self-Assembly: Size Mixing, Rods, Disks, Triangles, and Multiple Length Scales

**Brian A. Korgel**, University of Texas at Austin, Department of Chemical Engineering, Austin, TX 78712

Advances in nanomaterials synthesis have enabled a variety of size and shape monodisperse nanocrystals with characteristic dimension of ~10 nm and less to be obtained. These nanocrystals are excellent models for studying self-assembly since they readily disperse in a variety of solvents and are too small to be affected to gravitational settling. These nanocrystals have been observed to self-assemble into a wide range of different ordered structures, depending on the nanocrystal size, shape, and assembly process. Many of the fundamental forces that drive these self-assembly processes are known, however, experimental studies continue to surprise with many observations of unexpected complexity. This presentation will detail the fundamental underpinnings of nanocrystal self-assembly and will present a range of some surprising examples of self-assembly, including breath-figure templated inverse opals of nanocrystals, aligned nanorod "stripes", mixed monolayers of spherical and rod-shaped nanocrystals, and ordered arrays of triangular nanocrystals.

**ABSTRACT**

A fractal analysis is presented for the binding and dissociation (if applicable) of (a) different bradykinin concentrations (in nM) in solution to bradykinin B2 receptors immobilized on a resonant waveguide grating (RWG) biosensor surface (Feng et al., 2006), (b) mbetaCD cholesterol in solution to KeLa cells cultivated on a gold-coated prism surface (Ziblat et al., 2006), and (c) a calcium+FRET-based calcium biosensor employing troponin C. TN-XL fluorescence was observed in vivo in this case (Mank et al., 2006). Both single- and dual-fractal analysis were used. The dual-fractal analysis was used only when the single-fractal analysis did not provide an adequate fit. The fractal dimension provides a quantitative measure of the degree of heterogeneity present on the biosensor chip surface. The fractal dimension for the binding and the dissociation phase is not a typical independent variable, such as analyte concentration that may be directly manipulated. It may be considered as a derived variable. An increase in the fractal dimension value or the degree of heterogeneity on the biosensor surface, leads in general, to an increase in the binding rate coefficient. For example, for the binding of different bradykinin concentrations in solution in the 8-128 nM range (Feng et al., 2006), and for a dual-fractal analysis the binding rate coefficient, k2 exhibits a 6.57 order of dependence on the fractal dimension, Df or the degree of heterogeneity that exists on the biosensor surface. This indicates that, in this case at least, the binding rate coefficient, k2 is very sensitive to the fractal dimension or the degree of heterogeneity that exists on the biosensor surface.

# 188q Nanocomposites from An Amorphous Polyamide: Effect of Organoclay Structure on Morphology and Properties

**Youngjae Yoo** and D. R. Paul. Department of Chemical Engineering and Texas Materials Institute, The University of Texas at Austin, Austin, TX 78712

Amorphous polyamide (a-PA) nanocomposites were prepared by melt processing from three organoclays, $M_3(HT)_1$, $M_2(HT)_2$ and $(HE)_2M_1T_1$, to explore the effect of the organoclay structure on the extent of exfoliation and properties of these nanocomposites. Wide angle X-ray scattering (WAXS), transmission electron microscopy (TEM), and stress-strain behavior were used to examine the effects of organoclay structure on morphology and mechanical properties as well as the degree of exfoliation of the nanocomposites. For quantitative assessment of the structure of the nanocomposites, a detailed particle analysis from TEM images was made to provide various averages of the clay dimensions and aspect ratio. The results evaluated from different methods were generally consistent with each other. Nanocomposites based on the organoclays with one alkyl tail and hydroxyl ethyl groups gave well-exfoliated structures and high matrix reinforcement while nanocomposites from two-tailed organoclay contain a considerable concentration of intercalated stacks. Nanocomposites from the organoclays with one alkyl tail showed slightly better exfoliation and matrix reinforcement than those from the organoclays with hydroxyl ethyl groups. The organoclay structure trends for a-PA are analogous to what has been observed for nylon 6; this suggests that a-PA, like nylon 6, has good affinity for the pristine silicate surface of the clay leading to better exfoliation and enhanced mechanical properties with one-tailed organoclay than multiple-tailed organoclay. Furthermore, heat distortion temperatures were predicted from the dynamic mechanical properties of nanocomposites.

## 188r Degradable Nanoparticles as Alternative Ultraviolet Radiation Filters for Sunscreens

**Samuel C. Halim**[1], Ludwig K. Limbach[1], Tobias J. Brunner[1], Stefan Loher[1], Evagelos K. Athanassiou[1], Robert N. Grass[1], Neil Osterwalder[1], Norman A. Luechinger[1], Fabian Koehler[1], and Wendelin J. Stark[2]. (1) ETH Zurich, Zurich, Switzerland, (2) Chemistry and Applied Biosciences, ETH Zurich, Wolfgang-Pauli-Str. 10, ETH Hönggerberg, Zurich, 8093, Switzerland

The use of sunscreen to protect our skin against the damaging action of intense solar irradiation has become part of daily life particularly in the tropics. An increasingly controversial discussion, however, arises from modern sunscreen formulation's possible health risk. Widely used organic ultraviolet (UV) filters were reported to diffuse through intact human skin. Water streams containing elevated levels of these compounds affect the gender ratio of fish, snails and micro-organism in contaminated rivers, lakes and even seaside water. These health concerns have provoked development and application of inorganic UV filters to partially replace or minimize the use of organic filters with potential oestrogenic activity.

The rapidly increasing use of such nanoparticulate titania or zinc oxide pigments has initiated a debate about a possible penetration of these tiny particles through the human skin. The large production volumes of cosmetics directly entail the question whether the release of persistent inorganic nanoparticles into the environment may be harmful for our ecosystem. While both human health concerns and ecotoxicological investigations are under way, requiring several years to completion, the only timely alternative to a wait-and-see attitude is a proactive search for alternative sunscreen pigments circumventing above risks.

Here we show the one-step production of a new class of bio-degradable inorganic pigment nanoparticles with an UV radiation absorbance potential comparable to currently used titania. Our results suggest the substitution of currently used persistent inorganic UV filters in modern sunscreen formulations through such biodegradable compounds thus avoiding potential harm to both the environment and the human organism.

References

[1] T.J. Brunner, R.N. Grass, W.J. Stark, Chem. Commun., 13, 1384-6 (2006).

[2] S. Loher, W.J. Stark, M. Maciejewski, A. Baiker, S.E. Pratsinis, D. Reichardt, F. Maspero, D. Günther, Chem. Mater. 17 (1), 36-42 (2005).

[3] S. C. Halim, L. K. Limbach, T. J. Brunner, S. Loher, E. K. Athanassiou, R. N. Grass, N. Osterwalder, N. A. Luechinger, F. Koehler & W. J. Stark, in review (2008).

# 188s Static and Dynamic Properties of Molecularly Thin Perfluoropolyether Films

**Pil Seung Chung** and Myung S. Jhon. Chemical Engineering, Carnegie Mellon University, 5000 Forbes Ave., Pittsburgh, PA 15213

The ultimate goal of this paper is to find the best lubricant meeting all the requirements for ultra-small head-media spacing (less than 6.5 nm) in hard disk drive (HDD) systems. Newly synthesized lubricants (i.e., ztetraol multidentate (ZTMD)) [1] as well as an optimal blend ratio of the lubricants [2] can be a feasible and promising alternative for future HDD. Here, we examine the static and dynamic properties of nano-scale lubricant thin films of single component and binary mixture perfluoropolyether (PFPE) via molecular dynamics (MD) simulations.

The thin films were modeled via a coarse-grained, bead-spring model, which neglects the detailed atomic structural information while keeping the essence of the polymer molecular structures [3]. Lennard-Jones and finitely extensible nonlinear elastic models were used to describe potential energies among the beads in the system. Solid surface, which makes the films confined, is assumed perfectly smooth. With the total potential energy constructed, MD simulations were performed using the Langevin equation.

We examined the detailed conformation and dynamics of binary mixture films by analyzing the anisotropic radius of gyration and the self-diffusion coefficient as a function of chemical structures including functional groups, film thicknesses, and the blend ratio of the mixture. Our preliminary analysis shows that the film thicknesses and the diffusion processes are mainly dependent on the chemical structures of PFPE molecules and also found that a binary mixture monolayer provides more desirable properties in uniformity and mobility as compared to a single component case. By analyzing functional bead density, we observed that multilayer PFPE films with functional groups (e.g., ZTMD, Zdol, A20H) have layering structures. Furthermore, most monolayer single component lubricants have flat conformation. Due to strong interaction between functional groups and the surface, ZTMD exhibits more flat conformation, representing that monolayer ZTMD film becomes flatter, while it shows poor mobility. By mixing ZTMD and Z (i.e., PFPE without functional group), we found that both the film thickness and the mobility were enhanced for the specific blend ratio. We analyzed the diffusion mechanism connected with the film conformation and found that ZTMD molecules form the network structure to restrict the diffusion of Z molecules due to its strong intermolecular interaction. The static nano-structures and dynamic mobility changes were examined in conjunction with the optimal lubricant selection by systematically tuning the volume fraction.

Reference:

1. X.-C. Guo, B. Knigge, B. Marchon, R. J. Waltman, M. Carter, and J. Burns, "Multidentate functionalized lubricant for ultralow head/disk spacing in a disk drive," J. Appl. Physics, vol. 100, 044306, (2006).

2. P. S. Chung, H. Chen, and M. S. Jhon, "Molecular dynamics simulation of binary mixture lubricant films," J. Appl. Physics, vol, 103, 07F526 (2008).

3. Y.–T. Hsia, Q. Guo, S. Izumisawa, and M.S. Jhon, "The dynamic behavior of ultrathin lubricant films," Microsyst. Technol., vol. 11, 881-886, (2005).

# 188t Rheology and Tribology of the Nano-Scale Confined Polymeric Thin Films Using Molecular Dynamics

**Pil Seung Chung** and Myung S. Jhon. Chemical Engineering, Carnegie Mellon University, 5000 Forbes Ave., Pittsburgh, PA 15213

We examined the rheological and tribological properties of perfluoropolyether (PFPE) nanofilms to investigate the rheological and tribological properties of the polymeric thin films [1, 2] using molecular dynamics (MD) simulation [3].

The PFPE molecules were modeled via a coarse-grained, bead-spring model, which neglects the detailed atomistic structural information while keeping the essence of the molecular structure. Lennard Jones and finitely extensible nonlinear elastic models were used to describe potential energies among the beads in the system. Solid surfaces that confine the PFPE film are assumed perfectly flat. With the total potential energy constructed, MD simulations were performed using the Langevin equation.

We examined the shear deformation of confined, nano-scale PFPE films. By analyzing the viscosities and the relaxation processes as a function of wall interaction/separation and endgroup functionality, the nano-rheological response of PFPE films with and without functional endgroups was examined. We further examined the system with ultra-thin films to demonstrate solid-like deformation. The dynamic performance of nano-films by tuning the film thickness and the endgroup functionality was discussed in conjunction with the relaxation processes.

To examine the tribological fundamentals, we simulated the nano-mechanics (i.e., "compression" and "tension") of molecularly thin PFPE films. For "compression" process, functional PFPEs give higher clearance than nonfunctional PFPEs; while for "tension" process, a fluid bridge was more apparent for functional PFPEs. The normal stress profiles were calculated for both "compression" and "tension" processes, where the observed hysteresis phenomena indicate the irreversible nature of functional PFPE nano-mechanics. N-modes Maxwell model was then used to analyze the relaxation processes and we obtained a second relaxation process for N=2 case, suggesting the role of functional endgroups.

Reference:

1. T.E. Karis and M.S. Jhon, "The Relationship between PFPE Molecular Rheology and Tribology," Tribology Letters, 5, 283-286 (1998).

2. R.N. Kono, S. Izumisawa, M.S. Jhon, C.A. Kim, and H.J. Choi, "Rheology of perfluoropolyether lubricants," IEEE Trans. Mag. 37, 1827 (2001).

3. Q. Guo, P.S. Chung, H.G. Chen, and M.S. Jhon, "Molecular rheology of perfluoropolyether lubricant via nonequilibrium molecular dynamics simulation", Journal of Applied Physics, 99, No. 08N105 (2006).

4. M.S. Jhon, "Physicochemical Properties of Nano Structured Perfluoropolyether Films," Advances in Chemical Physics, 129, 1-79 (2004).

# 188u A Novel Heat Transfer Model for Ultrashort Laser Heating on Metals

**Dehee Kim**, Chemical Engineering, Carnegie Mellon University, 5000 Forbes Avenue, Pittsburgh, PA 15213, Sartaj S. Ghai, ExxonMobil Upstream Research Company, 3120 Buffalo Speedway, Houston, TX 77098, and Myung S. Jhon, Department of Chemical Engineering, Carnegie Mellon University, 5000 Forbes Avenue, Pittsburgh, PA 15213.

To overcome the technological limits of conventional magnetic recording systems, the heat assisted magnetic recording (HAMR) technology is being developed [1]. The HAMR technology utilizes the phenomenon of magnetic material property variation by tuning the temperature. Heating by a laser beam on the disk reduces the magnetic coercivity, and thus dramatically increases the aerial density. In this system, a nanoscale heat transfer process is initiated by laser heating, and the energy transport affects physical properties of the magnetic media as well as the stability and reliability of the system. Therefore, it is necessary to accurately estimate the heat conduction in the sub-continuum regime.

In addition to the HAMR application, similar concept was recently applied for micro-/nano-fabrication and material processing of thin metallic films as femtosecond laser heating has made technical innovation [2-4]. The thermal behavior of energy carriers, more specifically, electrons and phonons play a vital role in the theoretical analysis of sub-continuum thermal transport in metals.

The Boltzmann transport equation (BTE, represented in phase space) is used for the analysis by calculating the probability distribution function of both energy carriers (electron and phonon) inside the metals. These two BTEs for electrons and phonons should be solved simultaneously to examine microscale heat transfer in metals. The treatment of energy carriers as quasi-particles is simple yet powerful methodology, nonetheless, the formulation of BTEs becomes computationally intensive to solve. Furthermore, the coupling between the two BTEs for energy transport is still not understood very well, which creates difficulties in performing a full scale BTE simulation. The lattice Boltzmann method (LBM) is applied to complement BTE in this paper [5]. LBM is a technique that can be applied to solve the BTE in a simplified manner on lattices while maintaining the accuracy, as well as reducing the heavy computation tasks [5]. In our previous works [6, 7], LBM was developed and successfully applied to simulate the heat conduction of phonons. This method led us to develop a new LBM model to solve the energy transport of both electrons and phonons in metals. Here, we examine a thin gold film heating via ultrashort laser pulse and studied the electron-phonon coupling phenomenon [4].

For accurate prediction of the heat conduction in metals it is important to couple the two lattice Boltzmann equations properly. The Taylor series expansion and least squares based LBM (TLLBM) was adopted to couple two subsystems possessing different length and time scales [8]. The laser-induced heating problem was simulated successfully for the first time for metals using LBM. The numerical result is comparable to the widely used two-temperature models [4]. Although our LBM simulation has been validated only on gold films, it can be applied to other metallic films as long as the material properties are given. The material damage threshold by the hot thermal energy can be estimated using our present work for metals as well as for semiconductors and dielectric materials as shown in our previous work. Our LBM model can be a powerful tool in the investigation of thermal budget for the HAMR system, as laser-induced heating could affect the stability of the system including the substrate, lubricant film, and the air bearing.

REFERENCES


[1] R. E. Rottmayer et al., IEEE Trans. Magnet., Vol. 42, pp. 2417, 2006.

[2] T. Q. Qiu and C. L. Tien, Int. J. Heat Mass Trans., Vol. 35, pp. 719, 1992.

[3] J. K. Chen, J. E. Beraun, L. E. Grimes, and D. Y. Tzou, Int. J. Solids Struct., Vol. 39, pp. 3199, 2002.

[4] L. Jiang and H.-L. Tsai, J. Heat Trans., Vol. 127, pp. 1167, 2005.

[5] S. Succi, The Lattice-Boltzmann Equation for Fluid Dynamics and Beyond, Clarendon Press, Oxford, UK, 2001.

[6] S. S. Ghai, W. T. Kim, R. A. Escobar, C. H. Amon, and M. S. Jhon, J. Appl. Phys., Vol. 97, 10P703, 2004.

[7] S. S. Ghai, W. T. Kim, C. H. Amon, and M. S. Jhon, J. Appl. Phys., Vol. 99, 08F906, 2006.

[8] C. Shu, Y. T. Chew, and X. D. Niu, Phys. Rev. E., Vol. 64, 045701(R), 2001.

## 188v Fluorescent Gold Nanoparticle for Bio-Imaging

**Chang-won Lee**, Bioengineering, University of Utah, 122 S Central Campus Dr. RM 304, Salt Lake City, UT 84112 and Agnes E. Ostafin, Materials Science & Engineering, University of Utah, 122 South Central Campus Drive, Room 316, Salt Lake City, UT 84112.

Water soluble fluorescent gold nanoparticles have been developed for replacing commercial dye molecules and toxic quantum dots in many applications including target specific bio-imaging agents. Except for a few specialized organic molecules whose toxicity profiles are yet to be explored, this fluorescence gold nanoparticle can be the most promising material so far. It is smaller than 5 nm in size, less toxic than their CdSe counterparts and also has much less photo-bleaching property comparing to commercial dyes. Recipes to prepare gold quantum dots trapped in dendrimers to make them bright and water soluble have been reported [1]. We are developing an alternative strategy to make water soluble gold quantum dots, using thiol based capping molecules (Figure 1). This approach has the advantage of producing relatively monodispersed gold nanoparticles, with tailorable solubility, and which is covered with functionalizable carboxylic groups on the surface for the attachment of peptides, DNA, and/or other signal, therapeutic molecules.

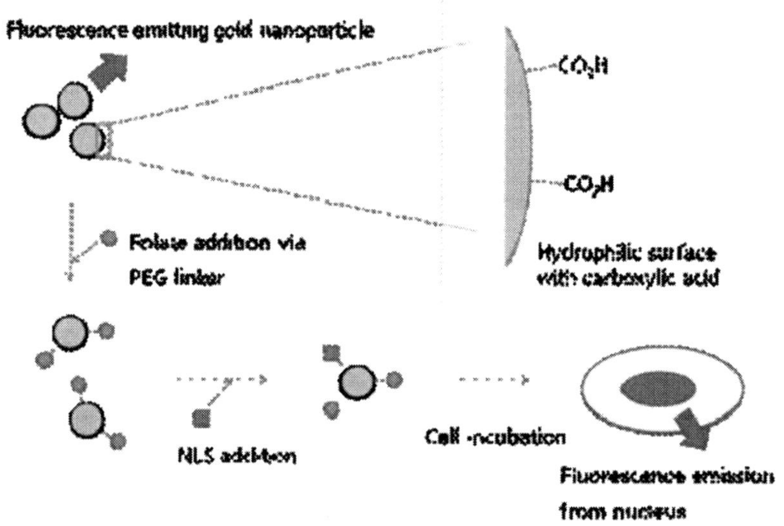


Figure 1. Scheme of gold nanoparticle for bio-imaging

In this work, we studied the emission spectra of gold quantum dots prepared using mercaptooctanoic acid. It has thiol functional groups to bind and stabilize gold surface and carboxylic acid groups on the other side for the anchoring point to the conjugation of other molecules and also provide water solubility of the particles. The TEM analysis showed the average size of the gold nanoparticle is in range of 1 ~ 5 nm in diameter. Large Stoke shift, with UV excitation and 620 nm emission, is monitored and it is one of the unique features for this gold nanoparticle. It is possible that the meta-stable state orbital has formed in between the nanoparticles HOMO and LUMO orbitals to lower the emission energy which eventually results far red-shifted emissions. The intensity of the emission is linearly proportional to the concentrations of the gold nanoparticle.

Changing the excitation and emission property of the gold nanoparticles by making alloy nanoparticles of gold and other transition metal atoms such as silver, platinum or palladium could be possible because the affinity to the electrons are different by each atoms which will result the change in orbital energy levels.

As a demonstration of the specific delivery application, folic acid has been conjugated on the surface using two different chemistries via PEG linker and further investigation will be performed with the cancer cells. Nucleus localization sequence (NLS) can be further attached to make the particles specifically delivered the nucleus of cancer cells as illustrated in Figure 1. Interestingly, upon the surface modification with folic acid via PEG linker, the emission color has been changed from 620 nm (red) to 460 nm (blue) without changing the excitation wavelength (Figure 2). It can also be explained as meta-stable state orbital of the gold nanoparticle is disrupted by the surface modification.

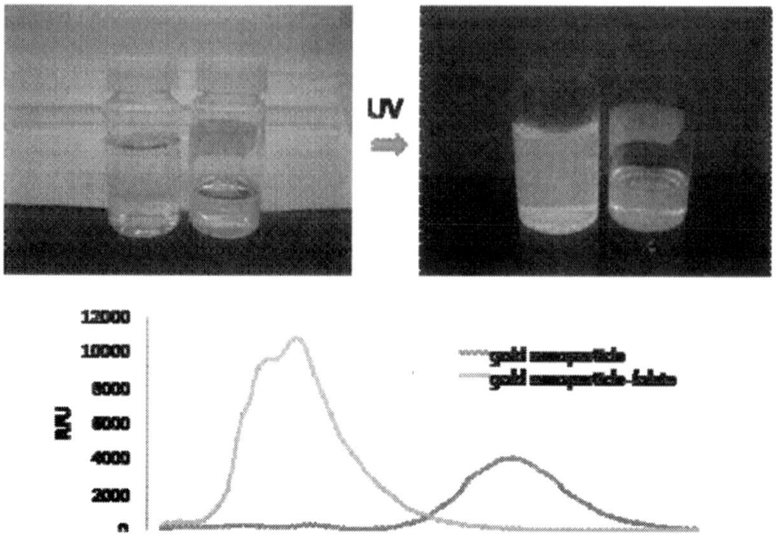


Figure 2. Pictures and emission plot @ 280 nm excitation of gold nanoparticles before and after folic acid attachment showing the change of the emission

Even though it is still in the on-going stage, we have successfully developed water soluble fluorescent gold nanoparticles and showed some of the possible modifications for bio-imaging applications. Further studies including the changing the emission-excitation properties of nanoparticles, conjugation of other targeting molecules, and folic acid-positive carcinoma incubation for the folic acid conjugated gold nanoparticles will be performed as future work of this project.

[1] Zheng et al., Physical Review Letters, 2004, 93, 77402-1~4.

## A simple methodology for producing super-hydrophobic composite films on various substrates

*P. N. Manoudis[1], I. Karapanagiotis[2], A. Tsakalof[3], I. Zuburtikudis[4], and C. Panayiotou[1]*

[1] *Department of Chemical Engineering, Aristotle University of Thessaloniki, 54124, Thessaloniki, Greece*
[2] *"Ormylia" Art Diagnosis Centre, Ormylia, Chalkidiki, 63071, Greece*
[3] *Medical Department, University of Thessaly, Larissa, 41222, Greece*
[4] *Department of Industrial Design Engineering, TEI of Western Macedonia, Kozani, 50100, Greece*

### Abstract

Various nanoparticles (silicon oxide, tin oxide and aluminium oxide) were mixed with poly (methyl siloxane) (Rhodorsil 224). The mixtures were sprayed on silicon substrates resulting in the formation of a rough, two-length-scaled hierarchical structure surface layer. We report that: (i) Super-hydrophobic composite films were produced with all the particles used, (ii) The effect of the particle's nature on the developed hydrophobicity was examined showing that the maximum static water contact angle is independent of the nature of the polymer-particle binary mixture, (iii) The effect of the particle concentration on the hydrophobic character of the composite film was also examined showing that the static contact angles measured on surfaces which were prepared from dilute dispersions, increase rapidly with particle concentration reaching a maximum value ($162 \pm 2^0$) and (iv) When the static contact angle is maximum, the contact angle hysteresis reaches some minimum values ($2$-$4^o$) providing evidence that the composites prepared with high particle concentration are super hydrophobic and exhibit augmented water repellency. The methodology was also tested on various substrates (glass, marble, concrete, aluminum, silk, wood). The static water contact angle measurements revealed that the substrate has no significant effect on the superhydrophobic character of the applied coatings.

### Introduction

Water repellency is important in many biological processes and technological applications. Prevention of snow sticking and metal corrosion, protection of monuments and self cleaning surfaces are only some of the applications where enhanced water repellency is required [1].

In nature, the leaves of lotus (*Nelumdo nucifera)* have been under thorough study showing that the high water contact angle is the result of the combination of the rough microstructure (cells) and the rough nanostructure (wax crystals) with a wax of low surface energy covering them [2-4]. The "lotus effect" has inspired the efforts for the creation of synthetic super-hydrophobic surfaces.

Within this framework the general strategy for the production of super-hydrophobic surfaces includes the use of materials with low surface energy and the creation of rough surfaces. The combination of surface roughness and low surface tension according to the Wenzel and Cassie-Baxter equations [5] enhances the hydrophobic character of a surface.

Most of the developed methodologies include the formation of roughness in the nano-scale, with plasma treatment, sol-gel or by picking suitable solvents for polymers [6-8].

In previous works we presented a simple methodology which was applied on a wide variety of surfaces [9,10]. The methodology included the addition of nanoparticles in polymeric solutions. The mixture was then sprayed on the substrate. Some of the advantages of the methodology are that: it is a low-cost and very simple method as it is a one step process and includes the use of some very common materials and that superhydrophobicity can be achieved by using different polymers.

In this work we follow the same methodology and evaluate the effect of the addition of the various particles on the hydrophobic character of the polymer/particle composite film. In particular various nanoparticles such as silicon oxide ($SiO_2$), aluminium oxide ($Al_2O_3$) and tin oxide ($SnO_2$) were used for mixing with a poly(alkyl siloxane) (Rhodorsil 224).

## Experimental

Poly(methyl siloxane) (Rhodorsil 224, Rhodia Silicones) which is a 7% wt. solution in white spirit, was used for mixing with various nanoparticles. The $\theta_s$ of the polymer on smooth substrate after spin coating was $103^0$. The nanoparticles were: (i) Silicon oxide ($SiO_2$) particles (fumed powder, Aldrich) with 7nm mean diameter, (ii) Silicon oxide ($SiO_2$) particles (fumed powder, Aldrich) with 14nm mean diameter, (iii) Aluminium oxide ($Al_2O_3$) particles (Alfa Aesar) with 40-50 nm diameter and (iv) Tin oxide ($SnO_2$) particles with 22-43 nm diameter. Rhodorsil-silica dispersions with particle concentrations 0.1, 0.3, 0.5, 1.0, 1.5 and 2.0 % w/v and Rhodorsil-tin/alumina dispersions with particle concentrations 0.5, 1, 2, 4, 6, 8, 10 % w/v were prepared. Mixtures were subsequently stirred vigorously for 20 min. The polymer-nanoparticles dispersions were immediately sprayed on clean Si wafer substrates. Silicon wafers with <100> orientation containing a native surface oxide were purchased from Siegert Consulting e.K. The spraying system was an airbrush system (Paasche Airbrush) with a nozzle of 733 μm diameter. The coatings were then annealed at 40 °C overnight in vacuum to remove residual solvent and kept at room temperature for 2-3 days. The morphologies of the film surfaces were investigated by Scanning Electron Microscopy (SEM, Jeol JSM 840A). Contact angle measurements were carried out by the sessile drop method using a Krüss DSA 100 (Krüss) apparatus and distilled water. Water droplets (2-4 μl) were delivered to different points of each specimen and from a height sufficiently close to the substrate, so that the needle remained in contact with the liquid droplet. Then, the delivery needle was withdrawn with minimal perturbation to the drop [11] and the image of the drop was captured immediately for contact angle measurement. The contact angle hysteresis was calculated by the dynamic sessile drop method. The advancing/receding contact angle was the maximum/minimum angle measured while the volume of the droplet was increased/decreased without increasing/decreasing the solid-liquid interfacial area. The reported contact angle values are averages of five measurements.

## Results & Discussion

### Surface morphology

In figure 1 SEM images of the surface morphologies formed by the Rhodorsil-particle mixtures sprayed on the substrate are presented. Various surface structures are formed on the substrate by varying the particle concentration (Figure 1, a-d). It is also shown that the morphologies formed are similar for all the particles used but there is a significant difference in particle content. The ratio of the surface area covered by the mixture increased with increasing the particle concentration. Apparently, the surface roughness of the treated substrate increased with increasing the particle content in the polymer-particle mixture. The particles were aggregated creating structures with diameter of about 20-40 μm (Figure 1, a-d). So by varying the concentration of the particles in the polymeric solution different surface morphologies in the micro-scale were formed.

We now turn our attention on the surface structure of the aggregates that protrude above the smooth polymer film. Figure 1 (e-h) show examples of the morphologies of aggregates formed in Rhodorsil-particle films deposited on silicon. It is demonstrated that each aggregate consists of a nanostructure which implies an augmented roughness at the nanoscale. In summary, figure 1 suggests that the randomly distributed protruded aggregates consist of further nanostructures, implying that a two-length-scaled hierarchical structure is formed in the surface. It is known that such structure exists in the lotus leaf which exhibits water repellent properties. In the following, the hydrophobic character and the water repellent properties of the surfaces, described above, are evaluated.

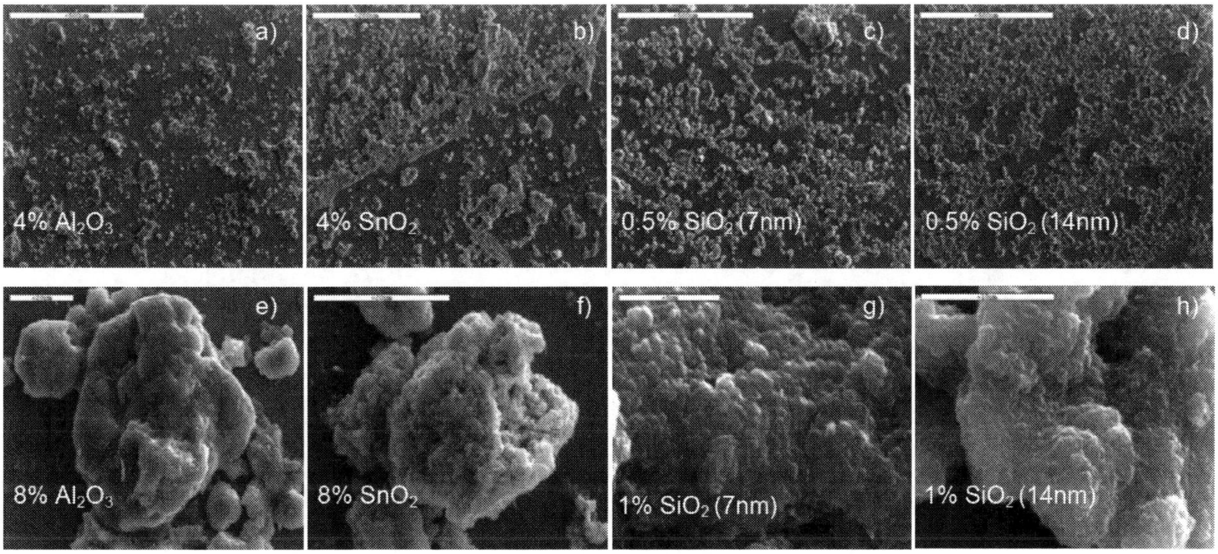


**Figure 1.** SEM images of Rhodorsil-particle films deposited on Si-wafer. The concentrations (w/v) of the particles in the dispersions which were deposited on the substrates are shown in the lower left corner of each image.

**Water contact angle**

Figure 2 shows water static contact angle ($\theta_s$) and contact angle hysteresis measurements, defined as the difference of the advancing ($\theta_a$) and the receding contact angle ($\theta_r$), as a function of particle concentration for siloxane-$SiO_2$ (7nm,14nm) - $SnO_2$ - $Al_2O_3$.

The results show a significant enhancement of the hydrophobicity for all the particles used. The increase follows the same tendency. For low particle content a significant increase of static water contact angle is observed. Beyond a "critical point" further addition of nanoparticles does not have a significant impact on the developed hydrophobicity. The static contact angle reaches a plateau of about $162\pm1^0$ for all the particles used. In a previous work, in PMMA/ particles mixture a maximum water contact angle of $154^0$ was reported. Therefore it is shown that the maximum water contact angle is determined by the surface tension of the polymer and not by the nature of the particles. As it is obtained from figure 2 there is a significant difference in the particle content where the plateau is reached. It is 0.5%, 1%, 4% and 6% w/v for silicon oxide 7nm, silicon oxide 14nm, aluminium oxide and tin oxide nanoparticles, respectively.

The variation of water contact hysteresis with particle content is more complicated. For low concentration of particles there is a gradual increase of advancing contact angle. The receding contact angle remains practically stable leading to an increase of water contact angle hysteresis. The hysteresis reaches some maximum values, which are $40\pm4^0$ depending on the particle used. The maximum hysteresis was measured at 1 % w/v for tin oxide and aluminium oxide nanoparticles and at 0.3% w/v for silicon oxide nanoparticles.

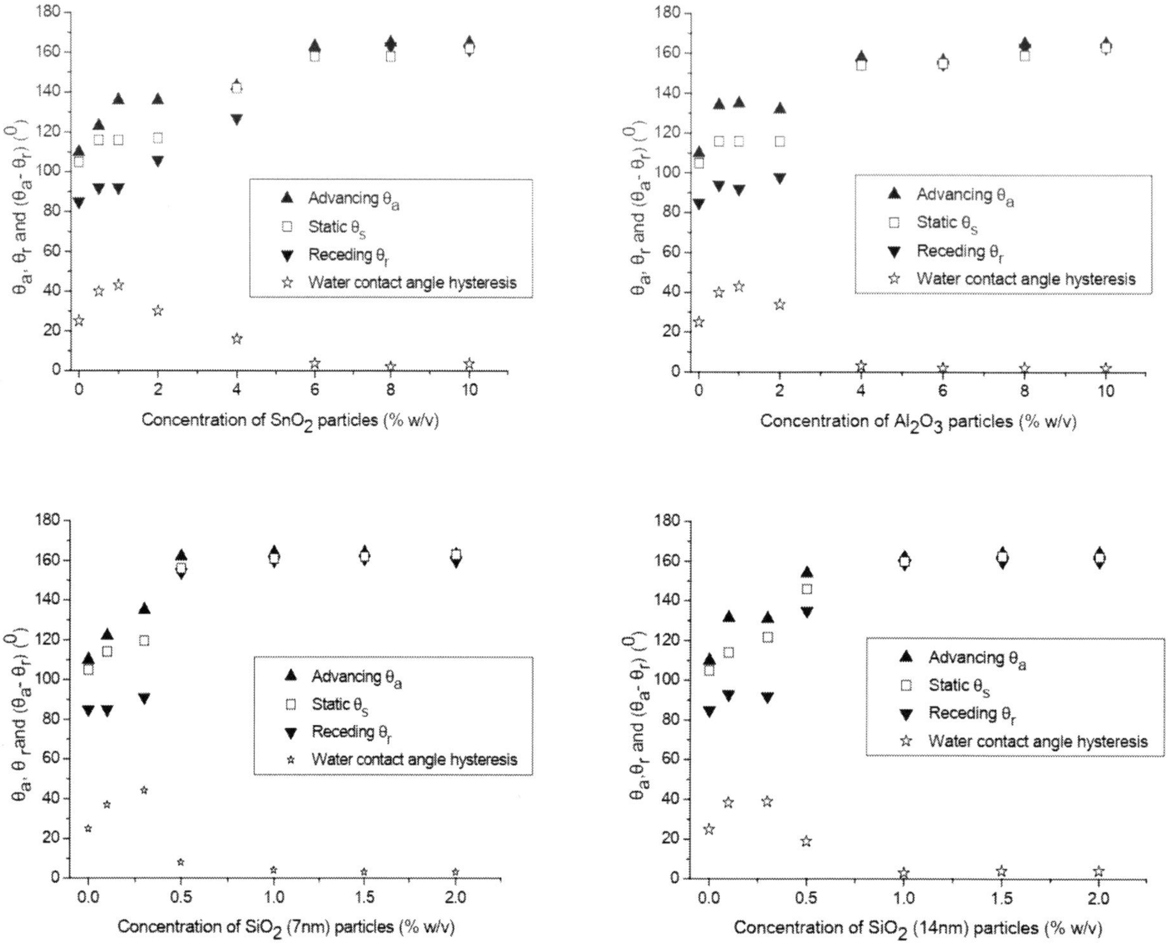


**Figure 2.** Static water contact angle ($\theta_s$) and contact angle hysteresis ($\theta_a$ - $\theta_r$) vs. concentration of particles.

As it was demonstrated in a previous work the effect of the particle concentration on the contact angle hysteresis ($\theta_a$-$\theta_r$) resembles to some extent the effect of roughness on wax surfaces [12,13]. In these works the water contact angle hysteresis first increases with roughness, then decreases rapidly and finally becomes constant. The rapid decrease of hysteresis can be attributed to the rapid transition of the surface from the Wenzel regime to the Cassie-Baxter regime.

In the case of Rhodorsil-particle dispersions the advancing contact angle increases but the increase is more rapid compared to the wax (Figure 2). The receding contact angle remains practically stable at low particle concentration before a significant increase. Finally at elevated particle concentration (>0.5% w/v for silica particles and >2%

for alumina and tin particles) hysteresis decreases rapidly and finally becomes constant. Polymer-particle composite films prepared from dispersions of high particle content correspond to the Cassie-Baxter scenario. This is evidenced by the enhanced $\theta_s$ values and the dramatic decrease in hysteresis that reflects a transition to a "composite" surface characterized by such a large roughness that a liquid cannot completely wet the crevices. The elevated hysteresis in the case of films prepared from dispersions of low particle content can be explained if we consider that water may fill the large, smooth areas that exist among the aggregates. In summary, the large, smooth areas are filled with water (Wenzel) but not the small nanocrevices that exist on the surface of the particle aggregates (Cassie-Baxter). This is a mixed regime.

The minimum hysteresis values reported were 2-4$^0$ depending on the particle used. Static water contact angle of over 150$^0$ and low contact angle hysteresis denotes a super-hydrophobic surface. So figure 2 provides evidence that composites prepared with high particle concentrations are super hydrophobic and exhibit augmented water repellency. This is also confirmed by figure 3. Figure 3 shows consecutive images of a water droplet bouncing on Si wafer covered by siloxane and 10% w/v $SnO_2$ nanoparticles. The droplet rolls-off and finally is removed from the substrate.

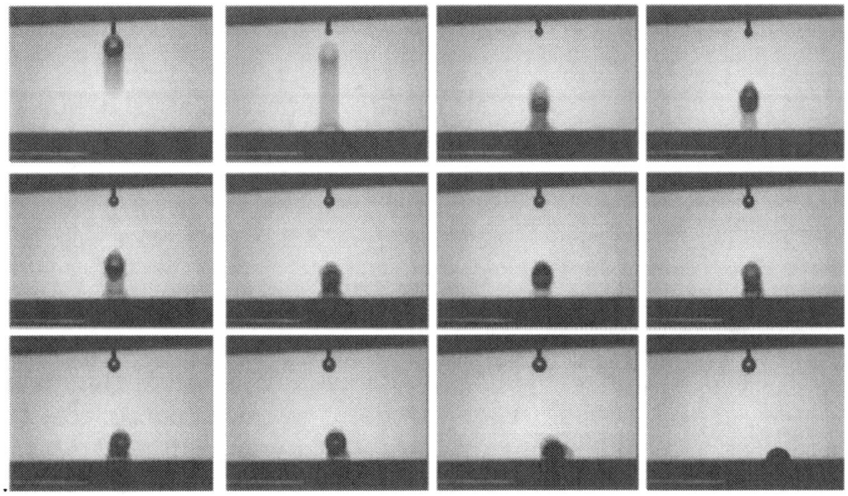

**Figure 3.** Water droplet bouncing on the substrate which was treated with the siloxane and 10% w/v tin particles.

### Application on various substrates

The spray technique has the advantage that it can be easily used to treat various surfaces. Various substrates were treated with Rhodorsil and silicon oxide particles (7nm) and static water contact angle measurements ($\theta_s$) on the treated surfaces of glass, silicon, concrete, aluminum, silk, wood, and marble were carried out. The obtained results are presented in Table 1:

**Table 1.** Static contact angle ($\theta_s$) of pure siloxane, siloxane-$SiO_2$ (2% w/v), applied on various substrates.

| Substrate | Static water contact angle ($^0$) | |
|---|---|---|
| | No particles | Silicon oxide particles (7nm) |
| Cement | 105 | 164 |
| Metal (Iron) | 136 | 160 |
| Fabric (Silk) | 106 | 162 |
| Wood (Plywood) | 130 | 162 |
| Marble (Pentelic) | 109 | 160 |
| Glass | 103 | 164 |

The obtained results show that: (i) The nanoparticles enhance the hydrophobic character of simple polymers applied on various substrate materials, (ii) The static contact angles measured for the polymer-particle composite films are nearly independent of the substrate material.

## Conclusions

A commercial siloxane composition was modified by the addition of nanoparticles (silicon oxide, tin oxide, aluminium oxide) and sprayed on different substrates - silica wafer, glass, wood, iron, silk. As a result rough, two-length-scaled hierarchical structure surface layer is formed. The maximum contact angle succeeded ($162\pm2^0$) was independent of substrate nature and nanoparticles chemistry. However, it was achieved at different concentrations of nanoparticles in modified composition and in particular at 0.5%, 1%, 4% and 6% w/v for silicon oxide 7nm, silicon oxide 14nm, aluminium oxide and tin oxide, respectively. At the same concentrations the very low contact angle hysteresis (2-4$^o$) was observed. Thus the modified siloxane composition enhance substantially the water repellence of the treated surfaces and render them super hydrophobic and self cleaning. The suggested modification is low cost and the modified composition can be applied for the most effective water protection of large surfaces such as monuments, iron construction and buildings.

## Acknowledgements

This research project (PENED) is co-financed by E.U.-European Social Fund (75%) and the Greek Ministry of Development-GSRT (25%). The financial support of Greek State Scholarship's Foundation to P. Manoudis is also gratefully acknowledged.

## References


1. Yildirim H., Erbil A., Demirel L, Avci Y. and Mert O., *Science*, 299 (2003), p. 1377.
2. Wagner P., Fürstner R., Barthlott W. and Neinhuis C., *J. of Exp. Bot.* (2003), *54, pp.*1295-1303.
3. Feng L. Li S., Li Y. Li H., Zhang L., Zhai J., Song Y., Liu B., Jiang L. and Zhu D., *Adv.Mater.* (2002), *14,* pp. 1857-1860.
4. Gao X. Jiang L., *Nature,* (2004), *432,*p.36.
5. Cassie, A. B. D., Baxter, S., *Trans. Faraday Soc.* (1944), *40,* pp. 546-551.
6. Tserepi A. D., Vlachopoulou M. -E. and Gogolides E., *Nanotechnology* (2006), *17,* pp. 3977–3983.
7. Zhang L., Zhou Z. Cheng B, DeSimone J. M. and Samulski E. T., *Langmuir* (2006), *22,* pp. 8576-8580.
8. Mahltig B., Böttcher H., *J. Sol-Gel Sci. Technol. (*2003), *27*, pp. 43-52.
9. Manoudis P., Papadopoulou S., Karapanagiotis I., Tsakalof A., Zuburtikudis I. and Panayiotou C., *J. of Phys.: Conf. Series,* 61, (2007), pp. 1361-1365.
10. Manoudis P.N., Karapanagiotis I. Tsakalof A., Zuburtikudis I. and Panayiotou C., *Langmuir*, (2008), ASAP Article, DOI: 10.1021/la801817e.
11. Murray M. D., Darvell B. W., *J. Phys. D: Appl. Phys.,* (1990), *23,* pp.1150-1155.
12. Quéré, D. *Phys. A* (2002), *313,* pp. 32-46.
13. Johnson R. E. Dettre R. H., *Adv. Chem. Ser.,* (1964), *43,* pp. 112 -135.

# ENHANCED THERMAL STABILITY OF BIODEGRADABLE POLY(3-HYDROXYBUTYRATE)/LAYERED SILICATE NANOCOMPOSITES

*I. Zuburtikudis[1*], S. I. Marras[1], K. Tornikidou[1,2], A. Tsimpliaraki[1], E. Panayotidou[1], G. Christofidou[1]*

*[1]Department of Industrial Design Engineering, TEI of Western Macedonia, 50100 Kozani, Greece*
*[2]Dentist, Private Dental Practice, I. Christidi 25, 50200, Ptolemaida, Greece*

## Introduction

Polymer nanocomposites are a class of materials that has already gained a big popularity among researchers both in academia and industry due to their exceptional properties, which are superior to those of the virgin polymers and of the conventional composites [1]. These hybrid materials consist of a polymer matrix in which inorganic particles with at least one dimension in the nanometer scale are homogeneously dispersed. In the case of polymer layered silicate nanocomposites, the high aspect ratio particles interact with the polymer matrix and improve substantially many physical properties of the pristine polymer including enhanced mechanical characteristics and barrier resistance, decreased flammability and increased biodegrability in the case of biodegradable polymers.

Recently, biodegradable and biocompatible polymers have received significant attention because they are environmentally friendly and are extensively used in biomedical applications. Poly(3-hydroxybutyrate) (PHB) is rapidly gaining recognition as one of the most promising biopolymers [2]. PHB is a linear, aliphatic polyester produced by different types of microorganisms. However, the poor thermal stability at temperatures slightly higher than its melting point restricts the application field of PHB.

In the present work, various loadings of montmorillonite organically modified by octadecylamine ($C_{18}$MMT) were dispersed in PHB by the melt intercalation method. The aim was to produce hybrid material with improved thermal properties over the pristine PHB. Morphological characterization of the composites was conducted by x-ray diffraction analysis (XRD). The effect of clay content on the thermal properties of the material was studied by differential scanning calorimetry (DSC) and thermogravimetric analysis (TGA).

## Experimental Section

### *Materials*

PHB was obtained from Aldrich, while sodium montmorillonite (NaMMT) (CEC=92.6 meq/100 g) was supplied by Southern Clay Products (Texas, USA). Octadecylamine ($C_{18}$A), chloroform ($CHCl_3$) and concentrated HCl were purchased from Sigma-Aldrich. All materials were used as received and without any further purification.

### *Preparation of the organophilic layered silicate.*

Organically modified montmorillonite ($C_{18}$MMT) was prepared through an ion-exchange reaction[3]. An amount of $C_{18}$A equal to 1.5 times the cation exchange capacity (CEC) of the inorganic material and an equivalent amount of HCl were dissolved in a 1:1 solution of ethanol and water at 75 °C. The inorganic material was dispersed in water at 75 °C and then, this aqueous suspension (3 wt%) was added to the alkylammine solution. The mixture was stirred vigorously for 24 h at the

* E-mail address: izub@teikoz.gr

aforementioned temperature. The precipitate was collected by filtration and washed with a hot 1:1 solution of ethanol and water until the $AgNO_3$ test verified the absence of chloride. Eventually, the cation-exchanged montmorillonite was dried and ground to a powder, which was kept in a desiccator for further use.

### *Preparation of hybrid materials*

Nanocomposites were fabricated with the aid of a microextruder-compounder (Minilab, ThermoHaake) at 175 °C and 130 rpm. The compounding time was 3 min. Constant nitrogen flow during nanocomposite preparation prevented thermal decomposition of the matrix. Eventually, nanocomposites with organoclay content ranging from 1 to 10 wt% were obtained.

### *Characterization*

The structure of the nanocomposite materials was investigated by XRD using a Rich. Seifert 3003 TT diffractometer and Ni-filtered CuKα radiation (λ=0.154 nm). The scanning range was varied from 2θ=2° to 10°.

The thermal properties of the produced hybrids were explored with the aid of a Scimadzu DSC-50 differential scanning calorimeter (DSC). Temperature scans were carried out at a heating rate of 10 °C/min under constant nitrogen flow of 20 $cm^3$/min. The samples were heated to 180 °C and remained at that temperature for 2 min in order to release their thermal history and, afterwards, they were quenched to -5 °C. The determination of the thermal characteristics took place during the second run.

Thermogravimetric analysis (TGA) of the hybrids was performed on a Shimadzu TGA-50 with a heating ramp of 10 °C/min up to 750 °C under both flowing nitrogen (20 $cm^3$/min) and static air. TGA runs in isothermal conditions were also conducted at 190 °C in air.

## Results and discussion

### *Organoclay morphology*

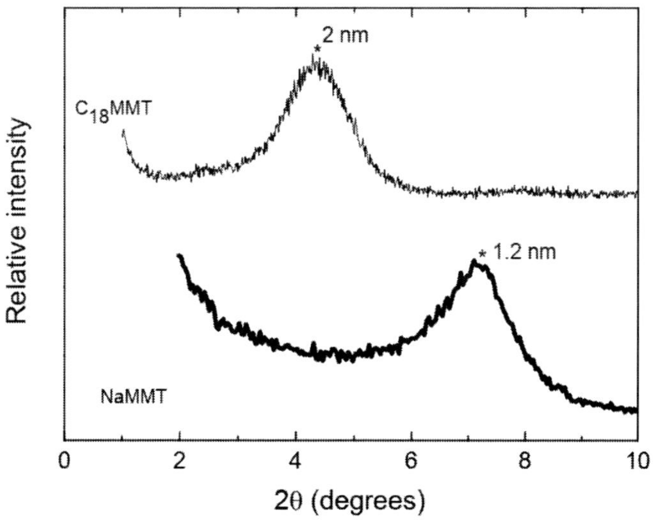


**Figure 1.** X-Ray diffraction (XRD) patterns of the natural and the organo-modified montmorillonite.

The basal spacing of both the natural (sodium) montmorillonite and the organically (octadecylammonium cation) modified montmorillonite were investigated by XRD and the results are presented in Figure 1. It appears that the basal spacing of the natural mineral is 1.2 nm. Insertion of octadecylammonium ions into the interlayer spaces of the natural clay led to an increase of this spacing from 1.2 nm to 2 nm indicating a bilayer arrangement of the alkylammonium molecules [4, 5].

***Structure of PHB hybrids***

Figure 2 illustrates the diffraction patterns of the organoclay and the prepared nanocomposites. The characteristic peak corresponding to the basal spacing of the inorganic material is shifted towards lower 2θ values indicating that polymer chains have diffused into the clay galleries expanding the clay structure. This means, that the morphology of these composites is mainly intercalated, even though, exfoliated clay platelets may coexist. The intensity of the characteristic peak of the clay ($Id_{(001)}$) increases as the percentage of the organo-modified clay in the hybrid material becomes bigger. This is because the number density of the intercalated clay particles in the polymer matrix increases, as the clay content arises, and many more x-rays are diffracted.

Unlike PLLA nanocomposites [6] where the interlayer distance was strongly dependent on the organoclay concentration, the d-spacing of PHB nanocomposites, is not significantly affected by mineral loading.

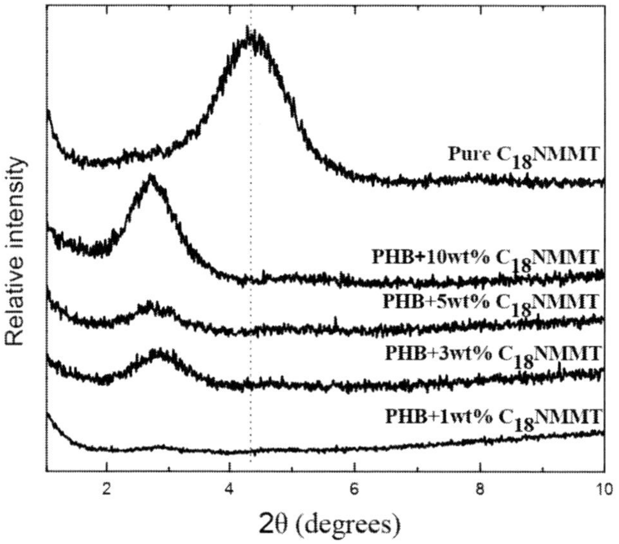


**Figure 2.** X-ray diffraction patterns of $C_{16}$MMT and PHB nanocomposites with various loadings of the organophilic clay.

***Thermal behavior of PHB hybrids***

Figure 3 describes the melting behavior of pure and nanocomposite PHB. It appears that the incorporation of the inorganic material affects the melting behavior of the polymer. Thus, an increase of the clay content causes a slight depression of the polymer's melting point.

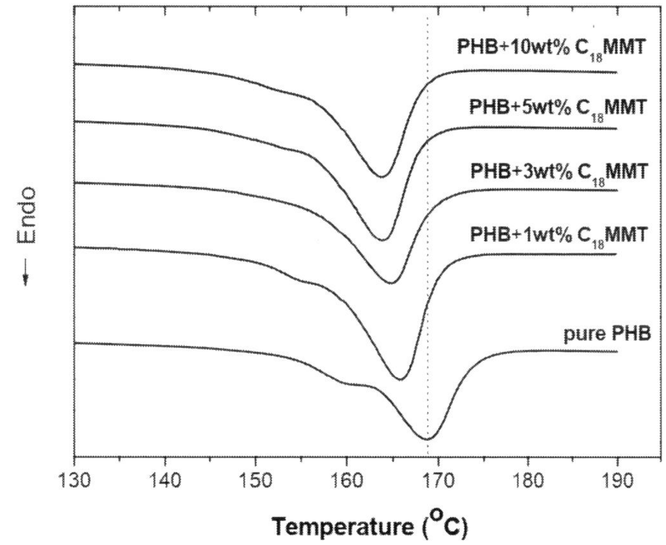


**Figure 3.** Melting endotherms of PHB nanocomposites with various organoclay loadings.

Figure 4a shows TGA data for pure PHB and its hybrids, which were taken in a $N_2$ atmosphere and in the temperature range of 150 °C, (where the samples are considered to be free of humidity), up to 750 °C. With a heating rate of 10 °C/min, PHB volatilizes completely in a single step beginning at about 220 °C. The introduction of organophilic inorganic material significantly improves the thermal stability of the polymer rising the temperature of decomposition initiation. This may be attributed to the labyrinth ('barrier') effect originated by the high aspect ratio clay platelets, which are dispersed in the polymer matrix delaying the escape of volatile degradation products within the nanocomposite [7].

Figure 4b presents the effect of the clay content on the temperature at which 2% of the material mass was lost, in the presence of nitrogen. It is clear that increase of the mineral loading enhances the thermal stability of the polymer. This in accordance with previous works [8].

The degradation properties of the polymer clay nanocomposites were also investigated under oxidative atmosphere and isothermal conditions (190 °C), as shown in Figure 5a. The organoclay addition significantly delays the polymer mass loss. As illustrated in Figure 5b, the nanohybrids, prepared by the incorporation of 10 wt% of organically modified clay, require almost five times the time that the pure polymer needs for a 40% mass loss. This greatly retarded material volatilization may be attributed to the formation of a carbonaceous-silicate char structure, which is produced on the surface of the nanocomposite during burning [9]. In the presence of oxygen, the intimate contact between the clay platelets and the polymer favors this char formation [10], which acts as an excellent insulator and mass transport barrier that delays the diffusion of oxygen from air to the polymer mass and slows down the escape of the volatile products generated during decomposition. Reduction in composite volume during volatilization improves this barrier property due to reassembly of the dispersed silicate layers [7].

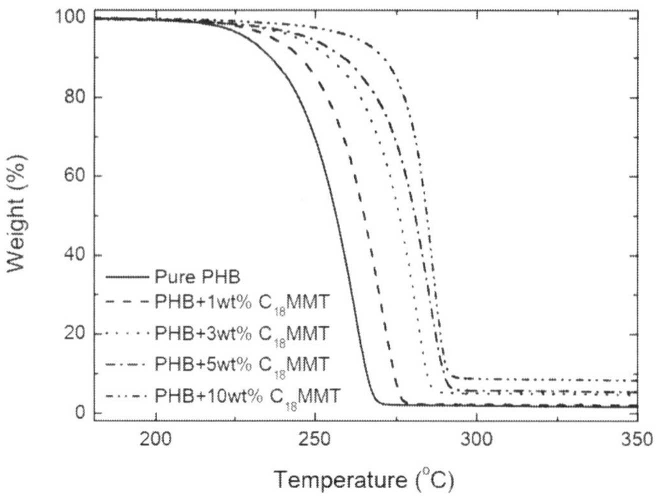


(a)

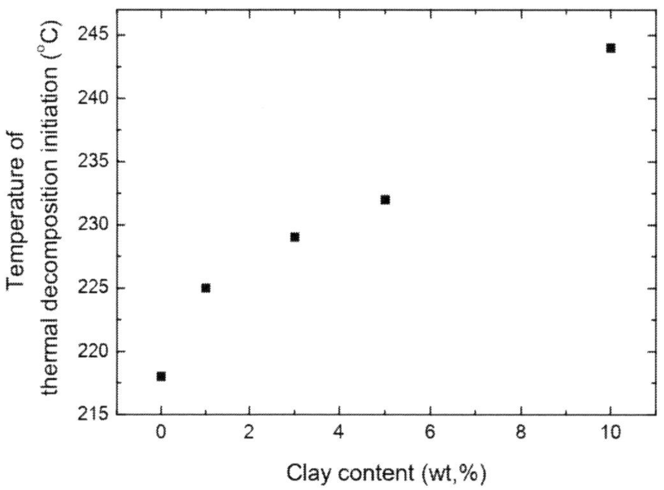


(b)

**Figure 4.** (a) TGA data for PHB hybrids prepared with various loadings of $C_{18}$MMT, in inert ($N_2$) atmosphere. (b) Effect of clay content on the onset decomposition temperature (estimated at 2% of polymer mass loss).

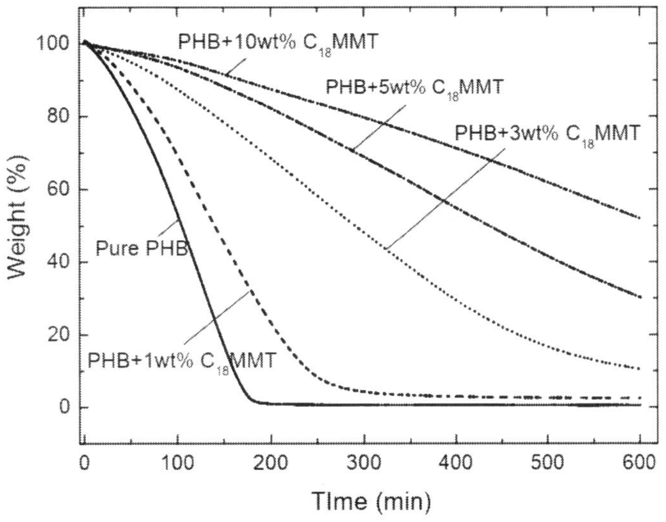


(a)

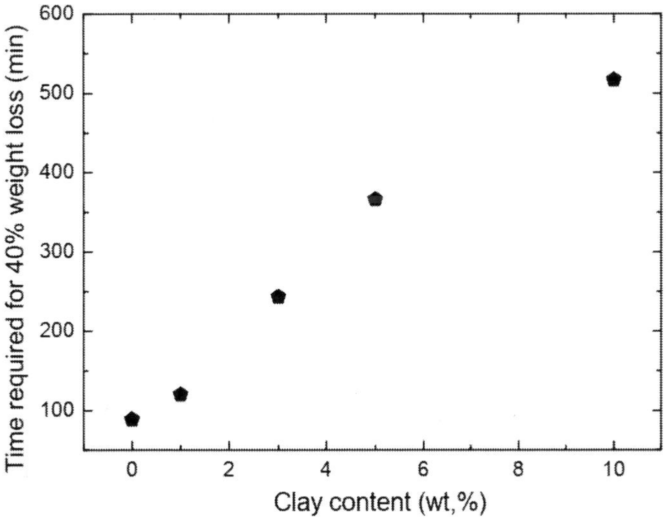


(b)

**Figure 5.** (a) TGA curves of PHB nanocomposites in oxidant atmosphere and isothermal conditions (190 °C). (b) Effect of clay content on the time required for 40% polymer mass loss.

## Conclusions

PHB nanocomposites were successfully prepared by the incorporation of various loadings of organophilic montmorillonite using the melt intercalation method. Constant nitrogen flow during the preparation process prevented thermal decomposition of the matrix. Investigation of the composite structure revealed that the basal spacing of the inorganic material was not significantly affected by the mineral loading.

The nanocomposites displayed enhanced thermal stability compared to pure polymer. The dispersed inorganic platelets assist the formation of a protective layer on the surface of the polymer, which acts as an insulator and a mass transport barrier retarding the decomposition rate of the polymer matrix. Moreover, the organomodified clay presence led to a slight depression of the polymer melting point.

*Acknowledgements*
This work was supported by the Region of Western Macedonia (Greece) through the PEP grant with MIS 105545 to I. Z.

***References***

1 Giannelis, E. P. (1996), "Polymer layered silicate nanocomposites," *Advanced Materials*, 8(1), pp. 29-35.
2 Ray, S. S., and Bousmina, M. (2005), "Biodegradable polymers and their layered silicate nano composites: In greening the 21st century materials world," *Progress in Materials Science*, 50(8), pp. 962-1079.
3 Marras, S. I., Tsimpliaraki, A., Zuburtikudis, I., and Panayiotou, C. (2007), "Thermal and colloidal behavior of amine-treated clays: The role of amphiphilic organic cation concentration," *Journal of Colloid and Interface Science*, 315(2), pp. 520-527.
4 Lagaly, G. (1986), "Interaction of Alkylamines with Different Types of Layered Compounds," *Solid State Ionics*, 22(1), pp. 43-51.
5 Hackett, E., Manias, E., and Giannelis, E. P. (1998), "Molecular dynamics simulations of organically modified layered silicates," *Journal of Chemical Physics*, 108(17), pp. 7410-7415.
6 Marras, S. I., Zuburtikudis, I., and Panayiotou, C. (2007), "Nanostructure vs. microstructure: Morphological and thermomechanical characterization of poly(l-lactic acid)/layered silicate hybrids," *European Polymer Journal*, 43(6), pp. 2191-2206.
7 Zanetti, M., Camino, G., Reichert, P., and Mulhaupt, R. (2001), "Thermal behaviour of poly(propylene) layered silicate nanocomposites," *Macromolecular Rapid Communications*, 22(3), pp. 176-180.
8 Maiti, P., Batt, C. A., and Giannelis, E. (2007), "New biodegradegradable polyhydroxybutyrate/layered silicate nanosomposites," *Biomacromolecules*, 8(11), pp. 3393-3400.
9 Gilman, J. W., Jackson, C. L., Morgan, A. B., Harris, R., Manias, E., Giannelis, E. P., Wuthenow, M., Hilton, D., and Phillips, S. H. (2000), "Flammability properties of polymer - Layered-silicate nanocomposites. Polypropylene and polystyrene nanocomposites," *Chemistry of Materials*, 12(7), pp. 1866-1873.
10 Berta, M., Lindsay, C., Pans, G., and Camino, G. (2006), "Effect of chemical structure on combustion and thermal behaviour of polyurethane elastomer layered silicate nanocomposites," *Polymer Degradation and Stability*, 91(5), pp. 1179-1191.

# 188y Synthesis and Characterization of Protogenic Liquid Crystals

**Supacharee Roddecha** and **Mitchell Anthamatten**. Chemical Engineering, University of Rochester, 206 Gavett Hall RC 270166, Rochester, NY 14627-0166

This paper presents a nanoscale approach to engineering anhydrous proton conducting materials. Fuel cells are electrochemical devices that convert stored chemical energy directly to electrical energy. They potentially offer a means to clean, reliable and efficient energy generation. The proton exchange membrane is the heart of a fuel cell serving to block fuel gases and electrons, while permitting proton transport. The majority of fuel cells employ perfluorinated ionomer membranes such as Nafion®. These membranes require hydration and are limited to low temperature (~80oC) operation. At low temperature, fuel cell catalysts are readily poisoned, lowering their activity. Imidazole has been recently investigated as a novel anhydrous proton conducting functional group, capable of high operating temperature [1,2,3]. Its amphoteric behavior can support Grotthuss-like proton transport. However, it is a liquid that cannot be easily incorporated into a proton exchange membrane. Also, imidazole functional groups must be concentrated and mobile (in a liquid-like state) to provide a platform for proton transport. To enhance mobility and overcome the difficulty in incorporating liquid imidazole, imidazole functionalized liquid crystals are considered. Our hypothesis is that liquid crystal ordering can be used to facilitate proton transport. The objective of this study is to determine whether liquid crystal ordering can promote proton conductivity. Two imidazole-terminated liquid crystal molecules were designed and synthesized (Fig. 1). The targeted liquid crystal molecules contain rigid diacylhydrazine core, a flexible hydrocarbon spacer and a terminal imidazole group. Together, the spacer and the core are designed to stabilize liquid crystal mesophase. In particular, smectic mesophase may be useful in confining imidazole groups into 2D planes to promote proton transport. Terminal imidazole- group are intended to promote Grothuss-like proton transfer. Moreover, its high thermal and chemical stability will be useful in fuel cell membrane. This poster will focus on the synthesis approach and characterization of liquid crystal molecules. References [1] M. Schuster, T. Rager, A. Noda, K.D. Kreuer, J. Maier Fuel Cells 2005, 5, 355. [2] G. Scharfenberger, W. H. Meyer, G. Wegner, M. Schuster, K.-D. Kreuer, and J. Maier Fuel Cells 2006, 6, 237. [3] S. Li, Z. Zhou, M. Liu, W. Li, J. Ukai, K. Hase, M.Nakahishi Electrochemica Acta 2006, 51, 1351.

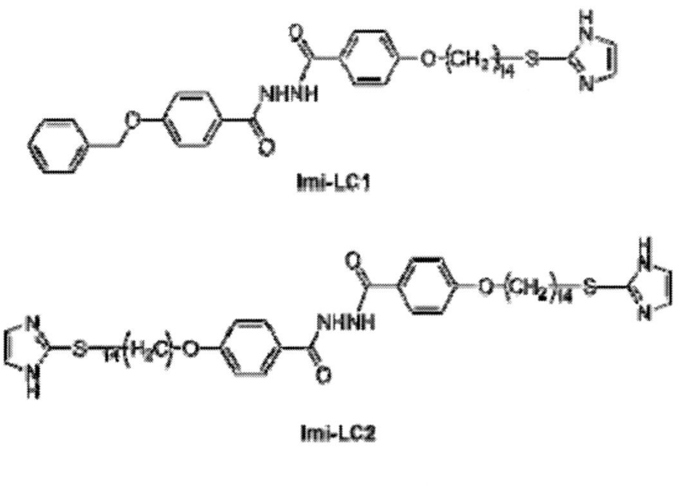


Figure 1. The proposed imidazole-terminated liquid crystal molecules

## 188z Self Assembly of Biomolecules at Nano Scales-Structure and Kinetics

**Weixian Shi**, Bristol-Myers Squibb Company, 1 Squibb Drive, New Brunswick, NJ 08901 and Ronald G. Larson, Chemical Engineering, University of Michigan, 2300 Hayward Street, 3074 H.H. Dow, Ann Arbor, MI 48109.

The self-organization of biomolecules, like DNA and proteins, are not only biologically important but also crucial to the understanding of general polymer behaviors. One of the powerful tools to visualize self-organized structures is atomic force microscopy (AFM). In this study, the nano complex formed by RecA protein and DNA molecules were investigated using AFM. The structure of complexes formed with and without the presence of single-stranded DNA binding (SSB) protein was found to be distinct from each other. More complicated but ordered assembly was seen when SSB was added in the preparation. SSB assists the assembly process between DNA and RecA protein by removing the secondary structure on DNA molecules which prevents continuous binding of RecA. The biological function of the complex was also evaluated using gel electrophoresis. The kinetics were determined and a preliminary mechanism was identified.

## 188aa Surface Coated Iron Particles for Magnetorheological Grease (MRGr)

**Alan Fuchs**[1], Joko Sutrisno[1], Faramarz Gordaninejad[2], and Husein Sahin[2]. (1) Chemical Engineering, University of Nevada, Reno, Reno, NV 89557, (2) Mechanical Engineering, University of Nevada, Reno, Mailstop 170, Reno, NV 89557-0136

Magnetorheological Fluid (MRF) is an intelligent material which has controllable rheological properties. MRF consists of magnetic particles and carrier fluid which are the dispersion medium for the magnetic particles. Recently, MRF has achieved significant attention for their excellent mechanical and rheological properties which can be controlled using an external magnetic field. MRF has properties similar to Newtonian fluids in the off-state condition. Magnetic particles create chain-like structures within the MRF when external magnetic fields are applied and fluid flow is then restricted. This fluid-like material then transforms into a pseudo-solid material. Generally, MRF consist of three different components: magnetic particles (iron), carrier fluid and additives. Carrier fluids act to suspend the iron particles and other components. Additives serve as stabilizers and protect the iron particle surface. Nanostructured polymers may be used as additives in MR fluids. This provides long operating life and prevents damage to the device during use. In addition, iron surface treatment prevents particle aggregating between the iron particles. MRF has several features including device non-abrasion, low settling, re-dispersibility, and a wide range temperature operation (typically -50oC – 150oC), chemical stability, high magnetic saturation, and excellent durability. Settling of iron particles in MRF is a challenge. There are several ways to reduce iron particle settling in MRF. By adding thickeners, a high viscosity can be achieved, which is a magnetorheological grease (MRG). This is one approach to reduce settling of iron particles. Additionally, appropriate additives provide a high yield stress and shear thinning behavior. MRGs have been characterized using shear rheometer. MRGs containing surface coated iron particles exhibit more stable rheological behavior than uncoated particles after oxidation testing. The oxidation test was used as a tool to examine the durability of MRGs. The iron particles with surface coating have constant magnetic permeability. It is shown by rheology of MRGs before and after oxidation testing in the on-state condition.

# 188ab Disperse Barium Oxide Nanoislands Synthesized on Aluminum Oxide by Successive Ionic Layer Deposition

**Thomas I. Gilbert** and Johannes W. Schwank. Chemical Engineering, University of Michigan, 2300 Hayward St, 3074 H.H. Dow Building, Ann Arbor, MI 48109-2136

Successive Ionic Layer Deposition (SILD) is an aqueous technique to deposit nanolayers of oxide materials on a substrate in an organized manner. With each SILD cycle submonolayers of metal hydroxides are strongly adsorbed on the substrate using precursor solutions of dilute aqueous metal salts. Varying the SILD synthesis conditions changes the chemistry and morphology at the nanometer scale.

Catalysis has always been a nanoscale science. The interface between the active metal catalyst and the support is believed to be a particularly active site in some catalytic systems. Increasing the dispersion of a catalyst on a support not only increases the overall catalyst surface area but also leads to a higher concentration of metal-support interactions. It has been reported that a dispersed phase of BaO on Al2O3 plays a major role in NOx emission control catalysis for lean burn engines. [1] However, conventional catalyst synthesis techniques tend to produce primarily a bulk-like phase of barium oxide as more catalyst is loaded onto the aluminum oxide support. SILD offers the opportunity to exclusively synthesize the dispersed phase of BaO on Al2O3 and study its properties.

In this poster, we describe how SILD was used to deposit disperse nanoislands of BaO on islands of Al2O3 supported on silicon wafer substrates. These nanomaterials were studied as model catalytic surfaces and characterized by techniques such as Atomic Force Microscopy, Scanning Electron Microscopy, and X-ray Photoelectron Spectroscopy.

REFERENCE

1. X. Chen, J. Schwank, J. Li, W.F. Schneider, C.T. Goralski, P.J. Schmitz, Appl. Catal. B Environ. 61 (2005).

## 188ac Effect of Crystallization Growth Inhibitors in the Synthesis of Sapo-34

**Surendar Reddy Venna** and Moises A Carreon. Chemical Engineering Department, University of Louisville, Ernst hall, Chemical Engineering Department, Belknap campus, Louisville, KY 40292

The silicoaluminophosphate molecular sieves have been the subject of great interest because of their high thermal resistance, mechanical strength, chemical stability, shape selectivity, chemical inertness, and uniform-sized pores of molecular dimensions. SAPO-34 with the composition of $Si_xAl_yP_zO_2$ (x = 0.01-0.98, y = 0.01-0.6 & z = 0.01-0.52) is a representative type of silicoaluminophosphate zeolite displaying chabazite structure with uniform microporous channels and cages. SAPO-34 with pore size of ~0.38 nm is an ideal material for carbon dioxide (kinetic diameter of ~0.33 nm) separation from light gases. Also, it is widely used as catalyst in methanol to olefin reaction in petrochemical industries.

Here, we present the synthesis of SAPO-34 crystals using various crystal growth inhibitors (PEG, $C_{12}E_6$ & MB) and several structure directing agents (TEAOH, DPA, Morpholine, TEA & CTAB). Crystal growth inhibitors play an important role in changing the alkalinity of the synthesis of gel and during phase transition of SAPO-34. They also interact with reactive sites of the precursors suppressing the crystal growth. The resultant SAPO-34 crystals were characterized by BET, XRD, SEM, Raman, FTIR, TGA and $CO_2/CH_4$ adsorption. The incorporation of crystal growth inhibitors during synthesis led to a remarkable decrease in crystal size from ~2 μm to ~0.5 μm. The surface area of the SAPO-34 crystals was significantly improved from ~450 $m^2/g$ to ~700 $m^2/g$. The crystallinity of SAPO-34 was confirmed by XRD and FTIR. These ~0.5 μm SAPO-34 crystals can be potentially used as seeds to prepare the thin membranes on porous supports to separate the $CO_2$ from $CH_4$ ¬to increase the energy content of the natural gas.

# 188ad Formation of Fractal Structures from Self-Assembling Dipeptide Derivatives

**Weiping Wang**[1], Siamrut Patanavanich[2], and Ying Chau[1]. (1) Department of Chemical Engineering, The Hong Kong University of Science and Technology, Clear Water Bay, Kowloon, Hong Kong, (2) Department of Materials Science and Engineering, Massachusetts Institute of Technology, Cambridge, MA 02139

Fractal structures displaying self-similarity provide the idea to produce materials incorporating multiple length scales simultaneously. Here we report the formation of fractal-like structures from self-assembling dipeptide derivatives, Fmoc-Phe-Tyr- phosphate (FPTP) and Fmoc-Phe-Tyr (FPT). Upon addition of alkaline phosphatase, FPTP solution is converted to FPT, and forms hydrogel with nanofiber network. When FPT solution is deposited on a silicon film, fractal structures are observed after drying from scanning electron microscopy image. More interestingly, FPTP which cannot form hydrogel in solution also assemble into fractal pattern on solid film. We study the effects of pH and building block concentration on the structure morphology. The fractal dimension of the patterns is measured using standard method. The formation process of fractal structures from diffusion-limited cluster-cluster aggregation is discussed. The present work will provide a molecular design concept for the construction of complex materials and patterns.

## 188ae Monolithic Tunnel Junctions for Single Molecule Detection

**Rahul Gupta** and Brian G. Willis. Chemical Engineering, University of Delaware, Colburn Laboratory, 150 Academy Street, Newark, DE 19716

In his classic talk on "There's plenty of room at the bottom" in 1959, Richard Feynman quoted several visionary statements on working at nanometer scale, which gave birth to the field of nanotechnology.1 He imagined devices that will involve manipulation of atoms and work on the principles of quantum mechanics. As a result, a lot of interest has developed on studying nanoparticles for sensor applications, which possess unique physical properties that do not exist in bulk materials. Nanosensors provide several performance advantages such as reduced size, higher sensitivity, and faster response time. Further, they can be self-assemble to provide economical solutions for designing logic circuits. They can provide unprecedented benefits in various fields, such as electronics, medicine, and chemistry. Most of the research on nanosensors has focused on nanometer scale materials such as quantum dots, carbon nanotubes, and fullerenes, for example, but less work has been conducted on molecule based sensor. The primary reason is the difficulty in making chemical contact to single molecules, which are only a few angstroms in length. However, single molecule sensors provide larger possibilities of control by tailoring the attachment chemical groups. Scanning tunneling microscopy (STM) studies on molecules have revealed interesting information, but the slow speed and economics do not favor mass production of STM-based sensors.

In this work, a novel design is presented where nanoscopic tunnel junctions are embedded in a thermally oxidized silicon wafer and used for single molecule detection. Molecules are adsorbed (i.e. trapped) under high electric field and detected using inelastic electron tunneling spectroscopy (IETS) at cryogenic temperatures. The peaks in IETS spectra provide critical information for the engineering of electrode-molecule interfaces and the study of vibrational features of molecules present in the tunnel junction (see figure). The design uses advanced reaction engineering principles including atomic layer deposition (ALD) and selective area growth to fabricate molecule size electrodes with the required 1-2 nm spacing. ALD has been selected because of its excellent control over the growth rate (typically $\leq$0.1 nm/cycle) and the electrode microstructure. This technique has the potential to fabricate parallel tunnel junctions with same electrode spacing at nanometer scale for practical molecule based sensors.

Successful nanofabrication of monolithic metal-vacuum-metal tunnel junctions has been achieved with demonstrated nanojunctions of 1-2 nm spacing. In-situ field emission and electron tunneling measurements are used to characterize the electrode properties and electrode spacing during the ALD of copper. We demonstrate that molecular adsorption into tunnel junctions can be precisely controlled using electric field and detection via IETS spectra. The IETS features are shown to distinguish between physisorbed and chemisorbed state of these molecules. Moreover, we demonstrate that the molecules can be desorbed and re-adsorbed reversibly without affecting the electrode properties, thereby demonstrating the reusability of these tunnel junctions for sensor applications. This work is a significant advance over previous nanofabrication designs that lacked sub-nanometer spacing and electrode structure control, and hence were unable to engineer nanoscopic tunnel junctions. Work is currently in progress to demonstrate parallel fabrication of these tunnel junctions for a viable sensor application.

1 http://www.zyvex.com/nanotech/feynman.html

## Confined nanoparticle solutions for nanotribological applications

*R. Chembeti and J.-C. Wang*, Missouri University of Science and Technology, Rolla, MO 65409-1230

Current advances in micro- and nano-electromechanical devices (MEMS and NEMS), micro and nanofluidic systems, and other nanotechnological areas all involve the tribological issues of friction, wear, and lubrication at the microscale or nanoscale. Under such nanotribological conditions, the lubricant films are molecularly thin and have been shown to possess properties that are significantly different from those of bulk fluids but not desirable for tribological applications. Recent experimental studies using surface force apparatus (SFA), atomic force microscopy (AFM), friction force microscopy (FFM), and quartz crystal microbalance (QCM) have confirmed the significant effects of the confining surfaces on the properties of nanoscopically confined fluid films and further suggested that nanoscale confinement can induce fluid molecules to form layered configurations, causing nanoconfined fluids to have very different properties than those of bulk fluids, where molecules possess no preferred orientation. Explicit evidences at the molecular level have come from computer molecular simulation studies using Monte Carlo (MC) and molecular dynamics (MD) techniques. Other theoretical approaches such as density functional theory and integral equation theory have also been extended to analyze nanoconfined fluids and confirm the formation of fluid layers under confinement.

It has now been well established that when the separation/spacing between the confining surfaces is larger than about ten molecular diameters, confined fluids behave much the same as bulk fluids in many aspects. As the separation decreases, molecular orientation and fluid configuration undergo changes and, more importantly, the isotropy between the perpendicular and parallel directions breaks down. The difference between the perpendicular pressure ($P_\perp$) and the parallel pressure ($P_{||}$) has been measured as the solvation force ($f_s$) by the surface force apparatus, where the confined fluid is open to the bulk reservoir under isothermal-isobaric condition and the parallel pressure as a result is taken to be the same as the bulk pressure. Interestingly, when symmetric molecules are confined under a constant bulk/parallel pressure, salvation/surface force oscillates as a function of surface separation with a periodicity that is equivalent to the mean molecular diameter [cf. Fig. 1(a)]. This behavior is not in accord with the expectations from conventional continuum theories but signifies the formation of layered configurations parallel to the confining surfaces by the confined fluid molecules. When a strongly layered confined fluid is sheared, instead of lubricated smooth motion, it exhibits yield stress, stick-slip motion [cf. Fig. 1(b)], and apparent shear viscosity that could be orders of magnitude higher than bulk values. These solid-like responses reflect the solid-like nature of the layered configurations, which could be linked to the peaked, repulsive surface forces, and can cause undesirably large friction and wear and consequently significant material and energy costs. Between two peaked surface forces and during the slip phase, nanoconfined fluids undergo configurational changes to become less layered, thereby exhibiting lubricated smooth motion and reduced friction and wear.

Alkanethiol-capped nanoparticles represent a recent technological breakthrough and have been touted as one of the most important ingredients for nanotechnologies. They are charge-neutral composites with an inorganic crystalline core capped in a dense shell of

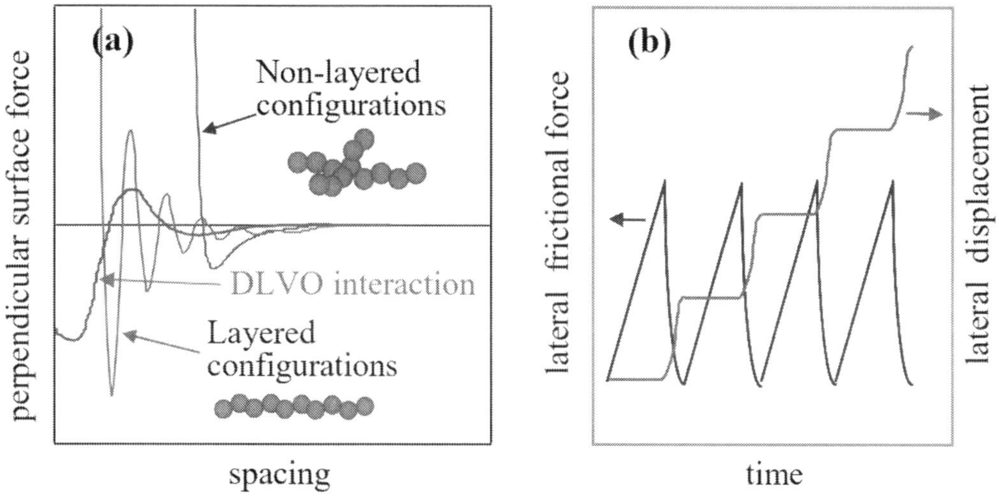


Figure 1. Schematics of (a) solvation/surface force profiles for confined fluids with layered or non-layered configurations and (b) stick-slip motion of confined fluids with layered configurations.

alkanethiol molecules, and have nowadays been routinely synthesized. They have good solubility in common non-polar solvents and are much larger in size than typical solvent molecules. To explore possible advantages resulting from good solubility and size mismatch of alkanethiol-capped nanoparticle, we perform MD simulations to study the nanotribological potential of confined liquid solution containing alkanethiol-capped nanoparticles. The model nanoparticle considered is hexanethiol-capped gold nanoparticle, $Au_{140}(SC_6H_{13})_{62}$, which is modeled as a soft sphere by using the Lennard-Jones (LJ) 12-6 potential. Separate atomistic simulations were performed in order to determine the LJ parameters for the gold-thiol nanoparticle ($\sigma_{gold\text{-}thiol}$=29 Å, $\varepsilon_{gold\text{-}thiol}$=10827.08$k_B$). For computational consistency and efficiency, spherical cyclohexane is chosen as the non-polar solvent and its LJ parameters are adopted $\sigma_{cyclohexane}$=5.466 Å and $\varepsilon_{cyclohexane}$=484$k_B$ from the literature. Since a relatively large portion of the envisioned nanotechnologies involve silicon (Si), we have considered the Si(111) surface as the confining surface, which has the diamond structure and a lattice constant equal to 5.43 Å. Following a standard approach, the Si lattice structure and energy results in $\sigma_{Si}$= 3.826 Å, $\varepsilon_{Si}$=202.45$k_B$ for its LJ potential. Taking hints from SFA, the thermodynamic state of the confined nanoparticle solutions is defined and controlled in our simulations by constraining the temperature and the parallel pressure ($P_{||}$) to prespecified bulk phase values at all times (300 K and 1 atm) via a recently developed *NhPT* MD method, which in effect is a virtual isothermal-isobaric reservoir method. The equations of motion are integrated with the fourth-order Gear predictor-corrector algorithm along with a time step of 5 fs. All confined nanoparticle solutions are first equilibrated at a separation of *h*=10 nm (100 Å) between two confining Si (111) surfaces and under continuous compression by decreasing the separation at a speed of 5 m/s.

The simulation studies of continuous compression of confined solutions containing 0, 1, 3, 5, and 10 nanoparticles were performed. As can be seen from Fig. 2 where the results from the case of 5 nanoparticles are shown, the employed *NhPT* method controls the parallel pressure well except when at the smallest separations the confined fluid is unphysically forced

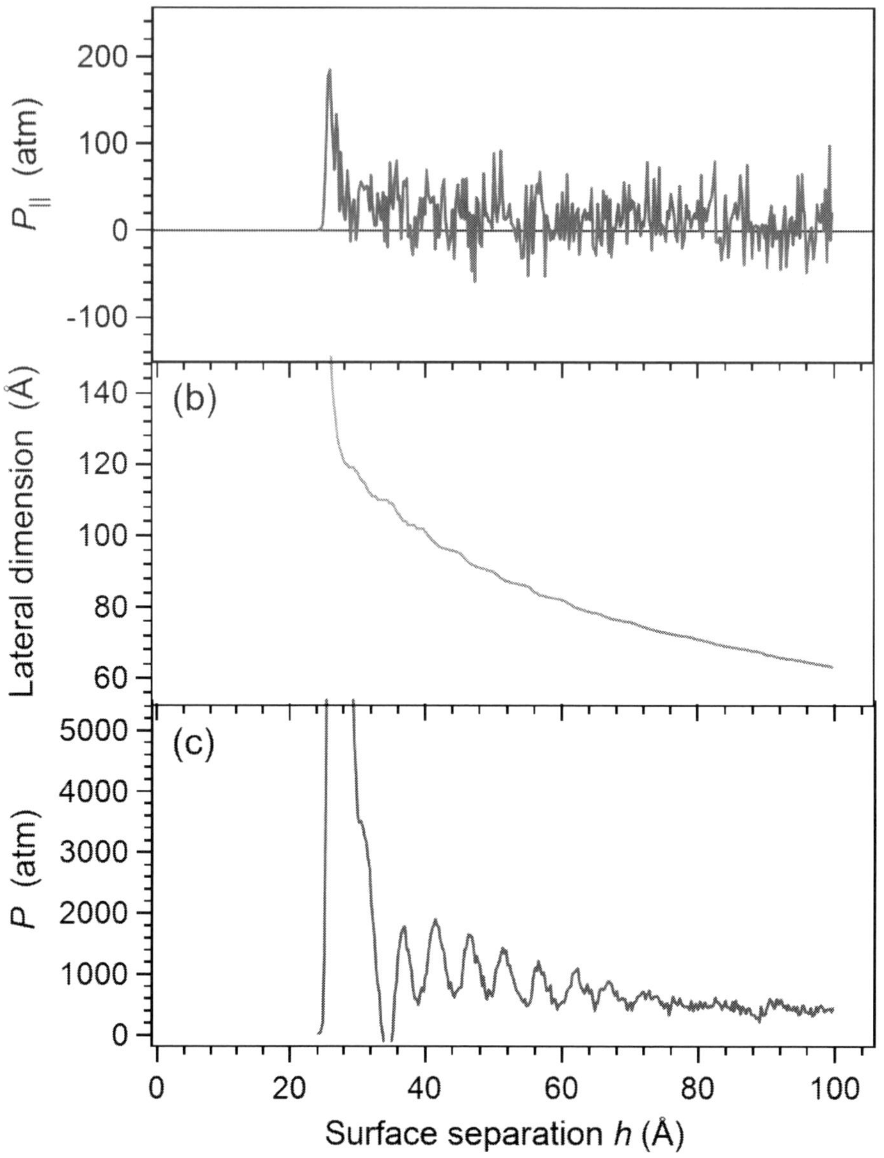


Figure 2. Instantaneous (a) parallel pressure, (b) lateral dimension, and (c) perpendicular pressure as a function surface separation during continuous compression.

to break apart by the imposed compression. Figure 3 shows the simulation results from continuous compression of different nanoparticles solutions. The data are instantaneous properties calculated and recorded every 1000 $\Delta$t's. Exhibiting the strongest force oscillations [cf. Fig. 3 (b)] is the pure cyclohexane fluid in nanoconfinement that can be understood to be most layered among all solutions studied. A closer examination of the force oscillations reveals that the periodicity is about the same as the cyclohexane molecular diameter, 5.466 Å. The connection between the surface force and the fluid configuration can be analyzed by comparing Figures 3 (a) and (b) together. It can be easily seen that the lateral dimension of the confined cyclohexane under continuous compression undergoes step-like expansion and the width of the steps is the same as the periodicity of the force oscillations. More specifically, minimum surface forces occur at separations that correspond to completion of step-like expansion and more disordered, less layered configurations. When compression continues, the less layered confined fluid are squeezed to become more layered and the surface force builds up again until the confined fluid could not sustain the strong pressure any more and sudden expansion occurs in the lateral directions.

When nanoparticles are added to the confined alkane fluid, they significantly reduce the magnitudes of the step-like expansion and force oscillations. In addition, nanoparticles bigger in size can now resist compression and strong perpendicular pressure at larger separations that are about the same as the size of the nanoparticles [cf. Fig. 3(a)]. These behaviors signify better nanotribological properties that could be provided by the presence of alkanethiol-capped nanoparticles in alkane/oil solutions.

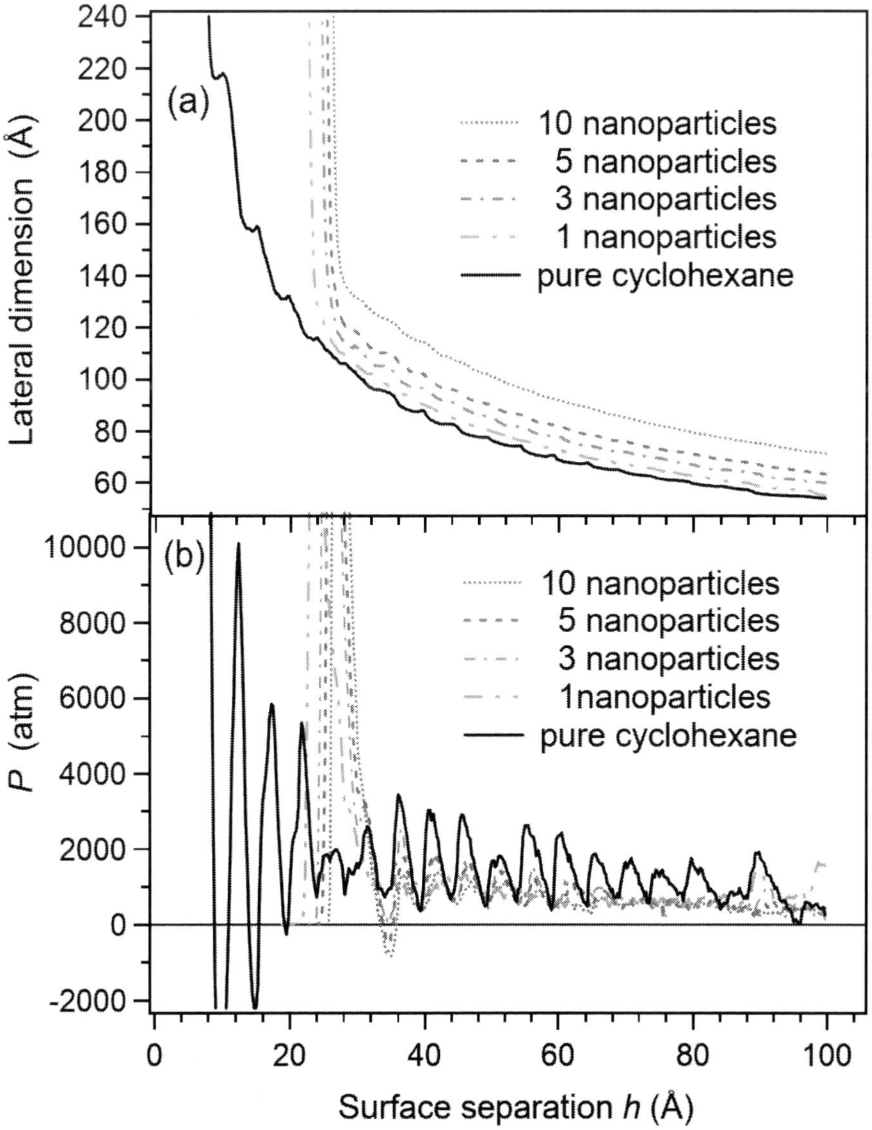


Figure 3. Instantaneous (a) lateral dimension and (b) perpendicular pressure as a function surface separation during continuous compression for different solutions.

## 188ag Comparing Mixing Quality of Nanosized Partcles by Environmentally Benign Mixing Techniques

**James V. Scicolone** and Rajesh Dave. New Jersey Center for Engineered Particulates, New Jersey Institute of Technology, Newark, NJ 07102

Mixtures of nanoparticles and nanocomposites, manufactured using nanoparticles as building blocks, have unique properties owing to small particle/grain size, and large interface area between individual nano-sized constituents. These properties have a great potential to improve performance of drugs, biomaterials, catalysts, energetic materials etc. However, to fully employ this potential, nanoparticles should be mixed at nanoscale level, which is a difficult task because they tend to form aggregates on the order of microns. Therefore, a major challenge in utilizing nanoparticles is to achieve their mixing at nano-scale level.

The objective of this study is to explore the homogeneity of nanoparticle mixtures obtained using environmentally benign dry mixing techniques. The mixing techniques studied in this research are Magnetically Assisted Impaction Mixing (MAIM), Fluidized Bed Mixing, Supercritical Stirring and Rapid Expansion of Supercritical Suspensions (RESS). The characterization of the resulting nanoparticle mixtures were conducted using Scanning Electron Microscope (SEM) equipped with Energy Dispersive X-ray Spectroscopy (EDS).

## 188ah Permeability Control of Nanoparticle-Assembled Capsules

**Hitesh G. Bagaria**, Dept of Chemical and Biomolecular Engineering, Rice University, 6100 Main Street, Houston, TX 77005 and Michael S. Wong, Departm of Chemical and Biomolecular Engineering, Dept. of Chemistry, and Center for Biological and Environmental Nanotechnology, Rice University, 6100 Main Street, MS-362, Houston, TX 77251-1892.

Recently, our lab has demonstrated a flexible, simple, safe and scalable method of synthesizing microcapsules called as nanoparticle-assembled capsules or NACs [1]. These capsules are synthesized by depositing negatively charged nanoparticles around cationic polyamine aggregates crosslinked by multivalent counteranions. As an example, commercial silica nanoparticles (Snowtex-O) have been deposited on polyallylamine aggregates crosslinked by either citrate or phosphate to form silica NACs [2-3].

We have successfully demonstrated the ability to encapsulate various water-soluble materials like cardiogreen [3] and sodium fluorescein [4] in these NACs. Depending on the application, the encapsulated material is either required to be completely entrapped in the capsule (e.g. photothermal therapy with cardiogreen) or be released at a controlled rate (e.g. drug delivery). The current technique to form NACs presents a porous surface that neither completely contains the encapsulated material, nor releases the material at a predesigned rate. In this study, we attempted to control the permeability of silica NACs by depositing silicic acid on the capsule surface. These studies that were performed with sodium fluorescein as the model encapsulated material will be presented here.

[1] R. K. Rana, V. S. Murthy, J. Yu, and M. S. Wong, "Nanoparticle self-assembly of hierarchically ordered microcapsule structures," Adv. Mater. 17, 1145-1150 (2005)

[2] V. S. Murthy, R. K. Rana and M. S. Wong, “Nanoparticle-Assembled Capsule Synthesis: Formation of Colloidal Polyamine-Salt Intermediates,” J. Phys. Chem. B 110, 25619-25627 (2006)

[3] J. Yu, M. A. Yaseen, B. Anvari, and M. S. Wong, "Synthesis of Near-Infrared-Absorbing Nanoparticle-Assembled Capsules," Chem. Mater. 19, 1277-1284 (2007)

[4] E. M. Tavera, S. B. Kadali, H. G. Bagaria and M. S. Wong, “Experimental and Analytical Modeling Studies of Diffusive Release from Multilayered Microcapsules,” in preparation

## 188ai Nanoscale Sensors for Multi-Modal Detection of Genotoxic Agents

**Daniel A. Heller**[1], Hong Jin, and Michael S. Strano[2]. (1) Chemical Engineering, Massachusetts Institute of Technology, 77 Massachusetts Avenue, 66-565, Cambridge, MA 02139, (2) 66-566 Department of Chemical Engineering, Massachusetts Institute of Technology, 77 Massachusetts Avenue, Cambridge, MA 02139

Reactive oxygen species (ROS) and active alkylating chemotherapeutics are important bioanalytes but difficult to measure selectively, in vivo, or in real time. A new nanoscale optical sensor platform under development can detect as well as identify these analytes in situ. Carbon nanotube-DNA complexes, formed by individually encapsulating nanotubes with oligonucleotides, respond in real time to these agents via multiple optical modes, giving each analyte a distinct optical signature. Single-walled carbon nanotubes are environmentally sensitive and emit tissue-transparent near-infrared photoluminescence. By their encapsulation in short strands of synthetic DNA, we introduce a selective handle for changing their emission, allowing detection of DNA-damaging oxygen species or alkylating agents, by red-shifts in emission energy as well as optical quenching, giving multiple detection modes which result in analyte specificity. The nanotube-DNA complexes exhibit uptake into mammalian cells via endocytosis without cytotoxic effects and transduce analyte activity information live and in situ.

## 188aj Integrated Plasmonic Lens Photodetector

**James A. Shackleford**[1], Richard Grote[1], Mark Currie[2], Jonathan Spanier[3], and Bahram Nabet[1]. (1) Electrical Engineering / Electrophysics, Drexel University, 3141 Chestnut Street, Philadelphia, PA 19104, (2) Optical Sciences Division, Naval Research Laboratory, 4555 Overlook Ave., S.W, Washington, DC 20375, (3) Materials Engineering, Drexel University, 3141 Chestnut Street, Philadelphia, PA 19104

The metal-semiconductor-metal (MSM) photodetector is an attractive alternative to other types of photodetectors due to its ease of fabrication. Typical MSM photodetectors are characterized not only by their low dark current but also by their high speed, which is a result of their small device capacitance. The biggest barrier preventing the proliferation of MSM photodetectors in commercial applications is their relatively poor responsivity, which is a result of several factors including reflection of incident light at the metal surfaces. It has been theoretically and experimentally shown, however, that it is possible to harness and guide such incident light by implementing a properly designed metal nanostructure at a metal/dielectric interface. Such light guiding systems are designed for very specific wavelengths and are capable of sustaining resonant collective electron oscillations at their design frequencies. These collective electron oscillations, or surface plasmon polaritons (SPPs), are a manifestation of incident light coupling to free electrons in a metal. It has been shown through FDTD simulation that the implementation of such a periodic corrugation nanostructure is capable of guiding light so as to concentrate the incident field about an aperture.

By implementing a properly designed plasmonic lens, it is possible to guide light that is normally reflected at the gold interface into the active area of the MSM photodetector. In this fashion, the responsivity of the MSM photodetector is increased without sacrificing the speed of the device. Here we present a fabricated MSM photodetector that exhibits an enhanced photocurrent by integrating a properly tuned nanoscale grating structure that serves to guide light into the active area of the device by coupling incident photons to surface plasmon polaritons (SPP). Through localized laser excitation and time response data we show that the responsivity of such an MSM device may be increased without sacrificing speed.

# 188ak Preparation of Anodic Aluminum Oxide Membrane for the Template Synthesis of Semiconductor Nanowires

Chang Hyun Ko, Sung-Youl Park, Kwang Bok Yi, **Jong-Ho Park**, and Jong-Nam Kim. Chemical Process Research Center, Korea Institute of Energy Research, 71-2 Jang-Dong, Yuseong-Gu, Daejeon, South Korea

Synthesis of nanostructured semiconductors have attracted much attention due to their distinctive electrical, optical and catalytic properties. Their properties mainly come from their high surface to volume ratio and quantum confinement effect. Many researchers have synthesized nanostructured semiconductors by various types of methods, such as, pyrolysis in solution and template synthesis, to find out specific properties. However, not only synthesis, but also fabrication of nanostructured semiconductors should be considered for the practical applications. Thus, template synthesis method seems to be superior to solution phase synthesis method because formation and fabrication of nanostructured semiconductor might be achieved simultaneously with proper templates located on adequate place. As a preliminary study for this purpose, anodic aluminum oxide (AAO) with membrane type, in which pores are directly connected and opened in both sides were prepared by anodization of aluminum plate using oxalic acid as electrolyte. Depending on the concentration of oxalic acid and anodization temperature, pore diameters of AAO membrane were controlled from 10 to 100 nm. In this pore, various types of semiconductors, such as zinc sulfide, zinc oxide, and silicon were selectively incorporated by gas phase deposition methods. Reducing the deposition rate of semiconductor is the critical factor for the formation of nanowires in the pores.

## 188al Ligand Exchange Using Alkosilanes on Oleic Acid-Cobalt Ferrite Nanoparticles

**Victoria L. Calero-DiazdelCastillo** and Carlos Rinaldi. Chemical Engineering, University of Puerto Rico, 636 Manuel Ramos, Mayaguez, PR 00682

Monodisperse cobalt ferrite nanoparticles have been synthesized by the thermal decomposition method. Controlling the temperature in the nucleation and growth stages between 150 - 250 °C and 300 - 330 °C respectively, monodisperse particles were obtained with various diameters. Particles with ~12 nm diameter and a geometric deviation of 0.11 were obtained when temperatures of 150°C and 300°C were used in the nucleation and growth stages. X-ray diffraction showed all peaks characteristic of the crystalline structure of cobalt ferrite. EDAX confirmed the incorporation of cobalt in the ferrite. Magnetic properties were determined using a SQUID magnetometer, at 5 K and 300 K. Some samples had superparamagnetic behaviour at 300 K. As synthesized., the particle surface is covered with oleic acid which does not permit suspending the particles in polar solvents. In order to suspend the particles in water, ligand exchange was accomplished using an alkoxysilane with carboxylic functional groups. FTIR confirmed the successful exchange between oleic acid and the alkoxysilane. A decrease in the relative intensity of vibration bands ascribed to $CH_2$ (2900-2800 $cm^{-1}$) was observed, which was attributed to exchange of the long chain of oleic acid by the short chain of the alkoxysilane. TEM shows the particles to be singly dispersed after exchange, which is an indication that the alkoxysilane provides sufficient steric and electrostatic barrier against agglomeration. The distance between particles in hexagonal arrays decreases from ~4 nm to ~2 nm, also confirming ligand exchange. The alkoxysilane-cobalt ferrite nanoparticles could be suspended in deionized water and the solution is stable for at least two months. COO-nanoparticles will be functionalized with Biotinyl-3,6-dioxaoctadiamine to use as a model sensor for avidin.

## 188am Smart Micro- and Nanostructured Hydrogels for Biomedical Applications Synthesized Via Pcμcp

**Hariharasudhan D. Chirra**, Department of Chemical and Materials Engineering, University of Kentucky, 177 F. Paul Anderson Tower, Lexington, KY 40506, J. Zach Hilt, Chemical & Materials Engineering, University of Kentucky, 177 F. Paul Anderson Tower, Lexington, KY 40506, and Dipti Biswal, Chemical and Materials Engineering, University of Kentucky, Lexington, KY 40506.

Surfaces that are spatially functionalized with intelligent hydrogels, especially at the micro- and nanoscale, are of high interest in the diagnostic and therapeutic fields. Though conventional methods of the semiconductor industry have been successfully employed for the patterning of hydrogels in devices, methods for fabricating precise 3-D patterns of hydrogels at the micro- and nanoscale over material surfaces are limited. Herein, polymerization controlled by micro contact printing (PC $\mu$ CP) was applied to synthesize a controlled array of environmentally responsive hydrogels (e.g., N-isopropylacrylamide) over material surfaces. XY control of the hydrogels was achieved using microcontact printing, and the Z/ thickness control was achieved using a variety of polymerization techniques such as UV photopolymerization and atom transfer radical polymerization (ATRP). The controlled growth and the responsive behavior of the hydrogels to external stimuli were characterized using optical microscopy, fourier transform infrared (FTIR) imaging, atomic force microscopy (AFM), and scanning electron microscopy (SEM). Using PC $\mu$ CP based techniques; it is possible to synthesize controlled responsive hydrogels over various surfaces for potential applications in the biomedical field.

# 188an Optical Tracking of Monodisperse Magnetite Nanoparticles

**Arjun Prakash**[1], Michail Stamatakis[1], Christopher J. Jones[2], J.T Mayo[2], Matteo Pasquali[1], and Vicki L. Colvin[2]. (1) Department of Chemical and Biomolecular Engineering, Rice University, 6100 S Main St, Houston, TX 77005, (2) Department of Chemistry, Rice University, 6100 S Main St, Houston, TX 77005

Magnetic separation using aqueous nanomagnetite dispersions have tremendous applications in environmental pollution, biotechnology and manufacturing processes[1,2].In this regard we have recently reported highly efficient Arsenic removal using low field magnetic separation[3]. This is achieved by utilizing the high surface to volume ratio of the nanoparticles, which is maximized on uniform dispersion. Therefore, understanding the aggregation of these particles is the key to explain the mechanism behind highly effective magnetic separation at low field. Despite being superparamagnetic, magnetite nanoparticles are known to aggregate even in the absence of a magnetic field owing to their dipolar interactions[4,5]. Here, we report diffusion measurements of nanoparticles under Brownian motion, to decipher the size of the particles and hence their degree of aggregation in the absence of a magnetic field by using Stokes-Einstein diffusion equation. An optical tracking system was used via which nanoparticles of 12nm core diameter were tracked using epifluorescence microscopy. Interestingly, the particles with concentration on the order of $10^{-7}$M were found to be monodisperse, in confirmation with their superparamagnetic behavior. The data is supported by corresponding Small Angle X-ray Scattering (SAXS) and Dynamic Light Scattering (DLS) analysis along with Cryogenic Transmission Electron Microscopy (Cryo-TEM) images. Also, DLS measurements under magnetic field showed that the particles aggregated with time irrespective of initial concentration, followed by re-dispersion into the medium on magnetic field removal without the loss of colloidal stability. This is in confirmation with a reversible aggregation phenomenon, on field application and subsequent removal to understand the separation efficiencies observed[3]. The results not only explain the trend observed by previous studies[4,6], but also provide an insight into the length scales of dipolar as well as van der Waals interactions between the magnetic nanoparticles.

Key words: Nanomagnetite, aggregation, diffusion.

References:

1. J. J. Hubbuch, D. B. Matthiesen, T. J. Hobley, O. R. T. Thomas, *Bioseparation* **10**, 99 (2001).

2. S. Bucak, D. A. Jones, P. E. Laibinis, T. A. Hatton, *Biotechnology Progress* **19**, 477 (Mar-Apr, 2003).

3. C. T. Yavuz et al., *Science* **314**, 964 (Nov, 2006).

4. K. Butter, P. H. Bomans, P. M. Frederik, G. J. Vroege, A. P. Philipse, *Journal of Physics-Condensed Matter* **15**, S1451 (Apr, 2003).

5. L. N. Donselaar et al., *Journal of Magnetism and Magnetic Materials* **201**, 58 (Jul, 1999).

6. E. B. H. Klokkenburg M., *Journal of Magnetism and Magnetic Materials* **306**, 85 (2006).

## 208a The Influence of Electric and Electromagnetic Fields on Water Self-Diffusion within Carbon Nanotubes Implanted In Lipid Membranes

J.M.Don MacElroy, **Jose-Antonio Garate**, and Niall J. English. Chemical and Bioprocess Engineering, University College Dublin, Engineering and Materials Science Centre, Belfield, Dublin, Ireland

Water self-diffusion is investigated by molecular dynamics (MD) simulation through armchair single walled carbon nanotubes (SWCNTs) implanted in 1-palmytoil-2-oleoyl-sn-glycero-3-Phosphatidylcholine (POPC) membrane patches. Four systems are investigated, each containing one of (5,5), (6,6), (8,8) and (11,11) CNTs with diameters of 6.89, 8.20, 11.04 and 15.02 Å and of varying lengths, oriented normal to the membrane. The CHARMM 27 potential has been employed, in conjunction with TIP3P water, with particle-mesh Ewald electrostatics. Equilibrium and non-equilibrium MD simulations are performed in the respective absence and presence of static electric and time-varying electromagnetic fields, applied both along the axis normal to the membrane and in the plane parallel to the membrane. Single-file diffusion is observed in the (5,5) and (6,6) cases, compared to classical diffusion in the larger pores. From an analysis of the molecular dipole moment distributions, time correlation functions, the number of water molecules present in the CNTs, and the hydrogen bonding characteristics of water inside the CNTs and at their mouth, we qualify the observed non-monotonic trends in the water molecule self-diffusion fluxes.

## 208b Mass Transport through Carbon Nanotube Membranes In Three Different Regimes: Ionic, Liquid and Gas

**Mainak Majumder**, Chemical and Biomolecular Engineering and Richard E. Smalley Instiute for Nano-scale Science and Technology, Rice University, 6100 Main Street, MS-362, Houston, TX 77005 and Bruce J. Hinds, Chemical and Materials Eng., University of Kentucky, 177 Anderson Hall, Lexington KY, KY 40506-0046.

Transport phenomena through the hollow conduits of carbon nanotubes (CNT) has been the subject of intense theoretical research due to the smooth graphitic structure of CNTs, the large van der Waals distances normal to the CNT wall and their similarity to biological membrane channels. We have experimentally studied transport processes over the large spectrum of ionic diffusion to pressure driven liquid and gaseous flow. Central to these studies is a membrane structure consisting of a high density (~ 1010/cm2) of aligned multiwalled carbon nanotubes with inner pore diameters (~ 7 nm) spanning a continuous polystyrene matrix allowing macroscopic measurement of transport through the structure. Several size-exclusion based control experiments including Au nano-crystal retention, separation of large from small molecule, ligand-gated 'on/off' ionic transport validates a membrane structure with open-ended carbon nanotubes as conduits for molecular transport.

Plasma oxidation step during the fabrication of the membrane introduces carboxylic acid groups at the CNT entrance, which provides electro-static 'gate-keper' effects on ionic transport. Capacitance measurements of the CNT membrane indicated a charge density of ~ 0.8 x 10-2 C/m2 at neutral pH. The ionic transport experiments are consistent with electrostatic hindrance effects at the entrances to carbon nanotubes (i.e a few nanometers of path length) and not along the entire path length. Diffusive transport of ions of different charge and size through the core of the CNT are therefore close to bulk diffusion expectations and allow estimation of the number of open pores or porosity of the membrane. Polar liquids such as water, ethanol, iso-propyl alcohol and non-polar liquids such as hexane and decane were observed to have pressure driven flow velocities four to five orders of magnitude higher than that predicted from the Hagen-Poiseuille equation. This suggests that the 'no-slip' boundary condition, is no longer valid between the fluids and the surfaces of the graphitic core. The observed flow velocities were close to those reported in biological membrane channels and in agreement with theoretical predictions. Flux of gases like N2, CO2, Ar, H2, CH4 scaled inversely with their molecular weight by an exponent of 0.4 suggesting Knudsen like transport through the CNT pores. However, the magnitude of the fluxes were observed to be one order of magnitude higher than predicted from Knudsen Diffusion calculations This is consistent with enhanced gas transport kinetics, expected for specular reflection inside smooth pores.

The transport studies indicate a membrane structure with exceptionally high mass transport kinetics, limited only by the functional molecules at the entrance to the CNT cores. The membrane structure has the potential to combine high selectivity with high permeability, not obvious with other membrane structures.

## 208c Crossover from Fickian to Single-File Diffusion of Fluids Confined In Carbon Nanotubes

**Ying-Chun Liu**[1], Joshua D. Moore[2], Thomas R. Roussel[2], Qu Chen[1], Qi Wang[1], and Keith E. Gubbins[2]. (1) Department of Chemistry, Zhejiang University, Hangzhou, 310027, China, (2) Department of Chemical and Biomolecular Engineering and Center for High Performance Simulation, North Carolina State University, Raleigh, NC 27695-7905

The self-diffusion of fluids in narrow pores is ubiquitous in many fields including material science, physics, chemistry, and biology[1-4]. It is therefore of practical interest to understand the properties that govern the diffusion mechanisms (e.g., ballistic motion, Fickian (3-D) diffusion, or single-file diffusion) of molecules confined in carbon nanotubes (CNTs) [5-7] Grand Canonical Monte Carlo (GCMC) and molecular dynamics (MD) simulations were used to compute adsorption isotherms, adsorption energies, and diffusivities of Lennard-Jones fluids and water confined in CNTs. We focus on the crossover from Fickian (3-D) to single-file diffusion in various helices of CNTs. We investigate the effects of fluid type, diameter, loading, and pressure.

We report that the transition between Fickian (3-D) and single-file diffusion strongly depends on the type of fluid, size of the CNT relative to the size of the fluid particle, and the loading of fluid in the CNT. Our results show a sudden decrease in the loading near a specific reduced diameter of the tube (which we define as the transition diameter). At diameters smaller than the transition diameter, the diffusion is purely single-file. However, at the transition, a decrease in the loading is found and is due to the tube diameter being too small to accommodate a second layer of fluid molecules. At diameters greater than the transition diameter, the particles start to pack into more than one layer, and the diffusion mechanism becomes Fickian (3-D). Our results are in agreement with those of Mon and Percus who studied hard spheres in cylindrical pores with hard walls [8].

The chirality of the CNT can affect the diffusivity of confined fluids in CNTs. For water at room temperature, the self-diffusion in zigzag CNTs is much slower than that in armchair CNTs at similar pore sizes. Calculation of PESs (potential energy surfaces) indicates that in armchair CNTs water diffuses in a spiral path along the axis of the tube. However, in zigzag carbon nanotubes more diffusion occurs in the $q$ direction of the tube, and the diffusion along the axis of the tube is slower in zigzag nanotubes compared to armchair nanotubes. For Lennard-Jones fluids we only see an effect of chirality when the temperature is low enough such that the kinetic energy is not enough to overcome the fluid-wall interaction energy.

1. V. Gupta, S. Nivarthi, A. McCormick, and H. T. Davis, *Chem. Phys. Lett.* **247**, 596 ( 1995).

2. Q.-H. Wei, C. Bechinger, and P. Leiderer, *Science* **287**, 625 (2000).

3. R. Allen, S. Melchionna, and J.-P. Hansen, *Phys. Rev. Lett.* **89**, 175502 (2002).

4. G. Hummer, J. C. Rasaiah, and J. P. Noworyta, *Nature*, London **414**,188 (2001).

5. C. Wei and D. Srivastava, *Phys. Rev. Lett.* **91**, 235901(2003).

6. O. G. Jepps, S. K. Bhatia, and D. J. Searles, *Phys. Rev. Lett.* **91**, 126102 (2003).

7. V. P.Sokhan,D. Nicholson,and N.Quirke, *J.Chem.Phys*, **120**, 3855 (2004).

8. Mon, K. K. and Percus, J. K., *J. Chem. Phys.*, **112**, 3457 (2000)

## 208d Pressure Gap In Carbon Nanomaterials? Effect of Temperature and Pressure on the Binding of Simple Molecules to Carbon Nanotubes

**Dmitry V. Kazachkin**[1], Yoshifumi Nishimura[2], Stephan Irle[2], Keiji Morokuma[3], Radisav Vidic[4], and Eric Borguet[1]. (1) Department of Chemistry, Temple University, Philadelphia, PA 19122, (2) Institute for Advanced Research and Department of Chemistry, Nagoya University, Furu-cho, Chikusa-ku Nagoya 464-8602, Japan, (3) Department of Chemistry, Emory University, Atlanta, GA 30322, (4) Department of Civil and Environmental Engineering, University of Pittsburgh, Pittsburgh, PA 15261

The interaction of simple molecules with single wall carbon nanotubes (SWCNTs) and carbon black was shown to depend on the experimental conditions. For example, acetone dosed to SWCNTs at 100 K and low pressure (~$10^{-6}$ Torr) has binding energy of ~75 kJ/mol. Increasing exposures at 100 K resulted in the filling of adsorption sites with lower energies. Adsorption of acetone at 300 K and high pressure (~7.6 Torr) results in acetone binding energies of ~120 kJ/ mol. Similar phenomena were observed for the interaction of 1,3-butadiene and n-butane with SWCNTs. Experiments suggest that these strongly bound species desorb as intact molecules. The adsorption under typical UHV conditions (100 K-$10^{-6}$ Torr) does allow for molecules to access the sites with highest desorption energy. Carbon black shows similar behavior to SWCNTs, but with significantly lower capacity for molecular uptake, consistent with its specific surface area and porous structure.

## 208e Atomic-Scale Analysis of the Interactions Between Atomic Hydrogen and Multi-Walled Carbon Nanotubes

**Andre R. Muniz**[1], Tejinder Singh[1], Michael J. Behr[2], Eray S. Aydil[2], and Dimitrios Maroudas[3]. (1) Department of Chemical Engineering, University of Massachusetts at Amherst, Amherst, MA 01003, (2) Department of Chemical Engineering and Materials Science, University of Minnesota, Minneapolis, MN 55455, (3) Department of Chemical Engineering, University of Massachusetts, 686 N Pleasant Street, Amherst, MA 01003

An atomic-scale analysis of the interactions of atomic hydrogen with multi-walled carbon nanotubes (MWCNTs) is presented, aiming to explore the structural changes undergone by MWCNTs when exposed to hydrogen plasmas. The analysis is based on a synergistic combination of classical molecular-dynamics (MD) simulations with first-principles density functional theory (DFT) calculations. The Adaptive Interatomic Reactive Empirical Bond Order (AIREBO) potential is employed in the MD simulations of H-MWCNT interactions and the resulting structural relaxations. The DFT calculations are performed within the generalized gradient approximation (GGA) and employ plane-wave basis sets, ultrasoft pseudopotentials, and supercell models. Parameters that have been varied in our analysis include nanotube diameters and chiralities, hydrogen concentration, and temperature.

Depending on the H dose and the resulting H coverage, the chemisorption of hydrogen onto the graphene walls of a MWCNT affects considerably its structure, leading to the deformation and local amorphization of the graphene walls to form a structure that consists of both $sp^2$- and $sp^3$-hybridized carbon atoms. The diffusion of hydrogen atoms on and between graphene walls and their reactions with the carbon structure are analyzed and the results of the analysis provide interpretations for the formation of nanocrystalline carbon phases (e.g., cubic and hexagonal diamond), which have been observed in our experiments upon $H_2$ plasma exposure of MWCNTs. In our theoretical study, special emphasis is placed on investigating systematically the possibility of formation of inter-shell $sp^3$ C-C bonds induced by atomic hydrogen. These localized inter-shell C-C bonds can act as a seed for the nucleation of nanocrystalline phases embedded into the MWCNTs. We present the results of a comprehensive protocol of DFT and MD calculations, which show that the resulting structures containing these inter-shell C-C bonds are stable and that seeds for the nucleation of different carbon phases (allotropes) can be formed, depending on the alignment between the graphene walls.

## 208f Characterization of Alkali Metal Doped Carbon Nanotubes

**Li Xiao**[1], J. Karl Johnson[1], Michael Buettner[2], and John T. Yates[2]. (1) Department of Chemical and Petroleum Engineering, University of Pittsburgh, 1242 Benedum Hall, Pittsburgh, PA 15261, (2) Department of Chemistry, University of Virginia, Charlottesville, VA 22901

Carbon nanotubes can be doped with alkali metals as a way to change the adsorption potential for other molecules on the inside and outside of the nanotubes. We present results from ab initio density functional theory calculations and temperature programmed desorption experiments that characterize the interaction of alkali metals with carbon nanotubes and the interaction of alkanes on alkali metal doped nanotubes. We investigate the interaction of Li atoms with carbon nanotubes (CNTs), both with and without defects. We find that Li atoms are highly mobile on defect free CNTs but are bound strongly to oxygen containing defect sites. Experiments show that the desorption of heptane is significantly altered by the presence of Li. We present calculations to probe the nature of the interaction between Li and alkanes within CNTs.

## 208g Facilitating Density-Based Electronic-Type Separation of Carbon Nanotubes Via Chemical Reactions: A Modeling Study

**Nitish Nair**[1], Woo-Jae Kim[1], Monica Usrey[2], and Michael S. Strano[3]. (1) Department of Chemical Engineering, Massachusetts Institute of Technology, Building 66-580, 25 Ames St, Cambridge, MA 02141, (2) University of Illinois at Urbana/Champaign, 104 RAL, 600 S. Mathews Ave., Urbana, IL 61802, (3) 66-566 Department of Chemical Engineering, Massachusetts Institute of Technology, 77 Massachusetts Avenue, Cambridge, MA 02139

A structure-reactivity relationship for electron transfer reactions of single walled carbon nanotubes (SWNT) has been derived and experimentally validated using 4-hydroxybenzene diazonium as a model electron acceptor. The formalism provides a mechanistic insight into electronically selective reactions. We have shown that covalently attached chemical groups can alter the densities of individual SWNT in a predictable and highly controllable manner, thus facilitating separation via ultracentrifugation. A hydrodynamic model is used to describe the motion of surfactant-suspended single walled carbon nanotubes in a density gradient, while being subjected to a centrifugal field. The number of surfactant molecules and functional groups adsorbed on each nanotube determine its effective density, and hence, its position in the gradient after centrifugation has been completed. Analysis of the spatial concentration distributions of unreacted CoMoCAT nanotubes suspended with 2 w/v% sodium cholate yielded 2.21, 2.36 and 2.74 surfactant molecules adsorbed per nanometer along the length of the (6,5), (7,5) and (8,7) nanotubes, respectively. A volume additivity model based is able to estimate the density difference between 4-hydroxy phenyl functionalized and non-functionalized HiPco SWNT as approximately 98.3 kg/m3, compared with 97.9 kg/m3 measured from density gradient centrifugation. We therefore conclude that chemical functionalization can provide an effective handle to separate out a particular SWNT from a typical diameter distribution.

## 245a Effects of Nanoparticle Size Polydispersity on Stability of the Double Gyroid Phase In a Tethered Nanoparticle System

**Carolyn L. Phillips**, Applied Physics, University of Michigan, 2350 Hayward Street, 3440 G.G. Brown, Ann Arbor, MI 48109, Christopher R. Iacovella, Chemical Engineering, University of Michigan, 2350 Hayward Street, 3440 G.G. Brown, Ann Arbor, MI 48109, and Sharon C. Glotzer, Chemical Engineering, Materials Science and Engineering, Physics, Macromolecular Science and Engineering, and Applied Physics, University of Michigan, 2350 Hayward Street, 3440 G.G. Brown, Ann Arbor, MI 48109.

Recent simulations have shown that aggregating nanospheres functionalized with polymer tethers can self-assemble to form the double gyroid phase previously seen in block copolymer systems [1,2]. Using Brownian dynamics simulations, we study the impact of nanosphere size polydispersity on both the static and dynamic properties of the double gyroid [3]. We applied the Voronoi Tessellation to determine the compactness of the nanoparticle component in the gyroid as a function of polydispersity. We show how this method can be used to quantify domain density in any microphased separated system. Related to the compactness of the gyroid, we also investigate local structure of nanoparticles as a function of polydispersity.

[1] Iacovella CR, Keys AS, Horsch MA, Glotzer SC., "Icosahedral packing of polymer-tethered nanospheres and stabilization of the gyroid phase", Physical Review E 75 (4), APR 2007

[2] Iacovella CR, Horsch MA,and Glotzer SC, "Local Ordering Of Tethered Nanospheres And Nanorods in the Double Gyroid Phase", preprint.

[3] Phillips CL, Iacovella, CR, and Glotzer SC, preprint

## 245b Ordering of Nanoparticles Mediated by End-Functionalized Triblock Copolymers

**Rastko Sknepnek**[1], Joshua A. Anderson[1], Monica H. Lamm[2], Joerg Schmalian[1], and Alex Travesset[1]. (1) Department of Physics and Astronomy, Iowa State University/DOE Ames Laboratory, A510 Zaffarano Hall, Ames, IA 50011, (2) Department of Chemical and Biological Engineering, Iowa State University/DOE Ames Laboratory, A510 Zaffarano Hall, Ames, IA 50011

Using molecular dynamics simulations we show that triblock copolymers, designed to have specific affinity for nanoparticles at the chain ends, can successfully mediate assembly of nanoparticle/ copolymer composites. In this talk, we will present a detailed investigation of the phase diagram of these nanocomposites as a function of both nanoparticle size and concentration. We find a rich phase diagram with a number of distinct ordered structures. Next to phases such as hexagons or gyroids, regularly observed in copolymer solutions and melts, we find a novel square columnar phase of two interpenetrating line-lattices of micellar cylinders and aligned nanoparticles. We argue that this phase is a realization of the packing problem of binary mixtures of disks. Using theory of minimal and constant mean curvature surfaces we study nanoparticle and copolymer density profiles and develop qualitative understanding of origin and stability of ordered structures. Further, based on the analysis of stretching of copolymers we argue that nanoparticles help stabilize gyroid order and drastically widen the region of phase diagram occupied by gyroids. We also propose experimental realizations of our model calculation. Our study suggests that combining nanoparticles with functionalized block copolymers can provide a simple and efficient tool for assembling novel materials with nanometer scale resolution.

Reference: Rastko Sknepnek, Joshua A. Anderson, Monica H. Lamm, Joerg Schmalian, and Alex Travesset: ``Nanoparticle Ordering via Functionalized Block Copolymers in Solution", ACS Nano, 2008, submitted.

## 245c Two-Dimensional Assembly of Aqueous Colloidal Metallic Nanowires at Surfaces

**Derek A. Triplett**[1], Lisa M. Dillenback[2], Benjamin D. Smith[2], Christine D. Keating[2], and Kristen A. Fichthorn[1]. (1) Dept. of Chemical Engineering, The Pennsylvania State University, University Park, PA 16802, (2) Dept. of Chemistry, The Pennsylvania State University, University Park, PA 16802

The assembly of metallic nanowires at surfaces is important in the construction of nanoelectronic circuits as well as in many other higher-order structures. The ability to understand and control the assembly of these anisotropic building blocks is fundamental to the creation of properly ordered structures. Here, we investigate the ordering of aqueous mixed-metal, 50% Au/ 50% Ag nanowires that possess a gold end and a silver end and are coated with 2-mercapto-ethanesulfonic acid (MESA). The nanowires settle from solution onto a glass cover slip and are visualized with an optical microscope. The wires form a smectic-like layer on the substrate and exhibit a slight preference for the gold ends of the wires to be aligned next to one another. The alignment of Au/Au, Au/Ag, and Ag/Ag ends is characterized using an order parameter that compares the observed alignment of the wires against a random distribution.

In an effort to understand assembly in this system and ultimately exercise control over it, we use canonical Monte Carlo (MC) simulations to model the ordering of the wires on the substrate. We account for van der Waals and electrostatic interactions between the wires. The MC simulations correctly reproduce the experimentally observed ordering and demonstrate that the preference for alignment of the Au ends stems from differences between the Hamaker constants for Au and Ag. We study how general differences between the Hamaker constants of the two materials, as well as differences in the zeta potentials can alter the value of the order parameter. We show that the propensity of the wires to exhibit a smectic vs. a nematic phase is governed by a balance between van der Waals attraction and electrostatic repulsion. If the zeta potential is sufficiently weak, van der Waals attraction leads to a smectic phase, as is observed experimentally. On the other hand, if the zeta potential is sufficiently strong, a nematic phase is predicted that mitigates electrostatic repulsion. The ordering of the nanowires is also shown to depend on their density and length.

## 245d Effect of Confinement and Internanoparticle Forces on the Nanoparticle Structuring and Ordering

**Younjin Min**, Chemical Engineering, University of California, Santa Barbara, Engineering II, Santa Barbara, CA 93106, Mustafa Akbulut, Chemical Engineering, Princeton University, Olden st., E-quad A-306, Princeton, NJ 08544, Yuval Golan, Department of Materials Engineering, Ilse Katz Institute of Nanotechnology, Ben-Gurion University of the Negev, Beer-Sheva 84105, Israel, Joseph A. Zasadzinski, Department of Chemical Engineering, University of California, Santa Barbara, CA 93106, and Jacob N. Israelachvili, Chemical Engineering, University of California at Santa Barbara, 3357 Engineering II, Santa Barbara, CA 93106-5080.

This presentation describes the study of the fundamental inter-nanoparticle interactions which govern the self- or directed-assembly of functionalized nanoparticles into hierarchical nanostructures, and the exploration of the conditions required for these processes by monitoring the structural evolution of nano-structured films. The effects of shape on nanowire interactions and shear-induced ordering were studied. Both the normal and lateral forces were sensitive to the particles' curvature. (i) No adhesion was observed between the confining surfaces, and the force profiles were short-ranged and mostly reversible for straight wires but longer-ranged and irreversible for curved wires. (ii) Structural phase transitions can occur in confined nanoparticle films. (iii) During shearing, straight wires order or align more easily than curved wires. Also the effects of humidity (water content or activity from 0 to ~0.98) on the frictional properties of surfactant-coated ZnS nanoparticles of various shapes, specifically, nanorods and nanowires were investigated.

## 245e The Role of Alpha-Helix Formation In the Self-Assembly of Protein-Analogous Micelles

**Mark Kastantin** and Matthew Tirrell. Chemical Engineering, University of California, Santa Barbara, Engineering 2, Room 3357, Santa Barbara, CA 93106-5080

Alpha-helical structure in a peptide chain has been shown to affect the geometry of self-assembled systems containing these molecules. In a related phenomenon, incorporation of a polypeptide into a self-assembled system has been shown to affect the helix-coil transition relative to the free polypeptide. This work presents a quantitative model of the interaction between self-assembly and the helix-coil transition in micelles formed from hydrophobic hydrocarbon tails and water-soluble polypeptides. Such structures are termed protein-analogous micelles and can exist in spherical, cylindrical, or flat geometries as dictated by thermodynamic considerations.

This work shows that alpha-helix formation is stabilized by the micellization process while the amount of stabilization increases with the strength of attraction between hydrophobic tails. Helix formation drives the micelle to form structures with less curvature than would be expected if the polypeptide existed in a random-coil state. A transition in protein-analogous micelles is expected from spherical to cylindrical micelles as helical content of the polypeptide headgroup increases. This prediction is verified experimentally for a protein-analogous micelle in which helical content is controlled using pH. These results aid the design of protein-analogous micelles to produce a desired shape.

# 245f The Significance of Sphere-to-Rod Transition Kinetics of Tri-Block Copolymer Micelles for the Synthesis of Mesoporous Silica Materials

Antonia G. Denkova[1], Eduardo Mendes[1], and **Marc-Olivier Coppens**[2]. (1) DelftChemTech, Delft University of Technology, Julianalaan 136, Delft, Netherlands, (2) Chemical Engineering, Rensselaer Polytechnic Institute, 110 8th Street, Troy, NY 12180

The sphere-to-rod transition kinetics of tri-block copolymer micelles are especially important for the synthesis of mesoporous silica materials, such as SBA-15, where these surfactants are used as templating agents [1]. Such a synthesis typically starts with spherical micelles, which, during silica hydrolysis and condensation, are converted into long cylindrical micelles. When the materials are calcined, the micelles are removed, resulting in a silica matrix with hexagonally ordered mesopores. It is highly probable that the kinetics of the sphere-to-rod transition versus the kinetics of silica hydrolysis and condensation will determine the final properties of the synthesized inorganic materials. Therefore, we have studied the kinetics of the sphere-to-rod transition of the tri-block copolymer P123 ($EO_{20}PO_{70}EO_{20}$) in aqueous solutions, using dynamic light scattering (DLS) and cryogenic electron transmission microscopy (Cryo-EM).

Sphere-to-rod transition is induced by a solvent jump, initiated by adding KCl and ethanol to an aqueous micellar solution. The growth process of the worm-like micelles depends on the experimental conditions and has two distinct regions that can be described as a slow initiation period, and actual growth to equilibrium. All growth curves exhibit a single relaxation time ($\tau_r$) that represents the lifetime of the micelles. The growth curves collapse into a master curve, when shifted by the relaxation time, indicating that the actual growth process of the micelles in all samples occurs through the same mechanism. The lifetime decreases with increasing surfactant and ethanol concentration. Additionally, as evidenced by Cryo-EM images, some of the formed micelles exhibit a caterpillar-like shape, in which several of the original spherical species can still be detected. These facts suggest that the micelles grow longer predominantly by random coagulation/fragmentation reactions involving species of different sizes [2].

Conclusions from this study allow us to enhance our understanding of the role played by salts and ethanol in the formation of micelle-templated mesoporous materials, such as SBA-15.

References:

[1] Zhao D.; Huo Q.; Feng J.; Chmelka B. F.; Stucky G. D. J. Am. Chem. Soc. 1998, 120, 6024

[2] Denkova A.G., Mendes E.,Coppens M.-O., submitted for publication.

## 245g Liposomal Nanoparticles Prepared by Nanoscale Directed Assembly for Delivery of Oligonucleotide Therapeutics

**Jingjiao Guan**[1], Megan Lynn Cavanaugh[2], Weibin Zha[3], and Ly James Lee[2]. (1) NSF Center for Affordable Nanoengineering of Polymeric Biomedical Devices, The Ohio State University, 1381 Kinnear Road, Suite 100, Columbus, OH 43212, (2) Chemical and Biomolecular Engineering, The Ohio State University, 125 Koffolt Labs, 140 W 19th Ave, Columbus, OH 43210, (3) Department of Chemical and Biomolecular Engineering, the Ohio State University, Rm. 125, 140 West 19th Ave., Columbus, OH 43210

Liposomal nanoparticles are clinically used for drug delivery and hold great promise for the delivery of emerging oligonucleotide therapeutics. Common to most techniques for liposomal nanoparticle preparation is the use of a self assembly process conducted at a scale much larger than the nanoparticles themselves. Such a large scale discrepancy usually leads to a limited control of particle size, structure, composition, and consequently delivery efficacy. Nanoscale directed assembly offers a much tighter control on the formation the liposomal nanoparticles at the same size scale as the particles. Here we report the use of an array of nanoscale wells to direct the assembly of lipids and oligonucleotides into nanoparticles with better controlled size and increased loading capacity characterized by AFM and cryoTEM.

## 278a Forum Award Lecture - One Dimensional Nanomaterials and Their Applications

**M. Meyyappan**, NASA Ames Center for Nanotechnology, NASA Ames Research Center, MS 229-3, Moffett Field, CA 94035

Carbon nanotubes(CNTs) have received much attention in the last decade due to their unique electronic properties and extraordinary mechanical properties. The ability to grow a variety of semiconductor, oxide and other inorganic materials in the form of nanowires with controlled properties and orientation also provides a competitive avenue for applications in logic, memory, data storage, sensors, instrumentation and others. Growth of the inorganic nanowires and carbon nanotubes by chemical vapor deposition follows a common basis of vapor-liquid-solid mechanism. This talk will focus on results from growth and characterization of these materials with a focus on application development in the above areas. Application examples will include carbon nanotube based chemical sensors, biosensors and instrumentation, phase change nanowire based random access memory, and other electronic devices.
The author acknowledges contributions from Cattien Nguyen, Bin Yu, Jeff Sun, Jing Li, Y. Lu, Prabhu Arumugam, Hua Chen and Jun Li.

## 278b Nanomaterials for Drug Delivery

**Samir Mitragotri**, Chemical Engineering, University of California, Santa Barbara, Santa Barbara, CA 93106

*Text Not Available.*

## 278c Colloidal Nanoparticles: Synthesis and Applications

**Darrell Velegol**, Department of Chemical Engineering, The Pennsylvania State University, University Park, PA 16802

How can we synthesize, classify, stabilize, and assemble nanoparticles? The wealth of nanoparticles recipes makes many types of nanoparticles -- though not all types -- feasible to synthesize. But what about processing issues? Synthesis reactions are often limited to producing a few grams of material at a time. Classifying nanoparticles by size is often a bottleneck problem, for instance in the dielectrics industry. Separating nanoparticles from assemblies of the particles is equally challenging. And whereas stabilizing 100 nm or larger particles is a well-known science, stabilizing 10 nm particles is often a delicate art, especially since the use of dispersants is often prohibited by the need for further surface chemistry functionalization. Does Chemical Engineering provide any solutions for the question of whether nanomotors, SERS, dielectrics, and other applications will become realities?

## 297a Controlled Aggregation of Iron Oxide Nanocrystals for Preparation of Magnetic Resonance Imaging Contrast Agents

**Marian E. Gindy**, Chemical Engineering, Princeton Unversity, E-Quad, Olden Street, Princeton, NJ 08544, Theodore Wolfson, Chemical Engineering, Princeton University, A301 E-Quad, Princeton, NJ 08544, Carlos R. N. Pacheco, Chemistry, Princeton University, Frick Lab, Washington Road, Princeton, NJ 08544, Victoria Calero, Chemical Engineering, University of Puerto Rico, PO Box 9046, Mayaguez, PR 00680, Carlos Rinaldi, Chemical Engineering, UPR-Mayaguez Campus, Mayaguez, PR 00680, and Robert K. Prudhomme, Princeton University, Dept of Chemical Eng Olden St, Princeton, NJ 08544.

Nanoparticle formulations of nanocrystralline materials present unique opportunities for improved medical imaging and diagnostics via optical, fluorescence, and magnetic resonance imaging modalities. The common requirements of these applications are biocompatibility, high payload of contrast agents, and precise control of particle size and functionality, which allow for selective localization of contrast agents at levels sufficient to provide enhanced detection in vivo. In this work, we present a novel technology for the preparation of composite nanoparticles (CNPs) based on iron oxide nanocrystals for application as magnetic resonance imaging (MRI) contrast agents. The process relies on controlled precipitation of components under diffusion-limited conditions to produce stable particles at high concentrations of encapsulated agents using amphiphilic block copolymers to direct self-assembly. Because the particles assemble spontaneously from solution by simultaneous desolvation of components, explicit surface functionalization of the nanocrystals is not required, and the method can be applied to a variety of nanocrystals that lack appropriate surface chemistry. Experimentally measured physical properties of the CNPs(particle size and size distributions) are well predicted by models for colloid coagulation in the diffusion limited regime. In addition, CNPs of various compositions exhibit a high relaxivity (r2 = 130-200 mM-1 s-1), rendering them very attractive as T2 contrast agents for MRI. Finally, the combined incorporation of iron oxide nanocrystals with drugs into an integrated system without reduction in relaxivity is demonstrated, enabling efficient and simultaneous drug delivery and imaging.

# IN VITRO TOXICITY AND INTRACELLULAR UPTAKE OF FLAME SYNTHESIZED IRON OXIDE NANOPARTICLES: AN ALTERNATIVE TO WET SYNTHESIS METHODS

*Kivilcim Buyukhatipoglu, Drexel University, Philadelphia, PA*
*Tiffany Miller, Drexel University, Philadelphia, PA*
*Alisa Morss Clyne, Drexel University, Philadelphia, PA*

## Abstract

Superparamagnetic iron oxide nanoparticles, including magnetite ($Fe_3O_4$), are widely used in applications such as hyperthermic malignant cell treatment, magnetic resonance imaging, targeted drug delivery, tissue engineering, gene therapy, and cell membrane manipulation. In the current work, superparamagnetic iron oxide nanoparticles were produced using a flame synthesis method, which provides significant advantages over other material synthesis processes. Flame synthesis allows control of particle size, size distribution, phase and composition by altering flame operating conditions and is further capable of commercial production rates with minimal post-processing of the final product materials. This study focuses on the interaction of flame synthesized iron oxide nanoparticles with porcine aortic endothelial cells and compares the results to those obtained using commercially available iron oxide nanoparticles. The materials characteristics of the flame synthesized iron oxide nanoparticles, including morphology, elemental composition, particle size, and magnetic properties, were analyzed by electron microscopy (TEM, ESEM, EDS), and Raman Spectroscopy. The data verified production of a heterogenous mixture of hematite and magnetite nanoparticles, which exhibit superparamagnetic properties. Monodisperse iron oxide particles of 6-12 nm diameter and aggregated clusters of these 6-12nm nanoparticles have been synthesized. Nanoparticle biocompatibility was assessed by incubating flame synthesized and commercially available iron oxide nanoparticles with endothelial cells for 24 hours. Both Alamar blue and Live/Dead cell assays showed no significant toxicity difference between flame synthesized and commercially available nanoparticles. Cells exposed to both types of nanoparticles maintained membrane integrity, as indicated by minimal lactase dehydrogenase release. Endothelial cells imaged by ESEM and confirmed by EDS demonstrated that uncoated flame synthesized nanoparticles are ingested into cells in a similar manner to commercially available nanoparticles. These data suggest that flame synthesized iron oxide nanoparticles are comparable to commercially available nanoparticles for biological applications. Flame synthesis has the advantage of a relatively simple synthesis process with higher purity products and lower time and energy manufacturing costs. Future work will include functionalizing the nanoparticle surfaces for specific biological applications, including specific cell targeting and bioactive factor delivery.

## Introduction

Recent developments in nanotechnology allow us to produce and functionalize nanoparticles for biomedical applications such as magnetic resonance imaging (MRI), hyperthermic treatment of malignant cells, targeted drug delivery, tissue engineering, and cell membrane manipulation [1-6]. Superparamagnetic nanoparticles, such as $Fe_2O_3$ and $Fe_3O_4$, are of particular interest for in vivo and in vitro applications because they do not show any magnetic behavior after the magnetic field has been removed [7].

In this study, iron oxide nanoparticles were produced by gas phase combustion synthesis, also called flame synthesis. Flame synthesis can be simpler and more economical than other material synthesis techniques such as sol-gel processing, chemical vapor deposition and wet chemical deposition. In particular, flame synthesis produces a continuous high yield of pure material with controlled particle size, size distribution, phase, and composition in a single step with no subsequent post-processing (see Wooldridge and Pratsinis for details) [8-10]. Interaction of flame synthesized iron oxide nanoparticles with endothelial cells was investigated, since endothelial cells directly interact with nanoparticles in the bloodstream. Cytotoxicity results were compared to those obtained using commercially available iron oxide nanoparticles.

## Methods

### *Nanoparticle Synthesis*

A laboratory scale combustion synthesis and sampling system was developed for direct synthesis of iron oxide nanoparticles. The combustion synthesis system was composed of three main components: the burner consisting of three concentric tubes, which was used to create a high temperature synthesis environment; a liquid precursor delivery system; and two particle sampling mechanisms to obtain discrete and bulk material samples via thermophoretic deposition. A schematic of the combustion synthesis system is shown in Fig. 1.

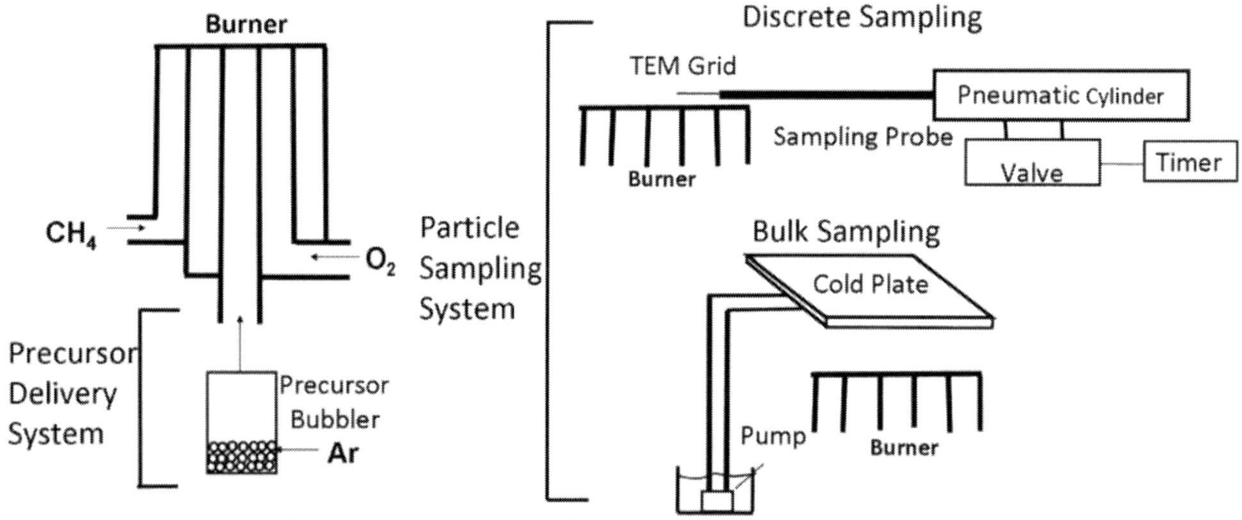


Figure 1: Schematic of combustion synthesis facility.

### *Nanoparticle Characterization*

Nanoparticle morphology, elemental composition and size were analyzed by transmission electron microscopy (TEM; JEM-2000FX). The iron oxide nanoparticle elemental composition was analyzed using a Renishaw RM1000 VIS Raman microspectrometer with 633 nm excitation wavelength.

### *Cell Culture*

Porcine aortic endothelial cells (PAEC) were maintained in low glucose Dulbecco's modified Eagle's medium (DMEM) supplemented with 5% fetal bovine serum, 1% penicillin-streptomycin, and

2% glutamine (Invitrogen). Culture media was changed every 48 hours and cells were used between passages 4 and 9. Confluent cells were incubated with media alone or media with 0.1 mg/ml iron oxide nanoparticles (flame synthesized or commercially available) for 24h. 10 ng/ml tumor necrosis factor-α (TNFα) was the positive control.

***Scanning Electron Microscopy for Cell Morphology***

Scanning electron microscopy (SEM) was used to visualize nanoparticle interaction with cells. PAEC were seeded on 8mm sterile glass coverslips and incubated with nanoparticles as previously described. Cells were fixed in 3% glutaraldehyde (EMS) at room temperature for 20 minutes, rinsed three times with phosphate buffered saline (PBS), and dehydrated in graded ethanol (Pharmco). Ethanol was then replaced with hexamethyldisilazane (HMDS; EMS), and samples were dried by overnight HMDS evaporation in a chemical fume hood. Samples were desiccated under vacuum, sputter coated with 0.75nm thick Pt/Pd layer to increase conductivity, and viewed under a Zeiss Supra 50VP SEM.

***Nanoparticle Cytotoxicity***

Alamar blue was used to measure cell proliferation and metabolic activity using an oxidation-reduction indicator. After 24 hours of cell nanoparticle exposure, 100µl of media from each condition was transferred into a 96 well flat-bottomed black assay plate. 10µl of Alamar Blue (AbD Serotec) was added to each well, and the well plate was incubated for 4h at 37°C. Fluorescence was measured at 535/590 nm in a GENios microplate reader.

The CytoTox-ONE Homogenous membrane integrity assay kit (Promega) measures release of lactate dehydrogenase (LDH) from cells with damaged membranes. 100 µl of media from PAEC incubated with nanoparticles for 24 hours was transferred into a 96 well flat-bottomed black plate. 100 µl of CytoTox-ONE reagent was added to each well, after which the plate was incubated at room temperature for 10 minutes protected from light. Fluorescence was measured at 560/590 nm in a GENios microplate reader.

The Live-Dead Assay (Molecular Probes) quantifies live and dead cells by measuring intracellular esterase activity (Calcein AM, green) and plasma membrane integrity (ethidium homodimer-1, red). PAEC seeded on sterile 13-mm round glass coverslips were incubated with nanoparticles for 24 h, washed with PBS, and treated with 150µl of 2µM calcein AM and 4 µM of ethidium homodimer-1 as per manufacturer instructions. Cells were incubated at room temperature for 45 minutes and imaged in an Olympus IX81 inverted fluorescent microscope.

## Results and Discussion

Two different size modes of iron oxide nanoparticles were synthesized using the flame synthesis method. Iron pentacarbonyl, $Fe(CO)_5$, was used as the liquid precursor and delivered to the central annulus of the reactor as saturated vapor entrained in Argon gas. All synthesis conditions were conducted under oxygen rich conditions with low equivalence ratios to minimize unwanted carbon contamination in the synthesized nanoparticles. An inverse diffusion flame configuration, with $O_2$ in the second annulus (flowrate: 3546 ml/min) and $CH_4$ in the third annulus (flowrate: 494ml/min), resulted in a high concentration of magnetite formation as verified by Raman spectroscopy. Representative TEM images of flame synthesized 6-12 nm iron oxide nanoparticles and small aggregated clusters of them collected at 5cm above the burner exit are shown in Fig. 2a. Raman spectroscopy of synthesized nanoparticles (Fig. 2b) showed a heterogeneous mixture of two common forms of iron oxide, hematite and magnetite, in the nanoparticles. The intense peak observed at 1320 $cm^{-1}$ is assigned to a two-magnon scattering which arises from interaction of magnons created on antiparallel close spin sites [11].

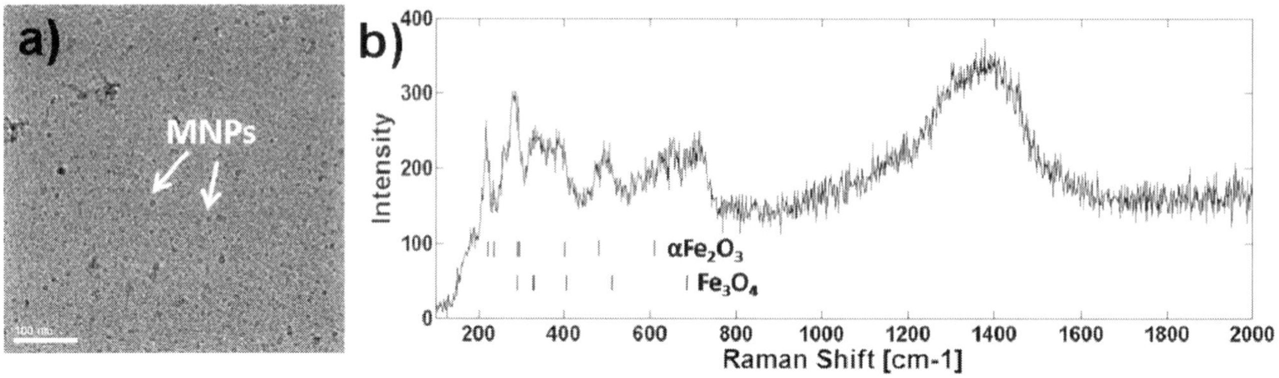


Figure 2: TEM images show monodispersed and aggregated clusters of 6-12 nm diameter iron oxide nanoparticles (a). Raman spectrum of flame synthesized iron oxide nanoparticles show magnetite, $Fe_3O_4$, and hematite, $\alpha Fe_2O_3$ content of the nanoparticles produced (b).

Endothelial cells remained attached and spread after 24 hours of exposure to 0.5 mg/ml nanoparticle solution. When incubated with nanoparticles (Fig. b, c), endothelial cells were less spread, more rounded, and appeared to have lost their typical cobblestone morphology, however no change in cell adhesion and total attached cell number was observed (Fig. 3 d)

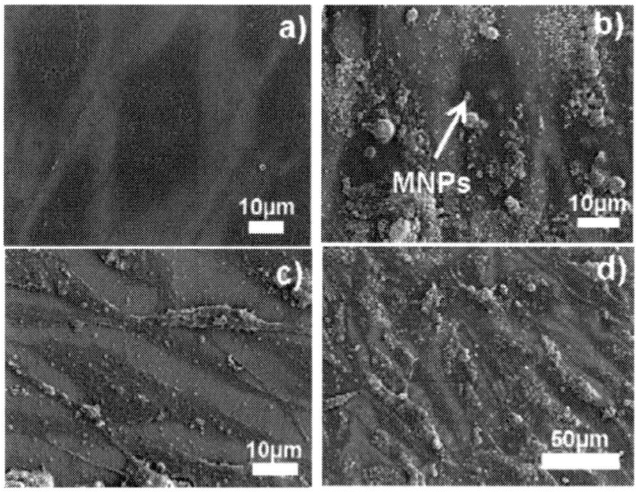


Figure 3: SEM images show that Individual cell spread less when incubated with nanoparticles (b, c). However there was no observed change in attached cell number and adhesion (d).

Flame synthesized superparamagnetic nanoparticles demonstrated low cytotoxicity comparable to commercially available nanoparticles. Endothelial cell viability, as measured by Alamar blue, did not change with nanoparticle exposure or with synthesis method (Fig. 4a). Endothelial cells demonstrated a slight loss of membrane integrity with purchased nanoparticles, as measured by LDH release, (Fig. 4b). However, no statistically significant changes in cell membrane integrity were observed in cells exposed to synthesized nanoparticles.

Endothelial cell viability with both synthesized and purchased nanoparticles was confirmed using a Live/Dead Cell Viability assay. When dead cell number was quantified using a microplate reader, cells exposed to both purchased and synthesized iron oxide nanoparticles showed slightly increased cell death over control cells (Fig. 4c). However, there was no statistically significant change in cell death for the two types of nanoparticles. Fluorescent images of representative cell regions showed no observable difference in live and dead cell density (Fig. 4 d, e, f).

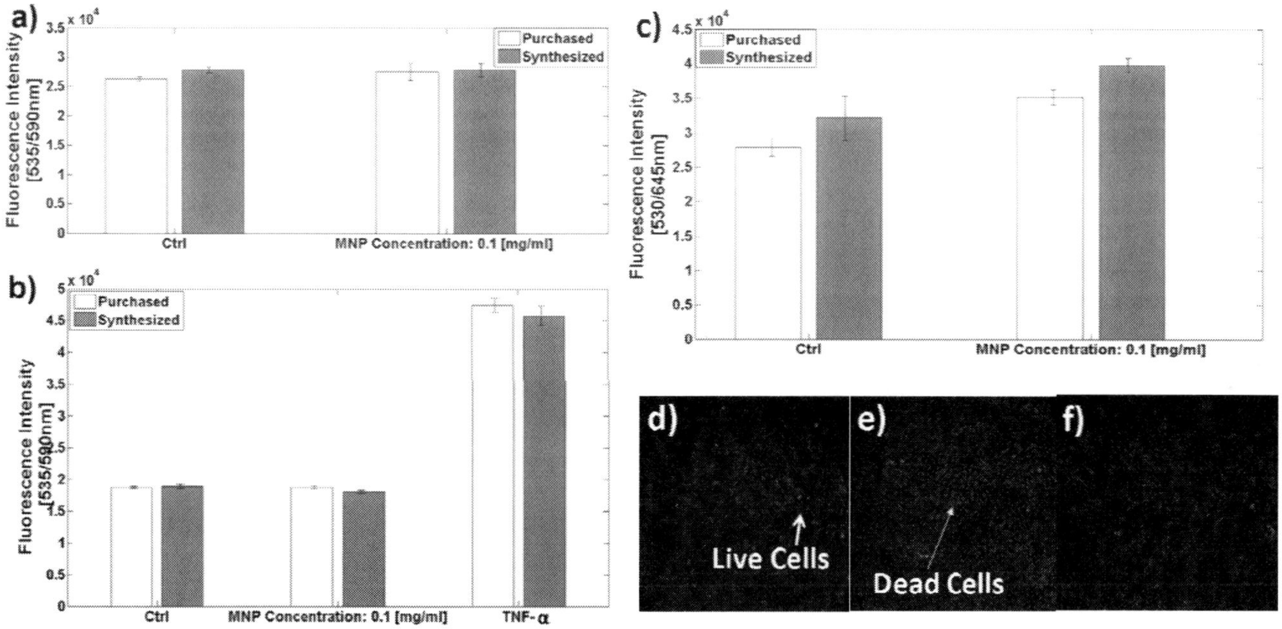


Figure 4: (a) Alamar blue, (b) LDH, (c) Live/Dead assays showed no significant difference in cytotoxicity between flame synthesized and commercially available nanoparticles. Live-Dead cell numbers are also compared with the data obtained from microplate reader by counting the cells at the fluorescent images: control cells (d), cells loaded with 0.1 mg/ml of purchased nanoparticles (e), cells loaded with 0.1mg/ml of synthesized nanoparticles (f).

## Conclusions

Iron oxide nanoparticles were synthesized using flame synthesis, which has the advantage of being a relatively simple synthesis process with higher purity products and lower time and energy manufacturing costs. Cellular uptake and cytotoxicity studies suggest that flame synthesized iron oxide nanoparticles are comparable to commercially available nanoparticles for biological applications. Future work will include functionalizing the nanoparticle surface for specific biological applications, including cell targeting and bioactive factor delivery.

## References


1. Halavaara, J. et al. (2002), "Efficacy of sequential use of superparamagnetic iron oxide and gadolinium in liver MR imaging," *Acta Radiol.*, 43(2): p. 180-5.
2. Bonnemain, B. (1998), "Superparamagnetic agents in magnetic resonance imaging: physicochemical characteristics and clinical applications. A review," *J Drug Target*, 6(3): p. 167-74.

3. Mitsumori, M., et al. (1996), Targeted hyperthermia using dextran magnetite complex: A new treatment modality for liver tumors. Hepato-Gastroenterology,"43(12): p. 1431-1437.
4. Scherer, F., et al. (2002), "Magnetofection: enhancing and targeting gene delivery by magnetic force in vitro and in vivo. Gene Therapy," 9(2): p. 102-109.
5. Dobson, J. (2006), "Gene therapy progress and prospects: magnetic nanoparticle-based gene delivery. Gene Therapy," 13(4): p. 283-287.
6. Gupta, A.K. and A.S.G. Curtis (2004), "Surface modified superparamagnetic nanoparticles for drug delivery: Interaction studies with human fibroblasts in culture," Journal of Materials Science-Materials in Medicine, 15(4): p. 493-496.
7. Gupta, A.K. and M. Gupta (2005), "Synthesis and surface engineering of iron oxide nanoparticles for biomedical applications," Biomaterials, 26(18): p. 3995-4021.
8. Wooldridge, M.S. (1998), "Gas-phase combustion synthesis of particles," Progress in Energy and Combustion Science, 24(1): p. 63-87.
9. Pratsinis, S.E. (1998), "Flame aerosol synthesis of ceramic powders," Progress in Energy and Combustion Science, 24(3): p. 197-219.
10. Kammler, H.K., L. Madler, and S.E. Pratsinis (2001), "Flame synthesis of nanoparticles," Chemical Engineering & Technology, 24(6): p. 583-596.
11. deFaria, D.L.A., S.V. Silva, and M.T. deOliveira (1997), "Raman microspectroscopy of some iron oxides and oxyhydroxides, " Journal of Raman Spectroscopy, 28(11): p. 873-878.

## 297c Synthesis and Characterization of Functionalized Magnetic Iron Oxide Nanoparticles

**Reynolds A. Frimpong** and J. Zach Hilt. Chemical and Materials Engineering, University of Kentucky, 177 F. Paul Anderson Tower, Lexington, KY 40506

The effectiveness of magnetic nanoparticles in various applications is enhanced through the use of suitable coatings. In addition to providing increased stability of the nanoparticles in different solvents or media, many of these coatings are designed to incorporate tailored functional moieties which yield magnetically responsive systems. In this work, magnetic iron oxide nanoparticles coated with citric acid, that show differences in their stability, magnetic properties and heating characteristics have been synthesized by the co-precipitation method. The nanoparticles obtained were further functionalized with poly(ethylene glycol) dimethacrylate based polymers using atom transfer radical polymerization. The nanoparticles were characterized using fourier transform infrared, transmission electron microscopy, x-ray diffraction, dynamic light scattering, thermo-gravimetric analysis, vibrational sample magnetometry, and magnetic induction heating.

# 297d Colloidal Stability of Polyether-Stabilized Magnetite Nanoparticles In Aqueous Media

**Richey M. Davis**, Virginia Tech, Department of Chemical Engineering, Blacksburg, VA 24061, William C. Miles, Chemical Engineering, Virginia Tech, Department of Chemical Engineering, Blacksburg, VA 24061, and Judy S. Riffle, Chemistry, Virginia Tech, Department of Chemical Engineering, Blacksburg, VA 24061.

Magnetic nanoparticles have many interesting applications in biological systems, particular as MRI contrast agents. In this study, we examine the effect of different polyether stabilizers with charged end groups adsorbed to magnetite nanoparticles and how particle interactions change based on the media in which they are dispersed. Some polyether stabilizers were found to be ineffective in phosphate buffered saline (PBS). This is due to the strong adsorption of phosphate to the iron oxide surface, as evidenced by highly negative zeta potentials over wide pH ranges and particle instability over time in PBS. Additionally, magnetite/polymer complex diameters were predicted with reasonable accuracy using an adapted version of the Density Distribution model by Vagberg, et. al., that is based on a model for star polymers. The model was then used to predict the half-life for flocculation for each complex depending upon its magnetite content. This half-life was then correlated to the complex diameter as measured via dynamic light scattering and the effect of particle clustering on magnetic resonance imaging processes – specifically T2 relaxivity - was examined.

# 297e Are Carbon Coated Metal Nanoparticles a High Magnetization Alternative to Oxide Based Beads?

**Robert N. Grass**, ETH Zurich, Zurich, Switzerland and Wendelin J. Stark, Chemistry and Applied Biosciences, ETH Zurich, Wolfgang-Pauli-Str. 10, ETH Hönggerberg, Zurich, 8093, Switzerland.

Ferrite particles and polymers loaded with iron oxides have been used in chemical and biological separation tasks for more than 50 years. Although metals (iron, nickel, cobalt & alloys) usually show highly enhanced magnetic properties, their use as magnetic micro- or even nano- particle has been hampered by their lack of oxidation-stability under most operation conditions. This limitation has however been circumvented by coating metallic nanoparticles with carbon coatings only 1-2 nm thick. [1]

We present advantages, possibilities and challenges of carbon coated metallic nanoparticles, which have recently been commercialized by TurboBeads Llc [2]. The high functional loading of these structures in combination with very responsive magnetic properties proposes applications in magnetic polymers, water purification, immunoprecipitation and chemical catalysis. These applications will be discussed in terms of chemical reactivity, dispersion stability and time of operation. We will use this discussion to highlight the opportunities as well as possible pitfalls of this novel technology in comparison with ferrite nanoparticles.

Figure 1 shows a photograph of an air-stable carbon coated nanopowder (left) and a TEM image of the same material (right) illustrating the core-shell structure of the material.

References: [1] R. N. Grass, E. K. Athanassiou, W. J. Stark, Angew. Chem. Int. Ed. 2007, 46, 4909.

[2] R. N. Grass, W. J. Stark, PCT Patent Application, Nov 2006.

## 297f Preparation of Biocompatible MAGNETIC COLLOIDS from Supercritical Fluid Extraction of Emulsion

**Marco Furlan**[1], Johannes Kluge[2], Marco Lattuada[1], Francesco Fusaro[3], Marco Mazzotti[2], and Massimo Morbidelli[1]. (1) Institute for Chemical and Bioengineering, Department of Chemistry and Applied Biosciences, ETH Zurich, Wolfgang Pauli Strasse 10, Zurich, 8093, Switzerland, (2) Institute of Process Engineering, Department of Mechanical and Process Engineering, ETH Zurich, Sonneggstrasse 3, Zurich, 8092, Switzerland, (3) ETH Swiss Federal Institute of Technology, Sonneggstrasse 3, Zurich, 8092, Switzerland

The preparation of biocompatible magnetic polymer nanoparticles of controlled size and uniform size distribution is of great interest for bio-applications, such as drug delivery. Such particles can combine the benefits of nanoparticles made of biodegradable and biocompatible polymers with the superparamagnetic properties conferred by magnetic nanocrystals encapsulated into them. However, the preparation of such particles is a challenging task. In this work, we present an innovative and clean process that allows one to prepare submicron monodispersed poly(lactic-co-glycolic) acid (PLGA) particles containing high loads of magnetite nanocrystals. First, hydrophobic magnetite nanocrystals are prepared through a coprecipitation reaction from iron chlorides solutions in alkaline conditions, with oleic acid used as capping agent. Then, magnetite nanocrystals and PLGA are first dispersed together in dichloromethane, and subsequently emulsified water in the presence of polyvinyl alcohol (PVA) as stabilizer. The organic solvent is finally removed by extraction with supercritical carbon dioxide $sCO_2$. This last step is accomplished by spraying the emulsion through a nozzle into a high pressure vessel containing the $sCO_2$. This step can be carried out both in batch or continuous operation modus. The result is a stable suspension of PLGA particles in water with a narrow size distribution. The initial composition of the organic phase determines both the magnetite loading as well as the size of the nanoparticles, which is however strongly dependent on the emulsification process and the corresponding droplet size before the extraction. The nanoparticles have been characterized by light scattering and transmission electron microscopy.

## 303a Self-Assembled Polyelectrolyte Films for Nonlinear Optics: Study of Film Structure, Composition and Kinetics of Film Formation

Akhilesh Garg and **Richey M. Davis**. Virginia Tech, Department of Chemical Engineering, Blacksburg, VA 24061

Organic materials exhibiting second-order non-linear optical (NLO) properties are a key to the development of advanced electro-optic modulators used in fiber-optic communications system. This research concerns a layer-by-layer technique used to fabricate films with useful second order NLO properties. A polycation - poly(allylamine) hydrochloride (PAH) - and a polyanion, poly{1-[4-(3-carboxy-4-hydroxyphenylazo)-benzenesulfonamido]-1,2ethandiyl} (PCBS) are self-assembled on a substrate by alternately adsorbing the two polymers. We study the film deposition process using a Quartz Crystal Microbalance (QCM), ellipsometry, and absorbance measurements.

The deposition of multiple layers of PAH and PCBS adsorption were studied at different concentrations using a silicon dioxide-coated substrate in a QCM. The results show a linear decrease in frequency with increasing number of adsorbed layer signifying a homogeneous growth of the self-assembled films. Data obtained from QCM were compared with thickness measurements from ellipsometric data and absorbance measurements to estimate the water content in these films. These findings will help us to understand this self-assembly at the molecular level in order to make a practical electro-optic modulation device.

## 303b Analyzing Photoresponsive Side Chain Dendritic Polyesters (SCDPE) Using Atomic Force Microscopy (AFM)

**Joshua J. Galgano**[1], David G. Rethwisch[1], and Alexei V. Tivanski[2]. (1) Department of Chemical and Biochemical Engineering, University of Iowa, 4139 Seamans Center, Iowa City, IA 52242, (2) Chemistry, Univerisity of Iowa, Iowa City, IA 52242

Polymers containing dendrimer wedges and azobenzene photoresponsive moieties were characterized using atomic force microscopy (AFM). These photoresponsive polyesters were polymerized by condensation of a diacylchloride azobenzene with a uniquely designed polyamidoamine (PAMAM) dendrimer wedge diol. AFM studies of individual chains supported on a mica substrate revealed surface morphology changes in response to irradiation. The polymer chains formed disc-like structures on the surface. The AFM studies indicated a decrease in the thickness of the polymer discs upon irradiation which correlated with the azobenzene photoisomerization studies carried out in polymer solutions. Future work will focus on converting the photoresponsive effects of azobenzene molecule into a physical mechanical response.

## 303c Theory of Polyelectrolyte Adsorption Onto Surfaces Patterned with Charge and Topography

**Nazish Hoda** and Satish Kumar. Chemical Engineering and Materials Science, University of Minnesota, 151 Amundson Hall, 421 Washington Ave. SE, Minneapolis, MN 55455

Mean-field theory is used to derive criteria for the adsorption of a weakly charged polyelectrolyte molecule from salt solution onto surfaces patterned with charge and topography. For flat surfaces patterned with periodic arrays of charged patches, the adsorbed layer thickness predicted using mean-field theory and that found by Brownian dynamics simulations are in quantitative agreement in the strong-adsorption regime. This corresponds to sufficiently small inverse Debye screening length, or to sufficiently large effective surface charge density or charge per chain segment. Qualitative agreement is obtained in the weak-adsorption regime, and for the case where surfaces are patterned with both charge and topography. For uniformly charged, sinusoidally corrugated surfaces, the theory predicts that the critical temperature required for adsorption can be greater than or less than the corresponding value for a flat surface depending on the relative values of the inverse screening length and the corrugation wavenumber. If the surface charge is also allowed to vary sinusoidally, then adsorption is predicted to occur only when the topography crests have a surface charge opposite to that of the polyelectrolyte. Surfaces patterned with rectangular indentations having charged bottoms which are separated by flat charged plateaus are investigated as well. Adsorption is predicted to occur even when the net surface charge is zero, provided that the plateaus have a charge opposite to that of the polyelectrolyte. If the charge on the plateaus and polyelectrolyte is the same, adsorption may still occur if electrostatic attraction from the indentation bottoms is sufficiently strong.

## 303d Engineering Robust Omniphobic Surfaces with Fluoro-POSS

**Anish Tuteja**[1], Wonjae Choi[2], Joseph M. Mabry[3], Gareth H. McKinley[2], and Robert E. Cohen[4]. (1) Department of Chemical Engineering, Massachusetts Institute of Technology, 77 Massachusetts Avenue, Bldg. NE-47, Room 583, Cambridge, MA 02139, (2) Mechanical Engineering, Massachusetts Institute of Technology, Building 3-252, 77 Massachusetts Avenue, Cambridge, MA 02139, (3) Edwards Air Force Base, RZSM, U.S. Air Force, Air Force Research Laboratory, 10 East Saturn Blvd, Edwards AFB, CA 93524, (4) Chemical Engineering, Massachusetts Institute of Technology, 77 Massachusetts Ave., 66-554, Cambridge, MA 02139

Superhydrophobicity, i.e. the ability of various surfaces to cause water droplets to bead up and roll off their surface, is a common observation in nature. In an attempt to develop bio-mimetic superhydrophobic surfaces, numerous researchers have investigated the superhydrophobic character of various natural surfaces, such as the lotus leaf and duck feathers. In their classic paper, Cassie and Baxter[1] noted, "The duck is generally regarded as having attained perfection in water-repellency, and it is usually taken for granted that the duck uses an oil or similar coating with larger contact angles than any known to man. In actual fact, the duck obtains its water-repellency from the structure of its feathers."

The synergistic effects of surface texture and relatively low surface energy enable duck feathers to trap pockets of air underneath droplets of a high surface tension liquid like water ($\gamma$ = 72.1 mN/m) and form a stable composite solid-liquid-air interface that resists wetting. However, the feathers natural surface chemistry and texture are insufficient to withstand the decrease in free energy arising from the spreading of lower surface tension fluids such as oils or alcohols. Thus, as is tragically evident after an oil-spill, low surface tension liquids like gasoline and rapeseed oil ($\gamma$ = 35.7 mN/m) completely wet the surface of a duck feather.

Designing and producing textured surfaces that can resist wetting by such low surface tension fluids has been a significant challenge in materials science, and no examples of such materials exist in nature. In this work we develop various surfaces possessing re-entrant texture[2] that can support composite solid-liquid-air interfaces, even with extremely low surface tension liquids such as pentane ($\gamma$ = 15.7 mN/m) - the lowest surface tension alkane in liquid state at atmospheric pressure. Further, to aid the systematic engineering of non-wetting surfaces, we develop four design parameters that allow us to provide an a priori estimation of both the apparent contact angles, as well as the robustness of the composite interface, obtained with a particular contacting liquid.

Based on these design parameters, we produce electrospun polymeric (poly methyl methacrylate - PMMA) fiber mats containing extremely low surface energy, fluoroalkyl-substituted silsequioxane molecules (FluoroPOSS).[2,3] Numerous fluoroPOSS molecules have recently been synthesized[3] in which the rigid 2 nm silsesquioxane cage is surrounded by different organic groups including 1H,1H,2H,2H-tridecafluorooctyl (fluorooctyl POSS) and 1H,1H,2H,2H-heptadecafluorodecyl (referred to as fluorodecyl POSS). The fluorodecyl POSS molecules were found to have the lowest surface energy amongst the various POSS molecules synthesized and were used exclusively in this study.

The low surface energy of fluorodecyl POSS molecules, as well as their small size, enables them to migrate (bloom) to the fiber-air interface during the electrospinning process. As a consequence, electrospun fibers (with radius ~ 500 nm) containing as low as 5 vol% POSS can display extremely high contact angles (> 150°) and support a robust composite interface with a wide range of polar and non-polar liquids. We suggest the term 'omniphobic', i.e. surfaces that resist wetting by all liquids, to describe such surfaces. In addition, we describe a simple 'dip-coating' process that enables us to provide a conformal coating of PMMA - fluoroPOSS blends on any surface that inherently possesses re-entrant texture, such as duck feathers or commercially available fabrics. The synergistic effect of roughness, re-entrant texture of the preformed substrate, and the low surface energy of fluoroPOSS-PMMA blends, imbues omniphobicity to the dip-coated surfaces, enabling them to support a composite interface with even the lowest surface tension liquids.

References:

1. Cassie, A.B.D. & Baxter, S. Wettability of porous surfaces. Trans. Faraday Soc. 40, 546–551 (1944).

2. Tuteja, A. et al. Designing superoleophobic surfaces. Science 318 (5856), 1618-1622 (2007).

3. Mabry, Joseph M. et al. Fluorinated Polyhedral Oligomeric Silsesquioxanes (F-POSS). Ange. Chem. Int. Ed. 47 (22), 4137-4140 (2008).

## 303e Patchy and Striped Phase Separation In Surfactant-Coated Surfaces

**Chetana Singh**, Department of Chemical Engineering, University of Michigan, Ann Arbor, MI 48109-2136 and Sharon C. Glotzer, Chemical Engineering and Materials Science and Engineering, University of Michigan, 2300 Hayward Street, Ann Arbor, MI 48109.

We perform dissipative particle dynamics (DPD) simulations to study phase separation of a binary mixture of immiscible surfactants adsorbed on a flat surface. Both patches and stripes have been reported for such systems in the experimental literature, and there is controversy regarding the preferred thermodynamic equilibrium phase. We previously showed that stripes form [1], rather than bulk phase separation, when the length mismatch between surfactants or the bulkiness of one of the tail groups is sufficient [2, 3]. Here we show that for critical and near critical mixtures, stripes are the preferred phase and patches are kinetically arrested patterns that form during phase separation, again when there is sufficient mismatch in surfactant length or bulkiness. We investigate generally under what conditions stripes or patches should form as a function of, e.g., degree of immiscibility, length difference, difference in tail-group bulkiness and surface coverage. Our simulations are in very good agreement with experiments on a wide variety of surfactant mixtures.

1. AM Jackson, JW Myerson, F Stellacci, "Spontaneous assembly of subnanometre-ordered domains in the ligand shell of monolayer-protected nanoparticles", Nature Materials, 3, 330-336 (2004).

2. C Singh, PK Ghorai, MA Horsch, AM Jackson, RG Larson, F Stellacci and SC Glotzer, "Entropy-Mediated Patterning of Surfactant-Coated Nanoparticles and Surfaces", Physical Review Letters 99, 226106 (2007).

3. RP Carney, GA DeVries, C Dubois, H Kim, JY Kim, C Singh, PK Ghorai, JB Tracy, RL Stiles, RW Murray, SC Glotzer and F Stellacci, "Size Limitations for the Formation of Ordered Striped Nanoparticles", Journal of American Chemical Society, (Communication), 130(3), 798-799 (2008).

## 303f Self-Organized Pattern Formation In Dewetting of Elastically Confined Thin Polymer Layer

**Danish Faruqui**, Chemical Engineering, Carnegie Mellon University, 5000 Forbes Ave, Pittsburgh, PA 15213 and Ashutosh Sharma, Chemical Engineering, Indian Institute of Technology, Kanpur, Indian Institute of Technology, Northern Labs, # 106, Kanpur, India.

We report various stages of self-organized, sub-micron, surface directed patterns in a thin polystyrene (PS) layer (thickness ~20nm) sandwiched between a silicon substrate and a cross-linked elastomeric layer (polydimethylsiloxane; PDMS, thickness ~20nm). The PS layer was swelled and dewetted by exposure to saturated cyclo-hexane vapor, owing to the lateral diffusion of the solvent vapors starting from the bilayer edges. Morphological evolution of the self-organized surface patterns was recorded both on the elastic PDMS surface and on visco-elastic PDMS-PS interface. The instability patterns could be aligned by placing a micro-stamp (pitch 1500 nm) in conformal contact on the surface of the PDMS-PS bilayer. The presence of the external micro-stamp on the PDMS layer directed the stress-driven patterns formed on the buried PDMS-PS interface, inducing anisotropic and regular surface pattern very similar to that on the master stamp. Such a control was found to be possible only when the order of magnitude of the characteristic length scale of the instability was similar to that of external stamp pattern. AFM scans of the top elastic PDMS surface and optical micrographs confirmed this contact-less transfer of master pattern uniformly over a larger area (cm2). The anisotropic surface pattern thus formed on the PDMS and PDMS-PS interfaces was employed to explore the subsequent stages of development of the self-organized sub micron structures formed by continued exposure of the bilayer to solvent vapors. Subsequent stages of self organization in this system refers to alignment of sub-micron structures in the grooves (width ~750 nm) of initial pattern followed by their morphological and topographical evolution in terms of shape, size, separation and aerial density. It was confirmed by the AFM scans of the top PDMS layer that the formation of such self-organized sub-micron structures occurs in the PS layer only. However the self organized micro structures along the tracks of the initial pattern were clearly visible in the optical micrographs of the bilayer samples. Thus the transparent nature of PDMS made it possible to record and investigate the later stages of self-organization where the self organized surface patterning occurred at the buried visco-elastic PDMS-PS interface only. The morphological evolution of these structures was studied in detail by characterizing their shape, size, average separation, number density along the grooves and aerial density. It is noteworthy that the periodicity of the corrugated pattern formed in the first stage of self organization remains undisturbed in the subsequent stages. Choice of materials, thickness of polymer films and certain key changes in experimental protocol made it possible to record different stages of self organization on a convenient time scale (~15 hrs).

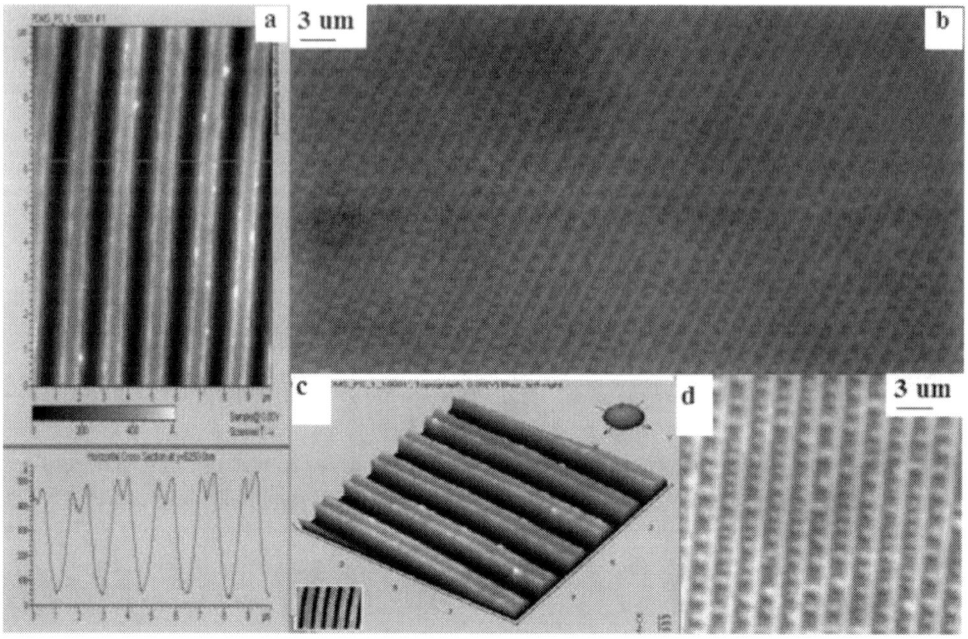


**Fig 1: Subsequent Self-organization at the visco-elastic, PDMS-PS, interface following the initial pattern replication in first stage.**

**a)** 10um x10um cross-sectional AFM scan of PDMS surface of the bilayer system after 3.5 hour of exposure to solvent vapor showing only micro trenches (formed in first stage) on the elastic PDMS Surface. **b)** Optical Micrograph of PS-PDMS bilayer system after 3.5 hour of exposure to solvent vapor (Track Width ~750nm). Showing alignment of sub micron structure formed on the buried PDMS-PS interface on extended swelling following the first stage of self organization. **c)** 10um x10um 3D AFM scan of PDMS surface of the bilayer system after 14 hour of exposure to solvent vapor still showing only micro trenches on the PDMS Surface. **d)** Optical Micrograph of PS-PDMS bilayer system after 14 hour of exposure to solvent vapor (Track Width ~750 nm). Showing alignment of half-spherical sub micron structure, eventually separated from one another, formed on the buried PDMS-PS interface on extended swelling following the first stage of self organization.

# Nanosculpting of Electrodeposited Multilayered Nanowires and Nanotubes

*Maoshi Guan[1] and Elizabeth J. Podlaha-Murphy[2]*

[1]*Department of Chemical Engineering, Louisiana State University, Baton Rouge, Louisiana*
[2]*Department of Chemical Engineering, Northeastern University, Boston, Massachusetts*

The electrochemical fabrication of compositionally modulated, multilayered nanowires, and to a lesser extent nanotubes, has been widely studied, but not the selective etching of them. We will present results on the electrochemical etching of one of the nanolayers that comprise the nanowires/tubes. In contrast to etching multilayered nanowires/tubes by a chemical route, an electrochemical method can result in novel intermediate structures ranging from bamboo-like wires and tubes to complete etching to form disks and rings. As a model system, nanolayers of Au and Co-rich CoAu were electrodeposited in a sandwich fashion into nanoporous alumina and polycarbonate membranes. The Co-rich layers were selectively etched potentiostatically and the current transients inspected. The current response indicated when the surface area of the nanowires/tubes is enhanced due to shape change under diffusion controlled dissolution and signaled when the layers were completely etched. Cyclic voltammetry was also employed to characterize and track the etching process. SEM confirmed the partially etched and completely etched nanostructures.

## *Introduction*

Nanoporous, metallic nanowires can provide unique properties and new applications with enhanced surface area and well-defined pore morphology. One, conventional fabrication method is to deposit an alloy and chemically dissolve a sacrificial element. For example, Ji and Searson [1,2] electrodeposited a gold-silver binary alloy nanowire using a nano-templating method. The nanowire was chemically etched to remove the silver and to create the nanoporous structure. Qin *et al.* [3] electrodeposited a layered alloy of Au/Ag and Au/Ni as a nanowire; the less noble element (i.e., Ag and Ni) were subsequently chemically etched. In this paper, we electrochemically etch a layered alloy nanowire in order to create intermediate structures and expand the technique to include nanotubes.

Electrochemical etching is applied to a collection of nanowires/tubes by means of an external circuit under potentiostatic conditions. The dissolution can be controlled by monitoring the applied current and hence charge, and it is carried out in more benign solutions compared to chemical etchants. In this study, the fabrication of Au/CoAu nanoscale bamboo-like structures is developed by etching nanowires and nanotubes in KCl electrolytes. The evolution of the shape change is monitored with time by examining the current during anodization and cyclic voltammetry at intermittent stages of etching.

## *Experimental*

The AuCo alloy and multilayered nanostructures were electrodeposited from a single, sulfite electrolyte containing 0.29 M $CoSO_4 \cdot 7H_2O$, 0.47 M $C_6H_8O_4 \cdot H_20$, 1.78 M KOH, 1.25 mM Au (Techni® Gold 25 E - a sulfite containing electrolyte), pH of 6.15. This non-cyanide-containing electrolyte was developed in our prior study [4]. To electrodeposit Au/CoAu multilayered nanowires into anodic alumina oxide (AAO) membranes a gold sputtering step was used to create a conductive layer on one end of the membrane followed by a two step galvanostatic, square-wave pulsed plating [5]. A gold layer was deposited at a current density of -2 mA/cm$^2$ and a cobalt rich Co-Au (96 wt % Co) alloy at -6 mA/cm$^2$. The deposition time was varied depending on the desired layer size. In order to release the nanowires from the template, the alumina template was dissolved in 2 M KOH with ultrasonication. Membranes with large pore diameters were used to create nanotubes by minimizing the amount of gold sputtering on the backside of the membrane.

The nanowires were then cleaned by sequential centrifugation and rinsed with deionized distilled water and ethanol. The electrode for etching the nanowires/tubes is shown in Figure 1. The nanowires/tubes were fixed on the electrode surface by a double sided adhesive conductive carbon tab (Electron Microscopy Sciences) typically used for SEM. The Co-rich layer of the Au/CoAu nanowires and nanotubes were electrochemically anodized in 2 M KCl under potentiostatic control at an overpotential of 0.04 V vs SCE.

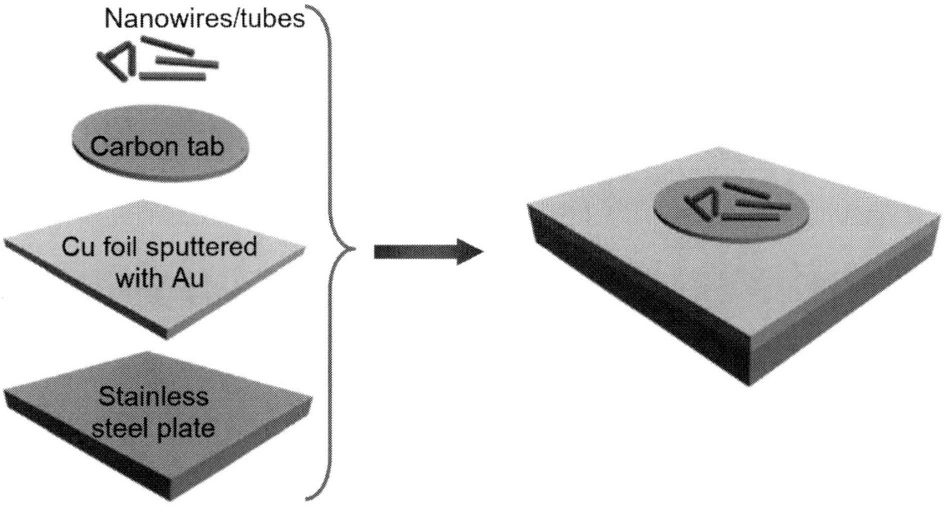


**Figure 1**. Schematic illustration of the working electrode used for electrochemical etching

A Hitachi S-3600N SEM was used to characterize the nanowires before and after electrochemical etching.

### *Results and Discussion*

Figure 2 (a-c) is a plane-view SEM image of a Au 25 nm / CoAu 50 nm multilayered nanowire during the process of electrochemically etching. Figure 2 (a) shows an array of nanowires and a single nanowire at higher magnification (inset) prior to etching. The gold layer was deposited for 180 s and the Co-rich layer for 33 s. The multilayer structure is observable. After partially etching the nanowire, Figure 2 (b), nano-bamboo like structures emerge from the preferential etching of the Co-rich layer. Images in different magnification scales are shown. After completely etching of the Co-rich CoAu alloy layers, Figure 4, uniform and parallel Au nanodisks are formed. The end stage of etching is the same result

**Figure 2 (a).** SEM images of Au 25 nm / CoAu 50 nm nanowire electrodeposited in 0.2 μm AAO membrane (as deposited)

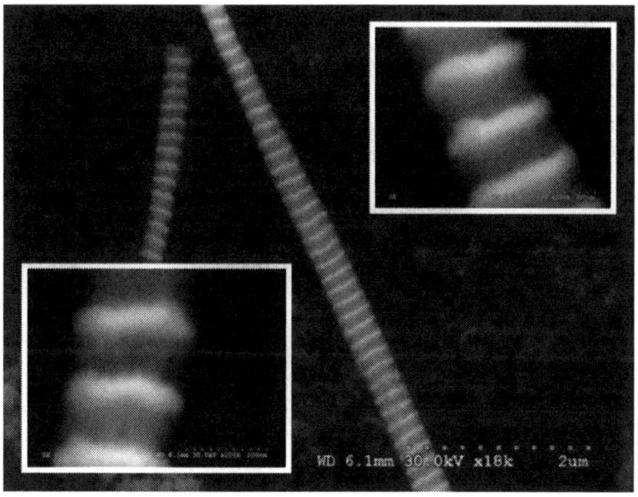


**Figure 2 (b).** SEM images of Au 25 nm / CoAu 50 nm nanowire partially etch electrochemically for 900 s (nanobamboo).

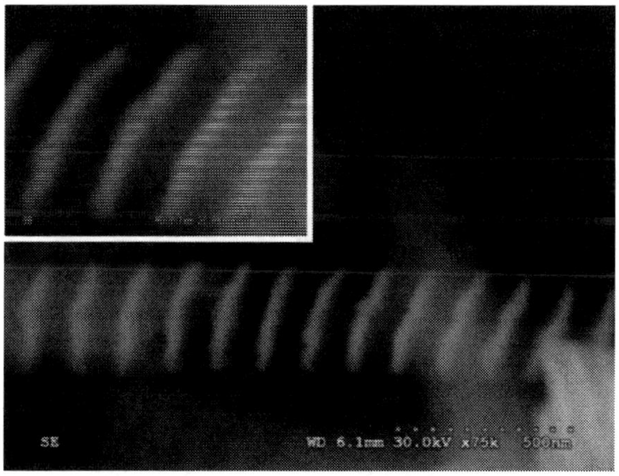


**Figure 2 (c).** SEM images of Au 25 nm / CoAu 50 nm nanowire completely etched

as chemical etching, but the intermediate structures are unique to the electrochemical etching method.

As etching progresses, there is a response in current and this current can be used to track the etching process. Also the total current response, or charge, $\int Idt$, should reflect the amount of mass etched. Nanowires with four different AuCo layer sizes were electrodeposited in AAO membranes with a pore size of 0.1 μm. Each set had the same total length of 12 μm and the same Au layer thickness of 25 nm, but different CoAu layer sizes: 20, 50, 100, 150 nm. The layer sizes were controlled by changing the deposition time. A constant volume of 60 μL of nanowire suspension was carefully transferred to the surface of the substrate. After the samples were fully dried, electrochemical etching experiments were tested. The current response as a function of time during electrochemical etching of four Au/CoAu nanowire electrodes is plotted in Figure 3. Each sample of a collection of nanowires with a fixed layer size exhibits a typical peak shape with a maximum. The current increases due to the larger surface area of the partially etched nanowires. As more and more material is removed the area decreases and hence the current decreases. The thicker the CoAu layer size, the higher value of the peak is achieved. The larger CoAu alloy layer size reflects more mass etched, consequently larger surface area is created during the etching process and a larger current arises. Finally, when the Co-rich CoAu alloy layers are completely etched away, there is no further material left for etching. The anodic current approaches zero. It is not always equal to zero because there is a distribution of current across the electrode and not every nanowire is etched exactly at the same time.

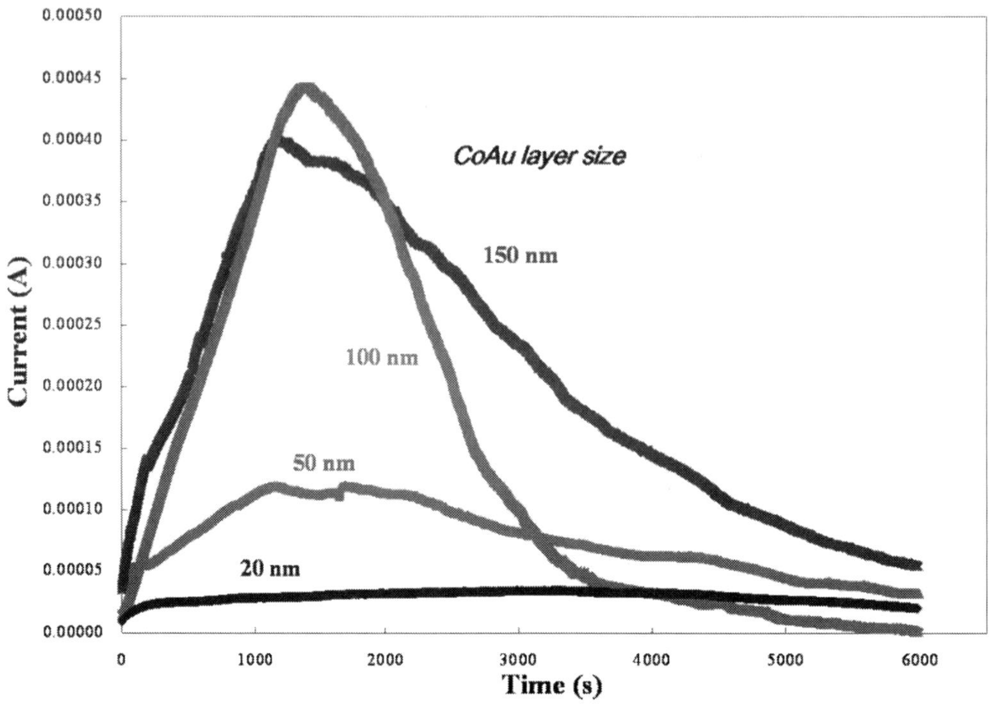


**Figure 3.** Anodic charges as the function of time during the process of the selective electrochemical etching of Au/CoAu nanowires on carbon electrodes with four different CoAu layer sizes including 20, 50, 100, and 150 nm

Table 1 is a summary of the charge measured (integrated current over time) during the evolution of etching the nanowires having different layer sizes. The maximum charge was found to be proportional to the etched layer size.

**Table I.** Ratio of CoAu layer sizes and ratio of charges @ 6000 s

| CoAu layer size (nm) | Ratio of CoAu layer size | Anodic charge (C) | Ratio of charge |
|---|---|---|---|
| **150** | 7.5 | 1.2509 | 7.2 |
| **100** | 5.0 | 0.8737 | 5.0 |
| **50** | 2.5 | 0.4597 | 2.6 |
| **20** | 1.0 | 0.1746 | 1.0 (reference) |

In a similar approach, nanotubes can also be electrochemically etched to create nanotube bamboos and nanorings. Figure 4 shows an SEM image of a nanotube partially etched. The Au layers were electrodeposited with the same conditions as in Figure 3 (25 nm) and the Co-rich CoAu layers were electrodeposited for 99 s to create layers of 150 nm. In order to readily deposit a tubular form, a large diameter membrane template was chosen, available commercially in polycarbonate filters. After deposition, the polycarbonate was removed by dichloromethane.

Figure 5 is a plot of the charge, determined from the measured current response, with time for three electrodes having different Co-rich CoAu nanolayers in the form of nanotubes. The gold layer was 50 nm and similar to the nanowire response in Table 1, the charge was proportional to the etched layer size.

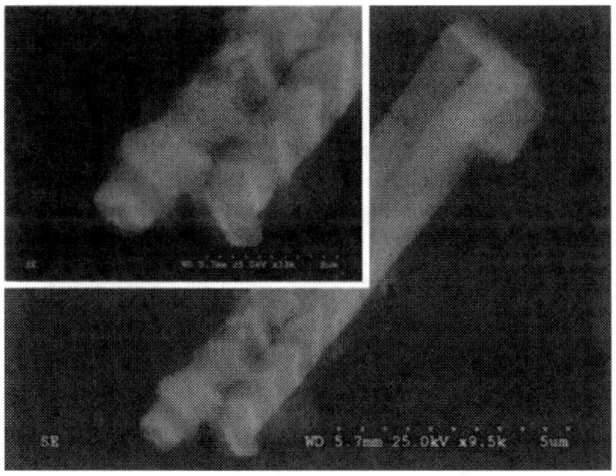

**Figure 4.** SEM images of Au 25 nm / CoAu 150 nm nanotubes with diameter of 1.0 μm after partial electrochemical etching of Co-rich CoAu layer

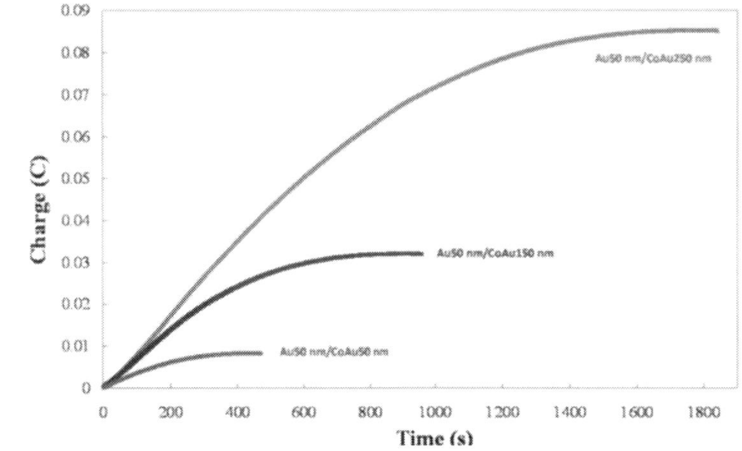


**Figure 5.** Anodic charges as the function of time during the process of selective electrochemical etching of Au/CoAu nanotubes on carbon electrodes with three different CoAu layer sizes: 50, 100, and 250 nm.

Cyclic voltammetry (cv) was used as a way to further distinguish changes in the etching process. The ferric/ferrocyanide redox couple is a reaction to characterize electrode surfaces since the reaction involves no deposit and thus no change in the electrode area. An electrolyte containing 0.001 M potassium ferricyanide, 0.003 M potassium ferrocyanide in 0.5 M NaOH where used to characterize a nanowire electrode. Figure 6 shows the reponse of the ferri/ferrocyanide reactions at different sweep rates (50, 100 and 150 mV/s) of the potential on an unetched nanowire electrode with layer sizes of Au 25 nm / CoAu 100 nm. For each sweep rate, in the forward scan, the main reaction at the working electrode was the reduction of the ferricyanide ion (near -0.6 V vs SCE).

$$Fe(CN)_6^{3-} + e^- \rightarrow Fe(CN)_6^{4-} \qquad (1)$$

When the potential becomes more negative than the potential of the main reaction, the current increases due to the side reaction of water reduction on the Au/CoAu nanowires and carbon substrate.

$$2H_2O + 2e^- \rightarrow H_2 + 2OH^- \qquad (2)$$

The applied potential was reversed and the potential was scanned in a positive direction. The large anodic peak (near 0 V vs SCE) is the oxidation of the ferrocyanide ion, which reflects the larger concentration of the ferrocyanide ion in solution.

$$Fe(CN)_6^{4-} \rightarrow Fe(CN)_6^{3-} + e \qquad (3)$$

The ending point of the scan does not return to its original value due to the capacitance effect of the double layer.

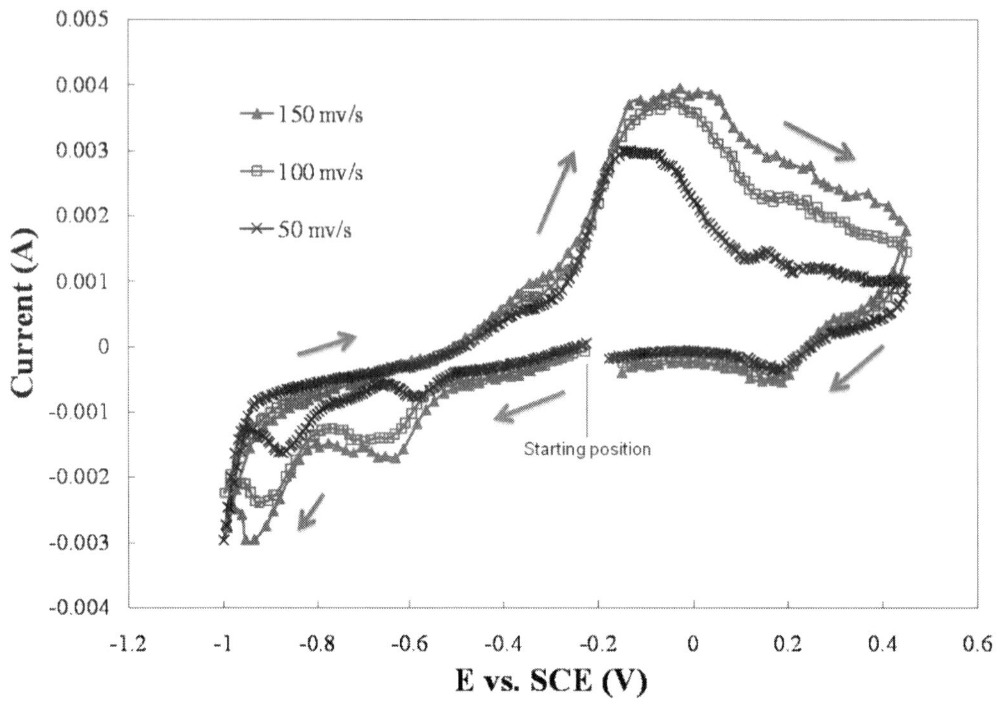


**Figure 6.** Cyclic voltammograms measured at three different sweep rates: 50, 100 and 150 mV/s in the ferricyanide/ferrocyanide solution on a carbon electrode with unetched nanowires Au 25 nm / CoAu 100 nm

Changes in the electrode nanolayers due to etching changed the current response and also changed the capacitance. Figure 7 is a set of cyclic voltammograms of a single size nanowire electrode (Au 25 nm / CoAu 50 nm) etched at different times in the region where ferricyanide reduction occurs. The etched nanowire response was more ohmic in nature reflective of an increase in the resistivity of the surface. During the forward scan (toward more negative potentials), the current increased in the region where ferricyanide is reduced (-0.6 V vs. SCE) only with large etching times and did not increase in a systematic way with etching time. However, the electrode capacitance increased with an increase of the etching time and could be a better measure to monitor the etching process. At the five different etching times: 600, 1300, 3500, and 6000 s, the capacitances were 0.6, 1.3, 2.2 and 2.5 mF, respectively. The increasing of capacitance as a function of etching time is largest in the beginning as the surface area is enhanced, but then changes less as the loss of mass of the electrode reduces the increase in surface area.

## *Conclusions*

Au/CoAu multilayers were electrodeposited into nanoporous templates to form either nanowires or tubes. The CoAu layer was subsequently etched potentiostatically and the current was monitored over time. The electrochemical etching technique was used to show that controlled, partial etching of nanowires/tubes can be achieved and bamboo nano-structures formed. The nanowires/tubes can be characterized by the charge measured during etching and the response of the electrode capacitance determined from cyclic voltammetry

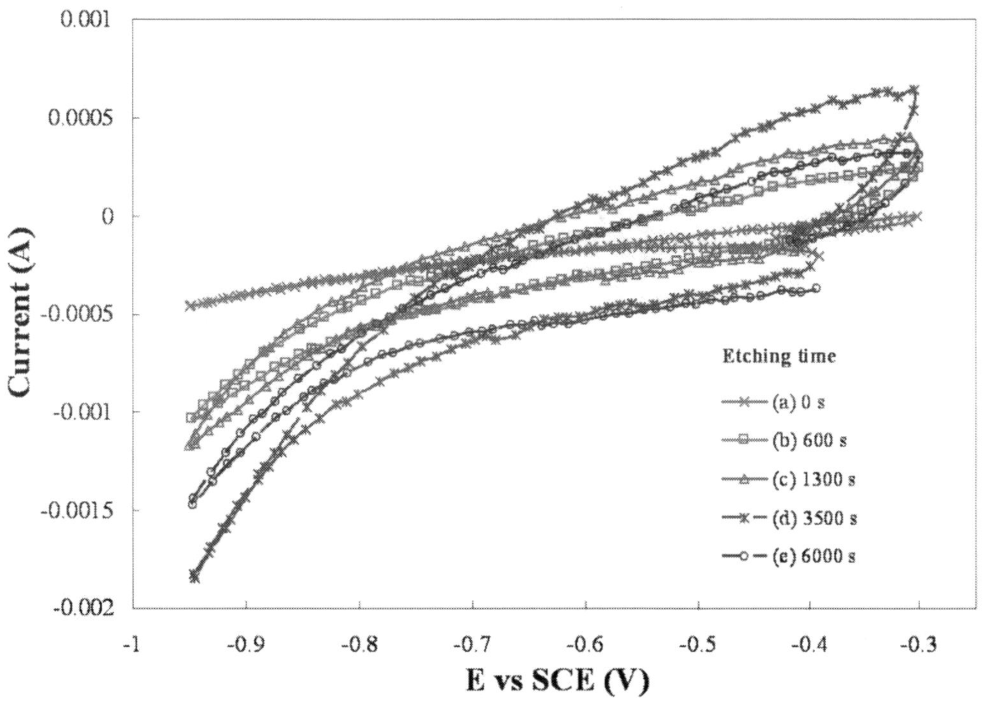


**Figure 7.** Cyclic voltammograms of Au 25 nm / CoAu 50 nm nanowires which were electrochemically etched with different etching time: (a) 0 s, (b) 600 s, (c) 1300 s, (d) 3500 s and (e) 6000 s

### *Acknowledgments*

The authors gratefully acknowledge NSF (CBET-0746467) and the Center for Biomodular Multi-scale Systems at LSU for partial support of this work.

### *References*

1. C.X. Ji and P.C. Searson, *Applied Physics Letters*, **81**(23) 4437-4439 (2002).
2. C.X. Ji and P.C. Searson, *Journal of Physical Chemistry B,* **107**(19), 4494-4499 (2003).
3. L.D. Qin, S. Park, L. Huang, and C.A. Mirkin, *Science*, **309**(5731), 113-115 (2005).
4. M. Guan and E.J. Podlaha, *Journal of Applied Electrochemistry,* **37** (5) 549-555 (2007).
5. M. Guan and E.J. Podlaha, "Electrodeposition and Electrochemical Etching of Au/CoAu Multilayered Nanowires" *ECS Transactions*, Meeting of The Electrochemical Society, Cancun, MX, Oct, (2006).

# 304b Templated Directed Electrodeposition of $SnO_2$-Nanotubes

**Min Lai**[1], Syed Mubeen[1], Jae-Hong Lim[1], Ashok Mulchandani[2], Marc A. Deshusses[1], and Nosang V. Myung[1]. (1) Department of Chemical and Environmental Engineering, University of California-Riverside, Riverside, CA 92521, (2) Department of Chemical and Environmental Engineering, University of California, Riverside, CA 92521

Templated electrosynthesis has widely been employed to prepare solids of defined dimension.[1] Here we illustrate how electrochemistry can be employed to manage the pH within the pores of a template and demonstrate that controlled growth of continuous $SnO_2$ nanotubes can be achieved. A gold electrode modified with a porous polycarbonate membrane was immersed in an aqueous tin dichloride solution. Electrochemistry was employed to control the local pH within the pores and drive a precipitation reaction. Removal of the gold and dissolution of the membrane yielded polycrystalline tin oxide nanotubes. The crystallinity of the as-deposited material was enhanced by annealing in ambient. It is proposed that the formation of nanotubes results from continuous side-wall precipitation along a reaction front. The nanotube size was controlled by using templates with different diameters and deposition time. Metal@$SnO_2$ core-shell nanostructures were fabricated after a further electrodeposition of metals within the $SnO_2$ nanotubes. We formed 1-D metal oxide composites by annealing as-prepared materials in air.

[1] J.C. Hulteen, C.R. Martin, J. Mater. Chem. 7 (1997) 1075.

## 304c Gold and Palladium Deposition on the Genetically Engineered Tobacco Mosaic Virus with Controlled Metal Ion Loading

**Jung-Sun Lim**[1], Sang-Yup Lee[2], James N. Culver[3], and Michael T. Harris[1]. (1) Chemical Engineering, Purdue University, Forney Hall of Chemical Engineering, 1137, 480 Stadium Mall Drive, West Lafayette, IN 47907, (2) Chemical Engineering, Yonsei University, 134 Sinchon-Dong, Seodaemoon-Gu, 120-749, Seoul, South Korea, (3) Center for Biosystems Research, University of Maryland College Park, College Park, MD 20742

Understanding the metal mineralization on the biotemplates has potential to design nanostructured materials for nano devices. The tobacco mosaic virus (TMV) has served as a model template to synthesize biologically inspired metal nanowires due to its unique tubular rod-shape. It has 4 nm/18 nm of inner/external diameter with 300 nm longitudinal length. Also, the genetic engineering of TMV is well established to modify it with active surface functionalities such as sulfhydry group. Researchers reported biotemplated metal nanowires with gold, silver, palladium, and platinum with various biotemplates including TMV. However, further progress of biotemplated metal nanowire synthesis is limited because few reports provided experimental supports elucidating the intermediates of metal deposition process.

As a model study, we investigate the gold(III) and palladium(II) ion uptake on the genetically modified tobacco mosaic virus (TMV1Cys). The gold(III) and palladium(II) ion loading on the TMV1Cys at equilibrium are measured using UV-Vis spectroscopy. Then, deposits of gold and palladium on TMV1Cys are prepared with the variation of metal ion loading under fixed TMV1Cys concentration to show the improvement of metal deposited morphology by increasing metal ion loading. Obtained samples are imaged with transmission electron microscopy (TEM) to visualize the metal deposition on the TMV1Cys.

## 304d Epitaxial Growth of Inp Nanowires on Silicon

**Li Gao**, R. L. Woo, and R. F. Hicks. Department of Chemical and Biomolecular Engineering, University of California, Los Angeles, Los Angeles, CA 90095

Semiconductor nanowires have potential applications in new and high-performance one-dimensional devices. Epitaxial growth of high-quality semiconductor nanowires on Si surfaces is ideal for utilizing semiconductor nanowires within the frame of mainstream silicon technology. We have achieved epitaxial growth of InP nanowires on Si(111) and Si(100) surfaces by metalorganic vapor phase epitaxy (MOVPE). The silicon surfaces are wet cleaned before growth. Indium droplets are deposited onto the silicon surface as the seeds for InP nanowire growth. By finely controlling the experimental conditions, InP nanowires prefer to grow in the <111> directions of the silicon substrates, which makes it possible to grow vertical InP nanowires on Si(111) surface. By restraining the growth of non-vertical nanowires, the growth of all vertical nanowires can be achieved on Si(111) substrate. TEM and PL studies of as-grown InP nanowires will also be presented.

# 304e Self-Catalysis Schemes for Group III-Antimonide Nanowires: Antimony Versus Group III Metal

**Chandrashekhar Pendyala**, Chemical Engineering, University of Louisville, Louisville, KY 40292 and Mahendra K. Sunkara, Department of Chemical Engineering, University of Louisville, 106 Ernst Hall, Department of Chemical Engineering,, University of Louisville, Louisville, KY 40292.

It has been recently reported that Group III-V compound nanowires could be synthesized using direct reaction schemes without the use of any foreign metal catalysts. Such direct reaction schemes are termed as "self-catalysis" schemes In these self-catalysis schemes, it has always been thought that the Group III metal droplet at tips lead the growth of III-V compound nanowires with subsequent dissolution and precipitation of Group V solutes [1, 2, 3].

We report first of its kind results with synthesis of III-antimonide nanowires, i.e., antimony (Sb) droplets can self-catalyze the growth of III-antimonide nanowires. Extensive experiments are conducted with GaSb system. This is a significant result in that one can actually use the same type of droplets (Sb in this case) to synthesize heterostructures involving different III-Sb compounds such as GaSb/InSb, etc. In addition, the use of Sb for tip led growth of nanowires could be advantageous for many other III-V compounds.

References:

[1] Eric A. Stach, et.al. Nano Letters, 3 (6), 867 -869, (2003)

[2] S. Vaddiraju, et.al. Nano Letters 5(8), 1625-1631 (2005)

[3] H. Li, et.al. Advanced Materials, 18(2), 216 (2006)

# 304f Synthesis of Metallic Nanostructures Using Mesoporous Templates and Supercritical $CO_2$ as Reaction Medium

**Candy S. Lin**[1], Frank Leung-Yuk Lam[1], Xijun Hu[1], Wing Yim Tam[2], and Ka M. Ng[3]. (1) Department of Chemical Engineering, The Hong Kong University of Science and Technology, Clear Water Bay, Kowloon, Hong Kong, China, (2) Department of Physics and Institute of Nano Science and Technology, The Hong Kong University of Science and Technology, Clear Water Bay, Kowloon, Hong Kong, China, (3) Department of Chemical Engineering, Hong Kong University of Science and Technology, Clear Water Bay, Kowloon, Hong Kong, China

The synthesis of metallic nanostructures has been, in the current decade, vastly investigated due to its extraordinary properties exerted at such small scales. These nanostructures can be used in many different industries including electronic, photonics and act as building blocks for future nanodevices. These techniques are developed to produce nanostructures suited to perform tasks according to the properties they possess at a particular scale range. One method developed to guarantee nanowire diameter is by using templates to mold out the necessary nanostructures. The chemical fluid deposition (CFD) process involves a metal precursor be dissolved in supercritical fluid and then reduced to deposit metal particles in the porous channels of the selected template.

Platinum nanostructures have been fabricated by CFD method using supercritical carbon dioxide ($CO_2$) as the reaction medium. Ordered platinum nanowires were molded out from a mesoporous silica template, SBA-15, with a pore channel diameter of 7 nm. Solubility study of the platinum precursor, platinum(II) dimethylcyclooctadiene, in the supercritical $CO_2$ was first carried out to obtain a phase diagram in order to optimize the operating parameters required to enhance the infiltration of metal precursor into the porous channels prior to the deposition process. High purity and uniform nanowire bundles resulted after template removal with large aspect ratio of 200:1 carried out at moderate operating temperatures and pressures.

## 317a The Assembly of Microwires from Monomeric Solution of Nanoparticles Using Dielectrophoresis

**Ryan P. Slopek** and James F. Gilchrist. Department of Chemical Engineering, Lehigh University, 111 Research Dr., Bethlehem, PA 18015

The self-assembly of colloidal particles in electric fields enables the one-step formation of microwires with unique structures and properties that have the potential to advance the fabrication of high quality photonic and electrical devices. Positive dielectrophoresis in an alternating current (AC) field was used to direct the growth of microwires between planer electrodes immersed in solutions of photosensitive acrylate monomer with a small fraction of conductive nanoparticles. The microwires were frozen in place via polymerization of the sample upon exposure to UV light. The growth rate and structure of the wires were characterized using optical microscopy and X-ray scanning microscopy. In addition, the effect of varying the particle concentration, particle composition, initiator concentration, and strength of the electric field on wire formation was examined.

## 317b Combined Electric and Magnetic Field-Driven Assembly of Magnetic Janus Particles

**Stoyan K. Smoukov**, Sumit Gangwal, and Orlin Velev. Department of Chemical and Biomolecular Engineering, North Carolina State University, 911 Partners Way, EB1, Raleigh, NC 27695

We will present results from the assembly of magnetic Janus colloids under combined (and sometimes competing) dielectrophoretic and magnetophoretic forces. While the dielectrophoretic interactions involve the electrostatic charges on the particle surfaces, the polarizability of the ionic atmosphere around them, and corresponding flows, the force exerted by the magnetic field can be independent of the dispersion medium and can be conveniently modulated in directions independent of the applied electric field. We have previously reported the assembly of metallo-dielectric Janus particles into 1D and 2D structures by controlling the magnitude and frequency of the AC electric field. By including magnetic interactions, the range of interactions, the types of possible static structures and the speed of dynamic manipulation are greatly expanded. These novel particles could serve as a tool for studying new physical phenomena and large-scale assembly, and have numerous potential applications, among which are bifunctional drug delivery agents and flexible displays.

# 317c Self-Assembly of Ditethered Nanospheres and Comparison with Triblock Copolymer Mesophases

**Christopher R. Iacovella**, Chemical Engineering, University of Michigan, 2350 Hayward Street, 3440 G.G. Brown, Ann Arbor, MI 48109 and Sharon C. Glotzer, Chemical Engineering and Materials Science and Engineering, University of Michigan, 2300 Hayward Street, Ann Arbor, MI 48109.

Linear and star triblock copolymers are known to self assemble into a wide array of complex morphologies. The use of polymer-tethered nanoparticles is a unique method to control the self-assembly of nanoparticles into phases reminiscent of these and other block copolymers. Here we present the results of Brownian dynamics simulations used to investigate the phase behavior of nanospheres functionalized with two tethers to form building blocks whose geometries are analogous to ABC triblock copolymers. In this analogy, the nanoparticle plays the role of the middle "block". We investigate the phase behavior under melt conditions as a function of both nanoparticle diameter and dihedral angle, the angle separating the two tethers. We find a variety of phases that resemble those of linear and star triblock copolymers, including the alternating diamond, alternating gyroid, and alternating orthorhombic tricontinuous lattices, tetragonal and [8,8,4] columnar phases, and a unique phase of NaCl ordered spherical micelles. We additionally explore the phase behavior (i) as a function of block fraction under melt conditions and show evidence of a transition from tricontinuous to lamellar structures, and (ii) as a function of volume fraction under selective solvent conditions for which we find the double gyroid, hexagonally packed cylinder, and BCC ordered micelle phases. We discuss under what conditions ditethered nanoparticles do and do not behave as more traditional triblock copolymers.

# 317d Theoretical Prediction of Branched-Tail Surfactant Micellization In Dilute Aqueous Solutions Using Computer Simulations and Molecular Thermodynamics

Shangchao Lin, **Jonathan D. Mendenhall**, and Daniel Blankschtein. Chemical Engineering, Massachusetts Institute of Technology, Cambridge, MA 02139

Surfactants possessing branched lipophilic domains, or tails, are frequently encountered in industrial formulations due to their higher aqueous solubility at low temperatures, relative to linear surfactants. For example, secondary alkyl sulfonates (SAS's) are frequently used in cold-water detergent applications. Other situations in which branched surfactants arise include emulsification applications (including such surfactants as Triton X-100 and Silwet L-77) and in ethoxylated surfactant mixtures, as by-products of linear surfactant synthesis. The prediction of the micellization properties of such branched surfactants, including the critical micelle concentration (CMC), micelle shape and size, and micelle composition (for mixed micelles), is in general useful to formulators, who often must use trial-and-error experimentation to develop surfactant formulations exhibiting a desired solution behavior. Molecular-thermodynamic (MT) modeling is a successful approach that has been used to theoretically characterize both single- and mixed-surfactant systems based solely on surfactant molecular properties and molecular thermodynamics. In recent years, Stephenson, Beers, and Blankschtein [J Phys Chem B, 111, 1063-1075, 2007] have demonstrated the utility of combining Molecular Dynamics (MD) simulation results with MT modeling of single-surfactant systems, creating a hybrid CS-MT model that uses detailed microstructural hydration information from the MD simulations to more accurately compute the free energy of micellization. We present the key modifications to the CS-MT modeling approach that we have made to further extend the range of applicability of the model to single, branched surfactant systems, including refinement of the counting of water contacts and development of an improved model for surfactant tail packing in a micelle core. We present results for a wide variety of branched surfactants, including two series of SAS surfactants; aromatic branched surfactants, including linear alkylbenzene sulfonates (LAS); and the emulsifiers Triton X-100 and Silwet L-77.

## 317e Molecular Modeling of the Self-Assembly of Amphiphilic Molecules Using Interfacial Statistical Associating Fluid Theory (iSAFT) Density Functional Theory

**Chris Emborsky**, Kenneth R Cox, Shekhar Jain, Adam Bymaster, Zhengzheng Feng, and Walter G Chapman. Chemical and Biomolecular Engineering Department, MS-362, Rice University, 6100 Main St., Houston, TX 77005

Amphiphilic molecules like surfactants and copolymers consist of two chemically distinct components, hydrophilic and hydrophobic. Due to the difference in the affinities of the two components with different solvents, these molecules self-assemble themselves into a number of microstructures in the presence of water and organic solvents. This self-assembly of amphiphilic molecules are utilized in a number of industrial applications ranging from energy (enhanced oil recovery) to medicine (drug delivery, controlled release).

We have developed a density functional theory known as interfacial statistical associating fluid theory (iSAFT) for complex inhomogeneous fluids. In the current work, we have applied iSAFT to study the self-assembly of amphiphilic molecules in solution. The effects of composition of amphiphilic molecules, solvent selectivity, and solvent amount on the self-assembly are investigated.

## 317f Elasticity of Chains of Paramagnetic Colloidal Particles

**Dichuan Li** and Sibani Lisa Biswal. Chemical and Biomolecular Engineering, Rice University, MS 362, Main 6100, Houston, TX 77005

A key obstacle in assembling colloidal structures is understanding the fundamental chemistry and physics of the assembly process. In this talk, we will focus on the influence of variable flexibility on the shape and morphology of linked colloidal chains. These multilink chains are comprised of magnetic colloidal particles which have been "glued" together with various linker molecules.

The flexural rigidity, or bending stiffness, is used to describe the chain's resistance to bending forces. We have previously shown that the properties of the linker molecules greatly influence the overall chain flexibility 1. In particular, by adjusting the length of the linker molecule, we can tune the overall flexibility of the chain. We have recently constructed a simplistic model using single linker molecule elasticity to estimate the overall flexural rigidity of the chain. The calculated results of particle chains linked by different linker molecules from the model are compared with the measured overall flexural rigidity values in Ref [1].

In this talk we will elucidate how the properties of the linker molecules affect the overall chain elasticity. Also, we will show that the overall chain flexural rigidity is a non-linear function of bending angle, instead of a constant independent of bending angle as in a linear homogenous elastic filament assumed by other researchers2-4.

1. Biswal SL, Gast AP. Mechanics of semiflexible chains formed by poly(ethylene glycol)-linked paramagnetic particles. Physical Review E 2003; 68: 021402/1-/9.

2. Goubault C, Jop P, Fermigier M et al. Flexible magnetic filaments as micromechanical sensors. Phys Rev Lett 2003; 91: 4.

3. Cebers A. Flexible magnetic filaments. Current Opinion in Colloid & Interface Science 2005; 10: 167-75.

4. Cebers A, Cirulis T. Magnetic elastica. Phys Rev E 2007; 76.

## 317g Mechanical Properties and Microstructure of Triblock Copolymer Network: DPD Simulation Study

**Yelena Sliozberg**[1], Jan Andzelm[1], Mark VanLandingham[1], John K. Brennan[1], Victor Pryamitsyn[2], and Venkat Ganesan[2]. (1) U.S. Army Research Laboratory, Aberdeen Proving Ground, MD 21005, (2) Chemical Engineering, University of Texas at Austin, 1 University Station, Code C0400, Austin, TX 78712

Amphiphilic triblock ABA copolymers undergo microphase separation in a middleblock-selective solvent that results in a formation of three-dimensional network. The polymer network consists of the junctions of the solvent-incompatible end blocks connected by the middle blocks. However, both end blocks of one chain could be incorporated in the same micelle core that leads to formation of the loop conformation of the middle block, excluding it from participation in the network. The loop/bridge ratio depends on several factors such as copolymer concentration, length of the blocks and addition of corresponding diblock copolymer. The main focus of this study was to explore different methods to control the loop/bridge ratio of a gel formed by association of ABA copolymer in B selective solvent to achieve the desirable mechanical properties of the material, described by a computational rheological experiment.

In this work, we employed both equilibrium and non-equilibrium mesoscale computer simulations of block copolymers in the explicit solvent. The dissipative dynamics method (DPD) was used with the non-equilibrium time dependent Lees-Edwards boundary conditions. We obtained dynamic shear storage (G′) and loss (G″) moduli for various ratios of the blocks length and polymer concentration. The behavior of a mix of triblock and diblock copolymers with different concentration of diblock was also investigated. We have observed formation of the polymer network with the diverse loop/bridge ratio. We have shown that reduction of relative length of the middle block leads to close association of the chains and causes the significant decrease in a number of the middle blocks participating in the polymer network because of the loop formation. This phenomenon manifests itself in reduction $G'(\omega)$ for low frequency region with comparison with the network with low loop/bridge ratio and ultimately leads to the principally different viscoelastic behavior of the system, where $G'(\omega)$ is proportional to $\omega$ showing an evident terminal zone. $G'(\omega)$ for the network with low loop/bridge ratio approaches the constant value. We obtained good agreement with other theoretical and experimental results. We extended this approach to predict viscoelastic properties of the block copolymer as function of polymer architecture and solvent selectivity.

# 338a Young Investigator Award Lecture - the Design of Nanoscale Polyvalent Therapeutics

**Ravi S. Kane**, Chemical and Biological Engineering, Rensselaer Polytechnic Institute, 4005 Center for Biotechnology, Troy, NY 12180

This talk will describe our efforts to interface biomolecules with nanomaterials and nanoscale scaffolds for designing potent therapeutics.

We are designing potent inhibitors of anthrax toxin based on the concept of polyvalency – the simultaneous binding of multiple ligands on one biological entity to multiple receptors on another. Since the major symptoms and death from anthrax are due primarily to the action of anthrax toxin, the toxin is a prime target for therapeutic intervention. We will describe the design and characterization of polyvalent inhibitors that are several orders of magnitude more potent than the corresponding monovalent inhibitors and can neutralize anthrax toxin in vivo. Studies relating the composition and structure of these inhibitors to their activity have shed light on fundamental aspects of polyvalent recognition, including the role of "statistical pattern matching". We will also discuss the structure-based design of potent oligovalent toxin inhibitors.

Next, we will describe the design of polyvalent inhibitors that target the cellular receptors for anthrax toxin. We used phage display to identify novel peptides that bind to the two cellular receptors for anthrax toxin, and NMR spectroscopy to characterize the peptide-protein interactions; the corresponding polyvalent inhibitors neutralize anthrax toxin both in vitro and in vivo. Receptor-directed therapeutics are particularly promising, because they may help counter the emerging problem of pathogen resistance to antimicrobial therapeutics. We will discuss applications of this concept for the inhibition of other important targets, such as HIV. Finally, we will discuss the ability to remotely control the activity of proteins adsorbed onto carbon nanotubes. We will discuss the mechanistic basis of this phenomenon, and its use to design polyvalent nanotube conjugates that selectively destroy anthrax toxin from a mixture of proteins.

## 338b Nanoporous Membranes: Fabrication, Analysis, and Applications

**Sankar Nair**, School of Chemical & Biomolecular Engineering, Georgia Institute of Technology, 311 Ferst Drive NW, Atlanta, GA 30332-0100

This talk will present an overview of our work on nanoporous membranes, highlighting the role of chemical engineering principles in the fabrication, application, and understanding of such membranes for a range of applications that are based on the control of molecular and energy transport. Of particular interest is the integration of fabrication strategies and structural characterization techniques with methods for transport property measurement, prediction, and control; thus allowing faster progress towards optimization and eventual application of these membranes. This approach will be illustrated with three membrane paradigms. The first is that of zeolite membranes, which will be discussed in the context of mass transport (for separations) and heat transport (for electronic and heat-pumping applications). The second paradigm is that of nanocomposite membranes, containing either nanoporous zeolite particles or nanoporous layered materials dispersed in a polymeric matrix. These will be discussed in the context of applications in controlling mass and ion transport. The third paradigm consists of membranes in which each nanopore is an individually addressable nanodevice. These membranes will be discussed in the context of applications in ultra-rapid biomolecule analysis.

# 338c Microplasma Techniques for Nanoparticle Synthesis

**R. Mohan Sankaran**, Chemical Engineering, Case Western Reserve University, 10900 Euclid Avenue, Cleveland, OH 44106-7217

Nanoscale inorganic materials have attracted a great deal of interest for their unique physiochemical properties and potential applications in catalysis, biochemical sensing, optoelectronics, and medicine. However, the precise synthesis and characterization of these materials remains a critical challenge to technological application. In this talk, atmospheric-pressure microplasmas are presented as an economically-viable, versatile, steady-state approach to the synthesis and processing of nanomaterials. Several examples of applications of our technique will be discussed including (i) non-lithographic fabrication of surface-enhanced Raman substrates [1], (ii) gas-phase nanoparticle synthesis [2], (iii) catalytic growth of carbon nanotubes [3,4], and (iv) colloidal nanoparticle synthesis.

References

1. W-H. Chiang et al., Appl. Phys. Lett., Vol. 91, 021501 (2007).

2. R. M. Sankaran et al., Nano Lett., Vol. 5, 537 (2005).

3. W-H. Chiang et al., Appl. Phys. Lett., Vol. 91, 121503 (2007).

4. W-H. Chiang et al., Adv. Mat., in press.

## 357a Characterization Methods for Magnetic Nanoparticle Systems

**Cindi Dennis**, NIST, Gaithersburg, MD 20899-8552

Magnetic characterization of thin film and bulk materials is well-established and -understood. To date, this knowledge and their techniques have been directly transferred onto the characterization of nanoparticles. Unfortunately, due to the fundamental differences in material behavior on the nanoscale, these characterization methods are often inadequate to quantitatively characterize nanoparticles, either as individuals or as collections. Our goal is to develop the fundamental measurement methods for quantifying the parameters of interest in magnetic nanoparticles. Amongst these parameters are the blocking temperature, anisotropy, interactions, and distributions of coercivity and interactions. For example, three alternative methods for measuring the blocking temperature will be presented that remove the ambiguity inherent in the "conventional" method (i.e. the peak position in the zero field cool magnetization vs. temperature measurement). First Order Reversal Curves (FORC) magnetometry is a technique being developed to quantify the interaction strength as well as the distribution in properties for a nanoparticle system. This model can be checked directly through small angle neutron scattering (SANS), although this technique is limited by available beam time and polydispersity of the nanoparticle system. Finally, measurement methods for determining both individual and collective anisotropy of nanoparticle systems using a combination of dispersion methods, temperature, and magnetic field in vector magnetometry will be presented. Each of the techniques will be described, along with initial experiments on cobalt and magnetite nanoparticle systems to demonstrate the information they can reveal.

## 357b On the Design of Stents for Implant Assisted Magnetic Drug Targeting

**Jan O. Mangual**, Armin D. Ebner, and James A. Ritter. Department of Chemical Engineering, University of South Carolina, Columbia, SC 29208

Ritter and co-workers [1-7] have been studying the possibility of using ferromagnetic stents to enhance magnetic drug targeting (MDT). In MDT, an external magnetic field is used to attract and retain magnetic drug carrier particles (MDCPs) at a specific site in the body. The additional use of ferromagnetic implants under an external field helps significantly to improve the collection of the MDCPs by more easily overcoming local convective forces and allowing collection farther from the source of the external magnetic field. However, the only stents that have been considered for use in an implant assisted (IA) MDT system have been those used in typical angioplasty procedures. But, this might not be the best design for MDT applications.

One of the main problems with the design of medical stents is that they are meant for vascular expansion not for collection. In the current form, the stent wires are against the wall and significantly away from the main flow of blood in the vessel, which could limit the ability of the stent to collect MDCPs that are not too close to the vessel wall. In addition the scaffold of stents for this particular approach may not be necessarily fully magnetic. The existence of non-ferromagnetic elements within the stent scaffold may help create stagnation zones that help improve collection at the ferromagnetic elements of the stent. This presentation will discuss the application of novel stent designs that are better suited for MDT applications.

In particular, these new stent designs will take advantage of having wires not just against the wall, but also more toward the center of the vessel to improve collection. Non-magnetic wires will also be used to create stagnation zones to slow the flow to also improve collection. This study will reveal the role of several design parameters on the performance of these new stent designs for IA-MDT including the relative location between the wires and the distance of the wires to the vessel wall, the fluid velocity and the strength of the applied magnetic field.

1. J. A. Ritter, A. D. Ebner, K. D. Daniel, and K. L. Stewart, "Application of High Gradient Magnetic Separation Principles to Magnetic Drug Targeting," J. Magnetism and Magnetic Materials, 280, 184-201 (2004).

2. J. A. Ritter, A. D. Ebner, K. D. Daniel, and K. L. Stewart, "Application of High Gradient Magnetic Separation Principles to Magnetic Drug Targeting," J. Magnetism and Magnetic Materials, 280, 184-201 (2004).

3. M. O. Aviles, A. D. Ebner, H. Chen, A. J. Rosengart, M. D. Kaminski, and J. A. Ritter, "Theoretical Analysis of a Transdermal Ferromagnetic Implant for Retention of Magnetic Drug Carrier Particles," J. Magnetism and Magnetic Materials, 293, 605-615 (2005).

4. A. J. Rosengart, M. D. Kaminski, H. Chen, P. L. Caviness, A. D. Ebner and J. A. Ritter, "Magnetizable Intraluminal Stent and Functionalized Magnetic Carriers: A Novel Approach for Non-Invasive Yet Targeted Drug Delivery," J. Magnetism and Magnetic Materials, 293, 633-638 (2005).

5. H. Chen, A. D. Ebner, A. J. Rosengart, M. D. Kaminski, and J. A. Ritter, "Analysis of Magnetic Drug Carrier Particle Capture by a Magnetizable Intravascular Stent. Part 2: Parametric Study with Multi-Wire Two-Dimensional Model," J. Magnetism and Magnetic Materials, 293, 616-632 (2005).

6. M. O. Avilés, A. D. Ebner and J. A. Ritter, "Ferromagnetic Seeding for the Magnetic Targeting of Drugs and Radiation in Capillary Beds," J. Magnetism and Magnetic Materials, 310, 131-144 (2007).

7. M. O. Avilés, A. D. Ebner and J. A. Ritter, "Ferromagnetic Seeding for the Magnetic Targeting of Drugs and Radiation in Capillary Beds," J. Magnetism and Magnetic Materials, 310, 131-144 (2007).

## 357c Hybrid Nanomaterials for Molecular Imaging and Cellular Engineering

**Gang Ruan**[1], Shuang Deng[1], Dhananjay Thakur[2], Ning Han[1], and Jessica O. Winter[3]. (1) Chemical and Biomolecular Engineering, the Ohio State University, 140 W 19th Ave, Columbus, OH 43210, (2) Chemical and Biomolecular Engineering and Biophysics, the Ohio State University, 140 W 19th Ave, Columbus, OH 43210, (3) Chemical and Biomolecular Engineering and Biomedical Engineering, the Ohio State University, 140 W 19th Ave, Columbus, OH 43210

The applications of nanoparticles based on semiconductors, iron oxides or other materials in biology and medicine have been extensively demonstrated. Herein we describe our efforts to synthesize nanomaterials combining two or more types of the above-mentioned materials, and to apply these nanocomposites for biological uses. Hybrid materials of fluorescent quantum dots and superparamagnetic iron oxide nanoparticles have been prepared by procedures of sequential crystal growth and physical encapsulation. An interesting finding is that the fluorescent and magnetic properties of the composite products are dependent upon the preparation procedures used. In addition to these fluorescent-magnetic nanoparticles, we have also explored the incorporation of other functionalities such as controlled release and enhanced cellular delivery.

The biocompatibility of our products has been confirmed in cellular studies. We are currently actively pursuing the use of the nanocomposites in basic biomedical research and clinical diagnosis, imaging and therapy. For example, combining the superparamagnetism of iron oxide nanoparticles and the bright fluorescence of semiconductor nanocrystals makes it possible to mechanically manipulate cells with the manipulation effect monitored in real time. Furthermore, encapsulating drugs along with magnetic and fluorescent nanoparticles into a polymer/lipid matrix could provide a 4-stage nanotechnology cancer treatment: the polymer/lipid nanoparticles find the tumor by active or passive targeting; magnetic nanoparticle-mediated MRI imaging locates the tumor deep in human body; fluorescent nanoparticles guide the physician to surgically remove the majority of tumor tissue; and finally the sustained release of anticancer drugs from the polymer/lipid matrix kills the remaining cancer cells.

## 357d Quantification of Non-Specific Binding of Magnetic Nanoparticles: Implication for Detection and Magnetic Cell Separation

**Jeffrey Chalmers**[1], Ying Xiong[2], Mei Shao[3], Xiaodong Tong[2], Sherif Farag[4], and Maciej Zborowski[5]. (1) Chemical and Biomolecular Engineering, Ohio State University, 140 W. 19th Avenue, Columbus, OH 43210, (2) Chemical and Biomolecular Engineering, The Ohio State Unversity, 140 W 19th Ave, Columbus, OH 43210, (3) The Ohio State University, 140 w19th Ave, Columbus, OH 43210, (4) Bone Marrow and Stem Cell Transplantation, Indiana University School of Medicine, 635 Barnhill Drive, Room 224G, Indianapolis, IN 46202, (5) Department of Biomedical Engineering, Cleveland Clinic Foundation, 9500 Euclid Ave, Cleveland, OH

The maturity of magnetic cell separation technology, as indicated by both the commercial success of a number of companies, and high number of citations indicating its use, also places increasing demands on magnetic cell separation performance. While a number of factors can cause sub-optimal performance, one of the major challenges can be non-specific binding of magnetic nano or micro particles to non-targeted cells. Depending on the type of separation, this non-specific binding can have a negative effect on the final purity, the recovery of the targeted cell, or both. In this work, we quantitatively demonstrate that non-specific binding of magnetic nanoparticles can impart a magnetization in cells such that they thee cells can be retained in a MACS separation column in such a way to significantly impact the purity of the final product. We also demonstrate that the nonspecific binding can negatively impact the recovery of desired cells. While highly variable, cells without specifically or non-specifically bound MACS beads can be retained in a MACS column, compromising the performance of the separation. How, this phenomenon is not easily reproducible. We also demonstrate experimentally, and through theoretical arguments that the number of MACS magnetic particles needed to impart a magnetization that can cause non-targeted cells to be retained in the column to be on the order of 500 to 1000 nanoparticles. This number was demonstrated experimentally with an instrument, cell tracking velocity, CTV, and it is demonstrated that the sensitivity of the CTV instruments for Fe atoms is several order of magnitude more sensitive than ICP-MS.

# 357e Magnetite Nanoparticle-Linked Immunosorbent Assay

**Lizeng Gao**, Jiamin Wu, Liangliang Cao, and Di Gao. Chemical and Petroleum Engineering, University of Pittsburgh, 1249 Benedum Hall, 3700 O' Hara street, Pittsburgh, PA 15261

Using chitosan-modified magnetite nanoparticles (CS-MNPs) as a replacement of enzymes in conventional ELISA configurations, a magnetic nanoparticle-linked immunosorbent assay is demonstrated. The CS-MNPs are synthesized by using a one-step solvothermal process, where chitosan in the reaction system is used both as a ligand and as a surface modification agent. The as-synthesized CS-MNPs have amine groups on their surface which provide good dispersibility in aqueous solutions and convenient sites for covalent linking of antibodies with the MNPs. They also possess catalytic properties that are able to catalyze color reactions in immunoassays, and magnetic properties with both a large enough saturation magnetization and a superparamagneticity that can be used to capture, separate, and enrich antigens prior to the assay procedure. By employing both the catalytic and magnetic properties of the CS-MNPs, a capture-detection immunoassay is developed, where antigens can be captured, separated, and enriched prior to the assay procedure. It is shown that such a procedure enabled by the intrinsic dual functionalities of MNPs may improve the detection limit of immunoassays.

# 357f In Situ Magnetic Separation and Immobilization of *Rhodococcus Erythropolis* LSSE8-1 for Dibenzothiophene Biodesulfurization

**Yuguang Li**, Jianmin Xing, and Huizhou Liu. Key Laboratory of Green Process and Engineering, National Key Laboratory of Biochemical Engineering, Institute of Process Engineering, Chinese Academy of Sciences, Beijing, China

In situ cell separation and immobilization method for dibenzothiophene biodesulfurization was developed by using superparamagnetic $Fe_3O_4$ nanoparticles. Magnetic $Fe_3O_4$ nanoparticles were synthesized by a coprecipitation method followed by modification with ammonium oleate. The surface-modified $Fe_3O_4$ nanoparticles were monodispersed in an aqueous solution, and the average particle size was about 20 nm with 50.8 emu/g saturation magnetization ($\delta_s$). After adding the magnetic fluids to the culture broth, the microbial cells of *Rhodococcus erythropolis* LSSE8-1 was adsorptive immobilized and then separated by external application of a magnetic field. The maximum adsorption amount was about 530 g dry cell weight/g particles. Transmission electron microscopy (TEM) analysis of the cells showed that the $Fe_3O_4$ nanoparticies were strongly absorbed to the surfaces and coated the cells.

The coated cells had distinct superparamagnetic properties, with the $\delta_s$ was 14.0 emu/g. Compared to the reusable limitation of the free cells, the coated cells not only had the same desulfurizing activity as free cells but could also be reused more than six times. The one-step magnetic immobilized LSSE8-1 cells exhibited good catalytic activity and repeated-batch desulfurization operational stability. The mechanism of cell adsorption on oleate-coated $Fe_3O_4$ nanoparticles was also discussed.

References

[1] J.J. Kilbane, Microbial biocatalyst developments to upgrade fossil fuels, Curr. Opin. Biotechnol. 17 (2006) 305-314.

[2] G.B. Shan, J.M. Xing, H.Y. Zhang, and H.Z. Liu, Biodesulfurization of dibenzothiophene by microbial cells coated with magnetite nanoparticles, Appl. Environ. Microbiol. 71 (2005) 4497-4502.

[3] H. Honda, A. Kawabe, M. Shinkai, and T. Kobayashi, Development of chitosan-conjugated magnetite for magnetic cell separation. J. Ferment. Bioeng. 86 (1998) 191-196.

## 362a Characterizing Interfacial Thermal Transitions In Polymer Nanocomposites

**Jason P. Killgore** and Rene M. Overney. Chemical Engineering, University of Washington, Box 351750, Seattle, WA 98195

Heated Tip atomic force microscopy (HT-AFM) allows for thermomechanical transitions to be locally determined, with nanoscale resolution. Here, HT-AFM is used in a scanning mode to activate a local transition at the polymer particle interface in a Poly(trimethylsilyl propyne) - Silica nanocomposite. Heated scanning ultimately resulted in debonding of the particle from the matrix. Consideration of the mechanism for debonding revealed both adhesive and deformative barriers. The existence of a significant deformative barrier suggests the technique is highly sensitive to the interfacial thermomechanical properties of the matrix. It was found that the impact force of the probe on the particle exhibits a stark transition at a temperature below that observed in the bulk material. The transition is attributed to enhanced mobility at the interface, which helps explain the enhanced properties of the resultant nanocomposite.

## 362b Application of Interfacial Statistical Associating Fluid Theory (iSAFT) to Tethered Polymers

**Shekhar Jain**, Adam Bymaster, and Walter G Chapman. Chemical and Biomolecular Engineering, Rice University, 6100 Main St., Houston, TX 77005

Tethered polymers have several industrial applications in colloid stabilization, adhesives, lubricants, modification of surface hydrophobicity, microfluidic devices, and biophysics. Two important aspects have major impact on the process and design of these applications: structure of tethered polymers at the surface/interface and force of interaction between the polymer tethered surfaces/interfaces. We extend our classical density functional theory (DFT) for complex inhomogeneous fluids known as interfacial statistical associating fluid theory (iSAFT) to investigate the structure and interaction force of these tethered polymers. Depending upon the number of polymer chains tethered to the surface (grafting density) and the quality of the solvent, these polymer chains can either stretch away from the surface or collapse onto it. And the force can be either attractive or repulsive. With iSAFT we will be able to explicitly account for the compressibility and difference in the segment sizes of the tethered and solvent polymer chains, which are missed by the previous scaling and self consistent field (SCF) theories for tethered polymers.

## 362c Forces Between Nanorods with End-Adsorbed Chains In Polymer Melts

**Amalie L. Frischknecht**, Sandia National Laboratories, PO Box 5800 MS 1411, Albuquerque, NM 87185-1411

Adsorbed or grafted polymers are often used to provide steric stabilization of colloidal particles. When the particle size approaches the nanoscale, the curvature of the particles becomes relevant. Here I use a classical density functional theory to study the polymer-mediated interactions between two nanorods. The rods are immersed in an athermal, melt polymer blend consisting of: 1) a small fraction of chains of length N=20 with "sticky" ends that are attracted to the rods with energy e/kT so that they form a polymer brush on the rods; and 2) a matrix of chains of length P which have no interactions with the rods. The structure of the brushes depends on the nanorod diameter, P, and e/kT. There is an attractive well in the force between the rods near contact, followed by a strong repulsion as the brushes are compressed. The depth of the well increases with increasing P. I will discuss the implications for experimental systems.

Sandia is a multiprogram laboratory operated by Sandia Corporation, a Lockheed Martin Company, for the United States Department of Energy's National Nuclear Security Administration under contract DE-AC04-94AL85000.

## 362d Fluids DFT Studies of Polymer Nanoparticle Thin Films

**Erin S. McGarrity**[1], Amalie L. Frischknecht[2], and Michael E. Mackay[1]. (1) Dept. of Chemical Engineering & Materials Science, Michigan State University, East Lansing, MI 48824, (2) Sandia National Laboratories, PO Box 5800 MS 1411, Albuquerque, NM 87185-1411

We study the phase behavior of polymer/nanoparticle blends near a substrate using a fluids density functional theory. Initially we model the blends as a mixture of hard spheres and freely jointed hard chains, near a hard wall. There is a first order phase transition present in these blends in which the nanoparticles expel the polymer from the substrate to form a monolayer at a certain nanoparticle concentration. The nanoparticle transition density depends on the length of the polymer, the nanoparticle diameter, and the overall bulk density of the system. The phase transition is due to both packing entropy effects related to size asymmetry between the components, and to the polymer configurational entropy, justifying the so-called "entropic push" observed in experiments. A layered state is found at higher densities which resembles that in colloidal crystals, in which the polymer and nanoparticles form alternating discrete layers. We show that this laminar state has nearly the same free energy as the homogeneously mixed fluid in the bulk and is nucleated by the surface. In addition to these athermal results, we will present preliminary data from calculations in which the polymers, particles, and substrate are attractive.

This work was performed in part at the U.S. Department of Energy, Center for Integrated Nanotechnologies, at Los Alamos National Laboratory (Contract No. DE-AC52-06NA25396) and Sandia National Laboratories (Contract No. DE-AC04-94AL85000).

## 362e Effects of High Concentration of Polymeric Stabilizers for Poorly Water-Soluble Drug Suspensions

**Kapil V. Deshpande**, Eric Jayjock, Frank Romanski, and Fernando J. Muzzio. Chemical and Biochemical Engineering, Rutgers University, 98 Brett Rd, Piscataway, NJ 08854

Delivery of poorly water-soluble drugs has been steadily moving toward developing stabilized particles of submicron size for improving uptake within the body. Due to the larger surface area of the smaller particles, stabilization using non-ionic surfactants is often impractical since a higher concentration is needed to provide full surface coverage. Instead, using biocompatible polymers that double as excipients has been suggested for stabilizing drug particle suspensions. This method raises many fundamental questions regarding the properties of polymer-drug suspensions. This work focuses on the effects of the interaction between the polymeric stabilizers and the poorly water-soluble drugs on the rheological properties of the system. We look at gel solutions of hydroxypropylmethylcellulose (HPMC) and sodium alginate in concentrations up to 10% by weight in water. The drug we are using is griseofulvin at concentrations up to 5% by weight. For this work, griseofulvin has been selected as the model drug since it is a BCS Class II, meaning that it has a low solubility and a high permeability through the intestinal cell layer. Suspensions with the different combinations of polymer and particle concentrations are tested for their rheological properties. In particular, we look at shear behavior and viscoelastic effects for evidence of changing polymer drug interactions by altering the respective concentrations. Temperature effects are also important and studied in detail. Particular interest is in determining the effect of the different concentrations on gelation temperature. This is done in conjunction with size analysis of the particles to determine the particle stability. This study can be used to determine the transition point for possible stabilization mechanisms. The possible interactions that can occur are polymer strands wrapping around the particles, a polymer matrix locking in the particles, and polymer and particles floating in suspension. With increasing polymer concentration it is expected that the interaction mechanism would be the particles locked into a matrix. From the rheological tests, we have constructed an envelope defined by a temperature, polymer concentration, and drug concentration range which is optimized for processing of these drug suspensions for film casting, gel capsule formation, etc.

# 362f Lipid Bilayer Curvature and Pore Formation Induced by Charged Linear Polymers and Dendrimers: The Effects of Charge Density, Concentration, Molecular Size and Shape

**Hwankyu Lee**, Lab. of Computational biology, National Heart, Lung and Blood Institute, National Institutes of Health, Rockville, MD 20850 and Ronald G. Larson, Chemical Engineering, University of Michigan, 2300 Hayward St., Ann Arbor, MI 48109.

Polyamidoamine (PAMAM) dendrimers, which consist of a central core, regularly branched monomeric building blocks, and many surface terminal groups, are considered to be good candidate nanoparticles for use as anti-tumor therapeutics and drug delivery due to their controlled mass, surface valency, and surface functionality. To investigate the effect on pore formation of the dendrimer properties (molecular size and shape, charge density, and concentration) and solution conditions (temperature and salt concentration), and to test the ability of coarse-grained (CG) models to predict experimentally measured dendrimer-bilayer properties, we have performed molecular dynamics (MD) simulations of differently acetylated G5 and G7 PAMAM dendrimers in DMPC bilayers with explicit water using the CG model developed by Marrink et al. (J. Phys. Chem. B. 2007, 111, 7812) with inclusion of long-range electrostatics. When initially clustered together near the bilayer, neutral acetylated dendrimers aggregate, whereas cationic un-acetylated dendrimers do not aggregate, but separate from each other, similar to observations from atomic force microscopy by Mecke et al. (Chem. Phys. Lipids. 2004, 132, 3). The bilayers interacting with un-acetylated dendrimers of higher concentration are significantly deformed and show pore formation on the positively curved portions, while acetylated dendrimers are unable to form pores. Un-acetylated G7 dendrimers bring more water molecules into the pores than do un-acetylated G5 dendrimers. These results agree qualitatively with experimental results showing that significant cytoplasmic-protein leakage is produced by un-acetylated G7 dendrimers at concentrations as low as 10 nM, but only at a much higher concentration of 400 nM for un-acetylated G5 dendrimers (Bioconj. Chem. 2004, 15, 774). At higher salt concentration (~500mM NaCl) or lower temperature (277 K), un-acetylated dendrimers do not insert into the bilayer, which again corresponds to the experimental results. To investigate the effect of molecular shape, we also performed MD simulations of multiple copies of poly-L-lysine (PLL) in DMPC bilayers. Membrane disruption is enhanced at higher concentrations and charge densities of both spheroidally shaped dendrimers and linear PLL polymers, in qualitatively agreement with experimental studies by Hong et al. (Bioconjugate Chem. 2006, 17, 728). However, larger molecular size enhances membrane disruption and pore formation only for dendrimers and not for the linear PLL. Despite more intimate electrostatic interactions of linear molecules than are possible for spheroidal dendrimers, only the dendrimers were found to perforate membranes, apparently because they cannot spread onto a single leaflet, and so must penetrate the bilayer to get favorable electrostatic interactions with head groups on the opposite leaflet. These results indicate that a relatively rigid spheroidal shape is more efficient than a flexible linear shape in increasing membrane permeability. These good qualitative agreements indicate that the effect on pore formation of the molecular size and shape, charge density, and concentration of large nanoparticles can be studied through coarse-grained MD simulations, provided that long-range electrostatic interactions are included.

## 363a Underlying Mechanisms of Gold Catalyzed Silicon Nanowire Growth: First Principles-Based Atomistic Modeling

**Soohwan Lee**, The University of Texas at Austin, 1 University Station C0400, Austin, TX 78712 and Gyeong S. Hwang, Chemical Engineering, Dept. of Chem. Eng.,The University of Texas at Austin, 1 University Station Stop C0400, Austin, TX 78712.

The growth direction, diameter, and surface structure of semiconductor nanowires can be controlled by varying process conditions and metal catalysts. The ability to manipulate their structural properties on the atomic scale makes semiconductor nanowires attractive for a variety of noble applications in electronics, optoelectronics, and sensors. Previous studies have suggested mechanisms underlying the metal catalyzed growth of Si nanowires, involving Si diffusion into a metal catalyst, eutectic Si-catalyst alloy formation, and Si precipitation at the catalyst-nanowire interface. However, many fundamental aspects regarding how synthesis conditions influence the nanowire growth and structure are still uncertain. In this presentation, based on first principles-based atomistic modeling we will focus on addressing the formation of silicon precipitates in a gold particle and the dependence of nanowire growth direction on the gold particle size. Our simulation results clearly demonstrate that silicon atoms preferentially remain near the gold particle surface, indicating that silicon clusters may form and grow in the surface region. The underlying reasons for the surface segregation of silicon will be discussed. We have also determined the interface structure and energetics between gold and silicon with different orientations [(111), (110), (211), and (100)]. The results provide good insight into the preferred growth of silicon nanowires in the (111) direction. We will also discuss mechanisms underlying the change of growth direction from the (111) to (110) with decreasing the gold particle size.

## 363b Dependence of Sensing Properties upon Metal Oxide Nanostructure

**Randy L. Vander Wal**[1], Gordon M. Berger[1], Gary W. Hunter[2], and Jennifer C. Xu[2]. (1) NCSER c/o NASA-Glenn Research Center, 21000 Brookpark Drive, Cleveland, OH 44135, (2) The NASA-Glenn Research Center, 21000 Brookpark Rd., M.S. 77-1, Cleveland, OH 44135

Applications of chemical sensors include environmental monitoring, automotive applications, emission monitoring, and aerospace vehicle health monitoring. The superiority of nanomaterials for sensor applications is clear. High surface area and controlled structure are the hallmarks. Each aspect is particularly relevant to sensors. Surface area is critical to reactive gas adsorption and translates into high sensitivity. Relative to micron sized grains or powders or layers or films, nanoparticles offer a 10 to 100-fold increase in each parameter. Controlled structure provides the reactive sites for adsorption and their effect upon the overall conductance. Nanomaterials often possess enhanced reactivity due both to a strained surface lattice and/or increased prevalence of step, edge, corner and terrace sites. Such activity further enhances sensitivity and potentially lowers operational temperature. Operation at lower temperatures saves power, extends operating lifetime and maintains reproducibility by preventing sintering-induced grain growth. Finally, lower operational temperature, combined with structure control can advantageously yield selectivity.

In recent years two competing approaches have been developed for synthesizing 1-dimensional forms of metal oxide semiconductors (MOSs): electrospinning and thermal-evaporative-condensation (TEC) synthesis. Each method produces 1-d sensor elements that can be incorporated into next generation sensors. Each synthesis method and product has attendant advantages and limitations. Apart from device fabrication and manufacturing issues, these two methods produce elements that differ primarily in their crystallinity. TEC synthesis can produce single crystalline nanowires. Electrospinning produces polycrystalline elements of the (as-spun) sol-gel fiber upon calcination. Though each form is a nanoscale material, an additional parameter, crystallinity, will have profound consequences upon the viability of each 1-d element for reactive gas sensing and devices incorporating them.

With a diameter approaching twice the nominal charge depletion layer thickness (~ 10's of nanometers, dependent upon temperature), the full conducting channel of a nanorod can be affected by adsorption of oxidizing gases, analogous to the electrical conductivity modulation in a field effect transistor. Although somewhat thicker, the polycrystallinity and cylindrical structure of a nanofiber potentially permits deeper penetration of the depletion layer into the structure. Essentially the same narrowing of the "effective" conduction channel occurs. Carrier depletion (or replenishment) throughout the "bulk" nanostructure will expand the sensor dynamic range by the virtue of oxidizing (reducing) adsorbates leading to full charge depletion (or replenishment) with corresponding near-zero or high conductivity, respectively. With the entire bulk of the material responsive to surface adsorbed species, sensitivity gains may be achieved relative to 2-d films where often only the outer surface layer is affected while the underlying portion remains unchanged (maintaining constant conductance).

While it would appear that the nanorod is the limit of the summation describing the nanofiber case, in practice the nanorod diameter is generally larger than twice the depletion layer thickness. Moreover the uniform crystallographic surface planes may not support a high surface concentration of defect sites or chemisorbed oxygen species. Alternatively in the polycrystalline nanofiber not all junctions may be accessible to adsorbates. Such spots would correspond to a "short". Moreover there could be a considerable variation in potential boundaries between grains, given the random orientation of single crystal nanoparticles with respect to each other. Therein experimental tests of the sensing performance of each material were conducted. The validity of such a comparison is facilitated by the one-dimensional morphology of each differently nanostructured metal oxide.

To further enhance the sensitivity of these morphologies to gases, catalyst nanoparticles were deposited upon the surface of the sensing elements. The utility of catalysts upon active oxide supports is well known in catalysis but has yet received little attention in gas sensors. Most approaches using catalysts of either noble metals or other transition metal doped the material. With a homogeneous distribution, most of the catalyst (whether atomically distributed or as nanophase particles) is inaccessible to the near-surface region, the region where active sensing occurs. Secondly, an atomic dispersion will alter lattice and grain structure, but not specifically create catalytic sites upon the sensor surface. The latter is the desirable structure from a catalysis viewpoint and is exploited for the 1-d sensing elements reported here. We note that this approach is frequently used in catalysis where the noble metal nanoparticles and/or the interfacial region between the particle and oxide support greatly accelerates the reaction compared to the bare oxide surface.

In general this particle coating will permit exposure of the underlying metal oxide support and most importantly establishes an interface between the particle and support oxide. These interfacial junctions will be self-polarized (forming a Schottky junction) by virtue of charge transfer, driven by the difference between the metal work function and oxide electron affinity. This interface is expected to be highly reactive for crystalline metal nanoparticles. Clearly this reactivity is critically dependent upon the nature of the oxide support material (composition, its crystallinity, chemisorbed oxygen sites and their binding energy, etc.) and availability of lattice oxygen from the semiconductor oxide surrounding the metal nanoparticle. Nevertheless such junctions are expected to be particularly reactive given the interfacial polarization. Additionally the catalyst nanoparticles will alter the free carrier charge density and band energy levels.

In summary, the central goal of this investigation was to operationally test the difference between these two nanostructured materials forms, with and without catalysts. In the fabrication of the prototype devices we gained practical knowledge of each synthesis method and realized caveats for integration of each element type towards commercial device manufacture. Advantages and limitations of each method towards practical sensor fabrication will also be presented.

## 363c Exploration of Lead-Free Nanosolders for Nanowire Assembly

**Fan Gao**, Subhadeep Mukherjee, and Zhiyong Gu. Chemical Engineering, University of Massachusetts Lowell, One University Ave, Lowell, MA 01854

Solders (low-melting point metals or metal alloys) are widely used in electronics industries. Lead-free solders are being implemented due to the toxicity of lead, even though several challenging issues still exist. In this research, nanoscale lead-free solders are proposed to not only address/overcome the problems in the lead-free implementation process in electronics industries, but also meet the high demand for nanotechnology applications including nanoelectronics. Nanowires have been researched extensively as key elements in sensors, nano-optics, and nanoelectronics (such as resistors, diodes, field-effect transistors). We show in this presentation our initial study in using tin (Sn)-based nanosolders (Sn-Ag, Sn-Ag-Cu, etc) to assemble nanowires into ordered structures. Electrodeposition in nanoporous templates is utilized to synthesize lead-free solders directly onto nanowires. The structures of lead-free nanosolders are characterized, and thermal reflow properties are studied in the lab-scale ovens, following typical reflow profiles. Assembling techniques such as self-assembly, together with lead-free nanosolders, are used to efficiently integrate multisegmented nanowires into ordered arrays with low contact resistance.

## 363d Maskless Electrodeposited Contact to Organic Nanowires

**Carlos Hangarter**, Youngwoo Rheem, Ashok Mulchandani, Wilfred Chen, Marc A. Deshusses, and Nosang V. Myung. Department of Chemcial and Environmental Engineering, University of California, Riverside, Riverside, CA 92521

Lithographic contact of assembled nanostructures is a bottleneck procedure increasing processing time and complexity with alignment requirements. Moreover, the seemingly benign lithographic processing conditions can be harsh environments for organic materials, subjecting the nanowires to high intensity ultra violet light, a polymer spin coat, and organic solvents that may permanently degrade, stress, or dissolve organic nanowire surfaces. This work investigates a post-assembly electrodeposition technique for mechanical joints and intimate contact of organic nanowires. Conducting polymer nanowires, PEDOT, P3HT, and PPy were dielectrophoretically assembled on prefabricated Au electrodes and the nanowire ends were embedded in ferromagnetic materials via electrodeposition. The high resistance of the polymer nanowire inhibited electrodeposition on the wire. The potential and different electrolytes were found to have a significant impact on the selective maskless deposition. Different dopants were also investigated to understand the polymer reduction during cathodic deposition of metal. A single dodecyl sulfate doped polypyrrole nanowire with maskless electrodeposited nickel contacts was shown to have improved sensitivity toward ammonia gas.

# 363e Molecular Dynamics Simulations on the Elongation Properties of Gold Nanowires in Benzendithiol

**Qing Pu**[1], Yongsheng Leng[1], Xiongce Zhao[2], and Peter T. Cummings[1]. (1) Department of Chemical Engineering, Vanderbilt University, Nashville, TN 37235-1604, (2) Center for Nanophase Materials Sciences, Oak Ridge National Laboratory, Oak Ridge, TN 37831

It is widely recognized that the contact geometry at organic molecule-metal interfaces has tremendous effects on electron transport through molecule-metal junctions. Understanding this fundamental issue directly correlates to the success of building molecular electronics devices. The current-voltage (I-V) characteristics for benzendithiolate (BDT) molecule sandwiched between two gold electrodes have been studied both experimentally and theoretically. However, a significant discrepancy in I-V values between theoretical predictions and experimental is still not well understood. Since in the quantum mechanical calculations for the electron transport, the geometry of the metal-molecule-metal contacts has to be assumed a priori, it is valuable to seek insights into the local bonding structures of BDTs on Au nanowires via classical molecular simulations to identify the detailed Au-BDT-Au contact molecular configurations. In this study, we combine grand canonical Monte Carlo (GCMC) with molecular dynamics (MD) simulations to investigate the dynamic elongations of gold nanowires in the presence of BDT molecules. We focus on the development of a simulation strategy to deal with the chemical adsorption and desorption of BDT molecules on Au nanowire by GCMC, followed by MD simulations for the elongations. Simulation results demonstrate that the existence of BDT molecules significantly affect the ductile elongations compared with those obtained in vacuum. The density of BDT and the number of nonbonded and bonded BDT molecules in the simulation box is monitored during the entire elongation process. In addition, the detailed contact geometry and elongation force variations are obtained. Our research result will help to resolve the original discrepancy between experimental and theoretical studies.

## 371a Multifunctional Polymer Structures: Enabling Technology for Miniaturized Devices

**Joseph L. Lenhart**, Sandia National Laboratories, PO Box 5800 MS 0888, Albuquerque, NM 87185

Polymer-based materials enable a broad range of applications from car bumpers to integrated circuitry. However, the current demands for device miniaturization coupled with enhanced performance, require novel polymeric structures with multifunctional properties. This talk will highlight several programs initiated at Sandia National Laboratories, where multifunctional polymers are being developed for emerging micro-device technologies, including 1) the design of extremely soft polymer gels; 2) the development of processable conductive polymers; 3) the investigation of high porosity polymeric scaffolds with controlled porous architectures, and; 4) the engineering of multifunctional polymeric composites. The polymer science has implications for a broad range of emerging micro-device technology including micro-electro-mechanical systems (MEMS), sensors, flexible electronics, biotechnology, three-dimensional micro-electronics packaging, and energy storage applications. The talk will discuss critical technical challenges, promising materials approaches, and initial examples of program success.

## 371b Biomolecular Sensing with Polyaniline-Poly(2-acrylamidomethylpropane sulfonic acid) Nanosystems

**Carolyn L. Bayer**[1], Alper A. Konuk[2], and Nicholas A. Peppas[2]. (1) Biomedical Engineering, The University of Texas at Austin, 1 University Station, CO400, Austin, TX 78712, (2) Chemical Engineering, The University of Texas at Austin, 1 University Station, CO400, Austin, TX 78712

Conductive polymers have been extensively researched for the creation of polymer electronics, such as thin film transistors, photovoltaics and organic light emitting diodes. In these devices, the processability and mechanical properties of conductive polymers are improved in comparison to traditional inorganic materials. Though less explored, these same properties encourage the use of conductive polymers in biomedical applications. In particular, conductive polymers may be chemically and physically tailored to be similar to established biomaterials, with the added feature of possessing unique electrical properties. The incorporation of these electrical properties in biomaterials presents opportunities for systems integration in biomaterials.

In this work, the conductive polymer polyaniline (PANI) complexed with poly(2-acrylamido-2-methylpropane sulfonic acid) (PAAMPSA), referred to as PANI-PAAMPSA, is used for biomolecular sensing. The development of PANI-PAAMPSA as an important material for polymer electronics was recently noted, and the inherent doping system of the material may be useful as an integrated biomolecular sensor. PANI-PAAMPSA is generated by a template synthesis process, where the PAAMPSA is dissolved in water with aniline monomer, allowing the aniline monomers to complex with the sulfonic acid groups of the PAAMPSA. Oxidative polymerization is initiated to form the PANI chain, which is doped by the dissociated acid functional groups of the PAAMPSA. The resulting material is highly conductive, and the processability of the conductive polymer is significantly enhanced. The premise of this research is that the introduction of biomolecules with ionic functional groups will disturb the complexation between the PANI and the PAAMPSA, leading to a decrease in dopant charge and therefore a decrease in the polymer conductivity.

Initial testing with the synthesized PANI-PAAMPSA material demonstrated its potential for use as a biomolecular sensor. To explore this effect further, it was desired to have a polymer film adhered to a device substrate, with electrodes for probing the film conductivity, which could be repeatedly exposed to solutions with various biomolecules, rinsed with water. To ensure a fast response time, a film of nanosized thickness was also desired. To accomplish these objectives, a self-assembled monolayer technique was used to adsorb the PANI-PAAMPSA to the device substrate. Silane treated device substrates were immersed in the PANI-PAAMPSA reaction solution, and the polymers adsorbed to the surface as the chains polymerized. This synthesis method has been previously described as an in situ polymerization method, however our methods differ in that we are using the template synthesis process to generate a film with PAAMPSA incorporated as a stable large molecule dopant, which does not easily rinse out with water or with a pH or ionic change in the environmental conditions, enabling the film to be used as a functional biomolecular sensor.

The produced films were characterized by profilometry, x-ray photoelectron spectroscopy (XPS), UV-VIS spectroscopy, and the electrical conductivity was measured using a 4 point probe technique. The films were characterized in their as-synthesized form, and upon exposure to a range of pH-controlled buffers. The films demonstrated a 3 order of magnitude change in conductivity when the pH was increased to 7.4, and the response time was less than 5 minutes. Additionally, upon rinsing with deionized water, the films recovered their conductivity.

The fabricated devices are highly sensitive and show promise as a platform for a biomolecular sensing device. The material's electrical properties, coupled with mechanical and physical properties ascribed to the PAAMPSA, enable the introduction of electronic properties into customizable biomaterials with integrated sensing properties.

This work was supported by the NSF-IGERT program grant DGE-0333080.

## 371c Iontophoretic Transdermal Drug Delivery System Using a Conducting Polymeric Membrane

**Qiuxi Fan**, Otto H. York Department of Chemical Engineering, New Jersey Institute of Technology, Newark, NJ 07102, **Kamalesh K. Sirkar**, Chemical Engineering, New Jersey Institute of Technology, Warren Street, Newark, NJ 07102-1982, and Bozena Michniak, Department of Pharmacology and Physiology, UMDNJ-New Jersey Medical School, Newark, NJ 07103.

This work investigated the application of a porous polyaniline (PANi) membrane as a conducting polymeric membrane as well as an electrode in an iontophoretic transdermal drug delivery (TDD) system. Model drugs studied were: caffeine (MW: 194.2), lidocaine HCl (MW: 270.8) and doxycycline HCl (MW: 480.1). The PANi membrane was first tested as a simple membrane between the donor and receptor solutions; it provided satisfactory permeation profiles; the observed flux values were well described by a simplified mass transport model. A mouse skin was then mounted beneath the PANi film; such a composite system also presented satisfactory permeation profiles. Iontophoretic TDD experiments were next performed using both Ag|AgCl electrodes and PANi|AgCl electrodes for comparison; a PANi anode replaced the Ag anode in the last set. For doxycycline HCl, the flux and the 24-hour accumulation from the PANi|AgCl set were 94.4±81.2 $\mu$ g/cm$^2$ hr and 2760±3980 $\mu$ g/cm$^2$ respectively; those from the Ag|AgCl set were zero. For lidocaine HCl, the flux and 10-hour accumulation from the PANi|AgCl set were respectively 43±15 $\mu$ g/cm$^2$ hr and 392±130 $\mu$ g/cm$^2$; the corresponding values from the Ag|AgCl set were 48±20 $\mu$ g/cm$^2$ hr and 348±78 $\mu$ g/cm$^2$. Porous polyaniline membrane appears to be capable of replacing the Ag part of Ag|AgCl electrode system; further such a membrane can exercise additional control over agent transport rate. Aqueous-organic partitioning system through the porous membrane of PANi was tested with this novel technique as well. Because of the rather low porosity of the synthesized PANi film, such a system did not yield a high permeation rate.

## 371d Delivery of Antibiotics and BMP-2 from Biodegradable Polyurethane Scaffolds

**Andrea E. Hafeman**, Bing Li, K. L. Zienkiewicz, and S. A. Guelcher. Chemical Engineering, Vanderbilt University, 211 Olin Hall, 2400 Highland Ave., Nashville, TN 37235

An urgent need for better bone healing and regeneration therapies results from the recognized limitations of current bone grafting procedures. Growth factors such as BMP-2 have been shown to enhance fracture healing, although infections can compromise healing even in the presence of BMP-2. We are developing a therapy for healing infected fractures using injectable, biodegradable polyurethane (PUR) scaffolds. These materials support osteoblast migration and proliferation, and degrade to non-cytotoxic products. In this study, we have developed PUR scaffolds that release BMP-2 and tobramycin for localized delivery. These biomaterials present potential clinical opportunities for enhanced treatment of infected fractures.

PUR foams were synthesized by reactive liquid molding of hexamethylene diisocyanate trimer and a hardener comprising a polyester triol, PEG 600, water, catalyst, stabilizer, and pore opener using previously reported techniques. Lyophilized tobramycin (8 wt%) and BMP-2 (0.001 wt%) with heparin (0.03 wt%) were incorporated separately into foams. In vitro release of tobramycin and BMP-2 in PBS at 37 °C was measured up to 30 days. BMP-2 release was measured by ELISA, and tobramycin by HPLC. The released substrates were evaluated for bioactivity: BMP-2 by alkaline phosphatase activity and Western blots, and tobramycin by Kirby-Bauer and time-kill assays with methicillin-susceptible S. aureus. The in vivo behavior of the scaffolds was evaluated by subcutaneous implantation and tibial plugs in rats.

22-27% of the BMP-2 released from the foams by 24 days, with a continuing upward trend, which suggests extended, ongoing release. 60-70% of the tobramycin eluted after 24 hrs, with 90% released by 5 days. Kirby-Bauer tests revealed that sufficient tobramycin diffused from the foams and was effective against the bacteria. Zones of inhibition (ZI) were larger than the minimum sensitivity levels of 15 mm: 8% tobramycin (34.3 ± 1.2 mm), and blank foams displayed no inhibition. Histological sections of the subcutaneous implants showed extensive scaffold biodegradation by day 21, with new tissue formation and a limited giant cell response confined to the material remnants. Cellular infiltration and new bone formation were also observed in the tibial plugs.

This biocompatible and biodegradable PUR scaffold shows promise as an effective therapy for bone fracture healing. Its injectable application and resilience promote thorough contact with the surrounding bone, and the material properties can be tuned for varied strength and elasticity. In vivo studies show extensive cellular infiltration and new bone formation with minimal inflammation. Furthermore, we can release growth factors and antibiotics from these scaffolds in a controlled manner. This scaffold therefore provides a surface supporting attachment of osteoprogenitor cells, as well as a delivery system from which bioactives can be released locally to enhance healing of infected fractures.

## 371e Purification of CD45+ Hematopoietic Cells Directly from Human Bone Marrow and Peripheral Blood Using a Flow-Based P-Selectin–Coated Microtube

**Srinivas Narasipura**[1], Joel Wojciechowski[2], Brian Duffy[1], Jane L. Liesveld[3], and Michael King[1]. (1) Biomedical Engineering, University of Rochester, Rochester, NY 14627, (2) CellTraffix inc, Rochester, NY 14623, (3) Leukemia/Blood and Marrow Transplant Program, James P. Wilmot Cancer Center, University of Rochester, 601 Elmwood Ave, Rochester, NY 14642

Recently, a flow-based selectin dependent method for capture and enrichment of specific types of blood cells (CD34+ hematopoietic stem and progenitor cells) from bone marrow and peripheral blood has been described. Using a similar approach of perfusing samples through P-selectin coated microtubes, here we show isolation of CD45+ blood cells directly from human bone marrow and plasma depleted peripheral blood to a high purity (90-97%). We observed that secreted P-selectin glycoprotein ligand-1 (PSGL-1) present in the blood plasma inhibited adherence of leukocytes onto P-selectin immobilized on the microtube surface. Hence an additional step of removing plasma from blood cells either by gravity-driven separation or centrifugation was incorporated prior to sample perfusion in the case of peripheral blood to achieve CD45+ cell purification. The process mimics a ubiquitous mechanism of cell trafficking in the body for recruitment of neutrophils during inflammation. The method is straightforward, rapid, and may represent a practical alternative to CD45+ cell enrichment procedures required for chimerism analysis following allogenic stem cell transplantation.

References:

1. Antin JH, childs R, Filipovich AH, et al. Establishment of complete and mixed donor chimerism after allogenic lymphohematopoietic transplantation: Recommendations from a workshop at the 2001 tandem meetings. Biol Blood Marrow Transplant 2001;7:473-485.

2. Kansas GS. Selectins and their ligands: current concepts and controversies. Blood 1996;88:3259-3287.

3. Narasipura SD, Wojciechowski JC, Charles N, et al. P-selectin- coated microtube for enrichment of CD34+ hematopoietic stem and progenitor cells from human bone marrow. Clin Chem 2008;54:77-85.

4. Charles N, Liesveld JL, King MR. Investigating the feasibility of stem cell enrichment mediated by immobilized selectins. Biotechnol Prog 2007;23:1463-72.

5. Wojciechowski JC, Narasipura SD, Charles N, et al. Capture and enrichment of CD34-positive haematopoietic stem and progenitor cells from blood circulation using P-selectin in an implantable device. Br J Haematol 2008;140:673-81.

# 371f Supported Biomembranes at Micro- and Nanoscale Interfaces

**Bin He**, Department of Chemical Engineering, City College of New York (of CUNY), T308 Steinman Hall, 140th St at Convent Ave, New York, NY 10031 and M. Lane Gilchrist, Chemical Engineering, City College and the Graduate Center of the City University of New York, Steinman Hall, 140th St @ Convent Ave, New York, NY 10031.

Membrane proteins make up ~30% of the genome of a typical eukaryotic cell, yet due largely to a strict requirement of a highly mobile, native-like biomembrane microenvironment, these molecules have not been widely built into functional materials. Membrane protein structure is often highly complex, typified by large, multi-subunit complexes that not only span the lipid bilayer but also contain large (>2 nm) cytoplasmic and extracellular domains that protrude from the membrane. Our current studies are focused on the implementation of a biomimetic approach to create solid-supported biomembranes interfaced with micro- and nanostructures.

To functionalize materials with active membrane proteins, the challenge is to build stabilized, supported biomembranes in which the substrate to biomembrane spacing can be controlled to accommodate larger membrane protein systems such as the ABC transporter proteins. In essence, our main objective is to create a biomimetic, surface-tethered artificial cytoskeleton where membrane-bound polymer tethered bioconjugates anchor the lipid bilayer at interfaces. Furthermore, the anchoring of biomembranes in this fashion could lead to assemblies that could hold up to shear, flow, and friction in challenging microenvironments such as found in microdevices, biomaterials, nanoporous membranes, and on micropipette surfaces.

Confocal fluorescence microscopy was utilized to analyze the tether-supported bilayer structures on silica microspheres, micropipettes, and in nanoporous anodic aluminum oxide (AAO) membrane filters with 200 nm and 100 nm pore diameters. We have initiated fluidity studies of the supported membranes using fluorescence recovery after photobleaching (FRAP) in these systems. Scanning and Cryo- electron microscopy has been used in concert with CLSM to characterize the micro or nanostructures formed. These systems have been employed as a new substrate used to functionally immobilize the yeast drug efflux pump PDR5 (as a GFP fusion), a member of the ATP binding cassette (ABC) transporter superfamily. These assemblies are being used to build highly miniaturized in vitro models of multidrug resistance for high-throughput screening and drug discovery. We furthermore aim to develop new functional materials and nanoscale imaging and diagnostic tools based on membrane protein active functions and molecular recognition.

## 398a Carbon Nanotube Electrochemical Photocathodes with Optical Rectenna Behavior

**Juan G. Duque**[1], Howard K. Schmidt[2], and Matteo Pasquali[1]. (1) Department of Chemical and Biomolecular Engineering, Rice University, 6100 Main Street, MS-100, Houston, TX 77005, (2) Chemical and Biomolecular Engineering, Rice University, 6100 Main Street, Houston, TX 77005

Graphitic and gold nanostructures are shown to produce rectified photocurrents without pre-bias when illuminated in the presence of aqueous electrolytes containing anionic surfactants. Graphite nanoplatelets, multi-wall carbon nanotube mats, single-wall carbon nanotube (SWNT) arrays and carbon black-loaded polymer all display this novel photovoltaic effect, and share a unique isolated photocurrent peak at ~300 nm. This feature apparently originates with their shared $\pi$-plasmon absorption. Rectified photocurrents are obtained only in the presence of anionic surfactants known to form self-assembled monolayers on graphitic surfaces. Carbon nanotube electrodes also produce minor photocurrents in the near infra-red. When coated with 5 nm of gold, single-walled carbon nanotube arrays generate similar photocurrents when treated with thiol-terminated anionic surfactant molecules. The Au-SWNT composite nanostructure displays a broad photocurrent spectrum spanning the entire visible region, along with a pronounced peak at ~400 nm. These features appear to arise from photoabsorption by gold nanorod plasmon hybridized with SWNT $\pi$-plasmons. A model explaining these results based on hot carriers traversing a rectifying tunnel barrier is presented and discussed. This mechanism is consonant with the concept of an optical rectifying antenna (rectenna), and could enable development of high efficiency photodetectors, photovoltaics and photochemical converters.

# 398b Effect of Pore Size and Rope Size on the Mechanical Properties of Vapor Grown Carbon Nanofiber – Unsaturated Polyester Resin Nanocomposites

**Vinod K. Radhakrishnan**[1], Matthew J. Kayatin[1], Robert L. Yuan[2], and Virginia A. Davis[1]. (1) Department of Chemical Engineering, Auburn University, 212 Ross Hall, Auburn, AL 36849, (2) Civil Engineering, Lamar University, P. O. Box 10024, Beaumont, TX 77710

The exceptional mechanical and thermal properties of vapor grown carbon nanofibers (VGCFs) and single-walled carbon nanotubes (SWNTs) have led to their extensive use as reinforcing agents in fabricating nanocomposites. However, difficulties in achieving good nanomaterial dispersion and interface interaction between the nanomaterial and polymer matrix have hindered the performance of the final nanocomposites. This has led to considerable research being focused on alternate routes of nanomaterial incorporation. Resin infiltration of the nanomaterial bucky paper is one such method. It involves layers of highly entangled sheets of nanomaterial, made by filtering nanomaterial dispersion in a solvent, infiltrated with resin usually assisted by vacuum. Many factors affect the properties of resin infiltrated nanomaterial bucky paper; most important of which are the pore size and rope size of the bucky paper that are largely dependent on mixing time and concentration nanomaterial dispersion.

A two level factorial design with mixing time of 2 and 4 days, concentration of 0.75 and 1.5 mg/ml, and thickness of 200 and 400 $\mu$m was employed to study effect of VGCF bucky paper pore size, rope size and thickness on the mechanical properties of unsaturated polyester resin infiltrated nanocomposite. Pore size of the bucky papers was measured on an autosorb pore size analyzer from Quantachrome while rope sizes were measured from microscopic images. The results of the study indicate that pore sizes and rope sizes indeed have a large effect on the properties of the final nanocomposite. Bucky papers should ideally have small rope sizes in order to allow for effective load transfers. On the other hand, small rope sizes result in small pore sizes which make resin infiltration difficult leading to the nanocomposite failure. This study helps in determining conditions to make ideal VGCF bucky papers that have pore sizes large enough to allow resin infiltration but also rope sizes small enough to allow good load transfer.

# 398c Electrochemical Double Layer Behavior on Vertically Oriented Cnt Electrodes

**Mainak Majumder**, Chemical and Biomolecular Engineering and Richard E. Smalley Instiute for Nano-scale Science and Technology, Rice University, 6100 Main Street, MS-362, Houston, TX 77005 and Bruce J. Hinds, Chemical and Materials Eng., University of Kentucky, 177 Anderson Hall, Lexington KY, KY 40506-0046.

Electrochemical double layer behavior of carbon materials is important in several applications such as super-capacitors, ionic separations and in responsive surfaces.

We have investigated the interfacial double layer behavior of vertically-oriented CNT electrodes encapsulated in a polymer matrix. These electrodes have been used as mass transport channels through the hollow interiors of the CNTs; however, they provide a interesting electrode materials with the edge-planes of the CNTs being exposed to the electrolytic solution.

Capacitance estimation, interpreted from electrochemical impedance spectroscopy measurements, indicated almost 4-5 times increase in functional density of carboxylic acid groups after electrochemical diazonium grafting chemistry. Potential of zero charge (PZC) measurements of these electrodes failed to provide a sharp transition associated with other atomically smooth surfaces, however, electrolyte screening effects were observed. Non-faradic impedance spectroscopy measurements of large macro-molecular tethers covalently attached to the electrodes indicated a potential dependent switching behavior. This switching behavior has been attributed to potential-dependent conformational change based on complementary faradaic impedance spectroscopy and equivalent circuit analysis. A similar potential-induced conformational change has been utilized for making voltage-gated CNT membranes.

# 398d Connecting Single Molecule Electrical Measurements to Ensemble Spectroscopic Properties for Quantification of Single-Walled Carbon Nanotube Separation

**Woo-Jae Kim**, Department of Chemical Engineering, Massachusetts Institute of Technology, 77 Massachusetts Avenue, building 66-580, Cambridge, MA 02139, Chang Young Lee, Chemical Engineering, Massachusetts Institute of Technology, 77 Massachusetts Avenue, NE47-577, Cambridge, MA 02139, and Michael S. Strano, 66-566 Department of Chemical Engineering, Massachusetts Institute of Technology, 77 Massachusetts Avenue, Cambridge, MA 02139.

We directly compare ensemble spectroscopic measurements to a systematic and statistically rigorous counting of individual SWNT devices using a high throughput electrical probe station. We derived simplified equation, which connects the absorption intensity ratio and the actual composition of metallic and semiconducting SWNT in SWNT mixture by only one coefficient, that is, the extinction coefficient ratio of two SWNT types. We have successfully quantified the % enrichment of each electronic type from separated SWNT samples for the first time, independently of potential spectroscopic artifacts, and report an extension coefficient ratio for metallic to semiconducting SWNT of $0.354 \pm 0.012$. This systematic classification of individual SWNT into metallic and semiconducting types also enables the assessment of defect generation during solution phase processing. We demonstrate that solution phase processing causes 2-3 orders of magnitude decrease in the field effect mobility distribution of the semiconducting SWNT.

# 398e Electronically Homogeneous, Photonic Semiconducting Single-Walled Carbon Nanotubes Bundles In Solid Phase

**Jae-Hee Han**[1], Ryuichiro Maruyama[2], Woo-Jae Kim[1], Chang Young Lee[1], Jong Hyun Choi[1], Daniel A. Heller[1], and Michael S. Strano[3]. (1) Chemical Engineering, Massachusetts Institute of Technology, 77 Massachusetts Avenue, NE47-577, Cambridge, MA 02139, (2) Solid State Ionics Lab., Advanced Material Laboratories,, Sony Corporation, Atsugi 2TEC 6F, 4-16-1 Okata, Atsugi, Kanagawa, Japan, (3) 66-566 Department of Chemical Engineering, Massachusetts Institute of Technology, 77 Massachusetts Avenue, Cambridge, MA 02139

The ability to synthesize and manipulate uniform bundles of electronically pure single walled carbon nanotubes (SWNTs) remains central to the advancement of these systems in nano-electronic applications. The assembly of separated SWNT into bundles (parallel aggregates) that are electronically homogeneous and of controlled size is currently beyond engineering capabilities. Using the advance of both density gradient[1,2] and dielectrophoresis methods[3], we have performed the first synthesis and measurements of electronically homogeneous bundles of SWNTs in solid state.[4] Several interesting properties are noted. We find that bundles of semi-conducting SWNT that are free of all metallic impurities emit photoluminescence (PL) in the near infrared, as anticipated by theory.[5] The emission is an important quality control metric, and allows for photophysical characterization of such systems. TEM image reveals a dielectrophoretically synthesized 36 nm bundle of (6,5)-enriched SWNTs bundle attached to the very end of a supporting chemically etched tungsten tip. A vertical conductance measurement characterizes the resistance with length and reveals non-ohmic, step changes in current at constant potential characteristic of ballistic conductance. A scattering length, $L$, of approximate 4 to 18 nm is calculated from the trace. We also find that such solid bundles show unprecedented significantly bright PL emission as imaged using an InGaAs camera at both 658 and 785 nm laser excitations, confirming a high degree of semiconductor purity. Controls with un-enriched samples show no emission. The $G/G_0$, where $G$ and $G_0$ are conductance and quantum conductance $2e^2/h \approx 1/13$ (k $\Omega$)$^{-1}$ respectively, ranges from 0.3 to 2.2 for a typical un-enriched SWNT bundle. High purity semi-conducting bundles exhibit significantly low ratios at about 0.002 to 0.004. Some interesting applications involving environmentally-related PL shift from these semiconducting solid bundles will be discussed.

References

1. Arnold, M. S.; Green, A. A.; Hulvat, J. F.; Stupp, S. I.; Hersam M. C. *Nature Nanotech.* **2006**, *1*, 60-65.

2. Kim, W. -J.; Usrey M. L.; Strano, M. S. *Chem. Mater.* **2007**, *19*, 1571-1576.

3. Lee, C. Y.; Baik, S.; Zhang, J.; Masel, R. I.; Strano, M. S. *J. Phys. Chem. B* **2006**, *110*, 11055-11061.

4. Han, J. H.; Maruyama, R.;, Kim, W. -J.; Lee, C. Y.; Choi, J. H.; Heller, D. A.; Strano, M. S. **2008** (submitted).

5. Graff, R. A.; Swanson, J. P.; Barone, P. W.; Baik, S.; Heller, D. A.; Strano, M. S. *Adv. Mater.* **2005**, *17*, 980-984.

# 398f Selective Enrichment of (n,m) Single-Walled Carbon Nanotubes Via Cosurfactant Extraction

**Yuan Chen**, Li Wei, and Bo Wang. School of Chemical and Biomedical Engineering, Nanyang Technological University, 62 Nanyang Drive, Singapore, Singapore

Highly selective enrichment of (6,5) and (8,3) SWCNTs (above 85% of the semiconducting tubes) was achieved

through multistep extraction by sodium dodecyl sulfate (SDS) and sodium cholate (SC) cosurfactant solution

from narrowly (n,m) distributed SWCNTs produced by the catalyst Co-MCM-41. A systematic change in

the chirality selectivity was observed when the weight ratio between SDS and SC varied in cosurfactant

solutions, with maximum enrichment selectivity for (6,5) tubes yielded at 1:4. Furthermore, surfactants were

washed away easily to produce “clean” SWCNTs. This observation sheds light on the possibility of obtaining

SWCNTs with the desired (n,m) structure via an easily scalable approach. No selectivity was detected when

using sodium dodecyl benzene sulfonate (SDBS)/SC cosurfactants, hence suggesting the need for further

exploration of various cosurfactant combinations for more effective extraction of different (n,m) species.

Reference:

ACS Nano VOL 1(4) 327–336 2007

Journal of Physcial Chemistry B 2008, 112, 2771-2774

Web Page: www.ntu.edu.sg/home/chenyuan

## 428a Kinetic Control of Self-Catalyzed Indium Phosphide Nano - Wires, Cones, and Pillars

**R. L. Woo**[1], L. Gao[1], Suneel Kodambaka[2], Kang L. Wang[3], Niti Goel[4], Mantu K. Hudait[4], and Robert F. Hicks[1]. (1) Chemical and Biomolecular Engineering, University of California, Los Angeles, Los Angeles, CA 90095, (2) Materials Science and Engineering, UCLA, Los Angeles, CA 90095, (3) Electrcial Engineering, UCLA, Los Angeles, CA 90095, (4) Intel Corporation

Nanowire research has gained tremendous momentum due to its potential applications in nano-electronics, photonics, solar cells, and thermoelectrics. Precise control of size, morphology, density, and uniformity is essential for realizing these commercial opportunities. Most indium phosphide nanowire research is prepared by vapor-liquid-solid (VLS) growth using gold catalysts. However, gold incorporation into the InP is unavoidable and can adversely affect optoelectronic properties. As an alternative, we have investigated the self-catalyzed growth of InP using liquid indium droplets in a metalorganic vapor-phase epitaxy (MOVPE) process. In this study, three distinct shapes were obtained depending on the growth conditions: nano- wires, cones, and pillars. The shape of nanostructures is determined by the relative rates of indium phosphide deposition via vapor-liquid-solid and vapor-phase epitaxial growth processes. At the meeting, we will discuss more details on the underlying mechanisms of the crystal growth process and the relationship between the deposition mechanism and the resultant shape of the nanostructure.

# A theoretical study of atomic wires and atomic junctions created on a Molybdenum disulfide surface

Diana M.Otalvaro,[1] Kian Soon Yong,[2] Ravi K. Tiwari,[1] Mark Saeys,[1] Christian Joachim[3,4]

[1]*Department of Chemical and Biomolecular Engineering, National University of Singapore*

[2]*Institute of High Performance Computing, Singapore*

[3]*CEMES, CNRS, 29 rue J. Marvig, 31055 Toulouse Cedex, France*

[4]*Institute of Materials Research and Engineering, 3 Research Link, Singapore 117602*

## Introduction

The powerful drive for enhanced computational power is swiftly driving the electronics industry towards the atomic realm, where clusters of atoms, molecules and even single atoms act as the primary units for fabricating wires, switches, transistors and storage devices. Distinct from classical electronic devices, molecular and atomic electronics are governed by quantum mechanical principles, and thus understanding the transmission properties of these devices at the nano scale poses a scientific challenge. Experimentally, the scanning tunneling microscopy (STM) offers the possibility of manipulating and visualizing single atoms and molecules with atomic precision, and has facilitated the controlled fabrication of atomic structures.[1 2] One such structure that has attracted much attention are atomic wires fabricated on the surface of a solid substrate; these wires pose a pathway to access molecules and atomic structures that can ultimately act as quantum logic gates. Moreover, recent experiments have shown that small metal islands deposited on surfaces can be manipulated by the STM; [3] these metal clusters can then act as nano-electrodes, and can be used to contact exactly one atomic wire thus offering the possibility to characterize the electrical properties of the wire through them.[4]

Dangling bonds created by removing atoms from a surface has been shown to possess promising conductance properties. These surface structures might replace molecular or oligomer wires, which are difficult to align reproducibly on the surface and difficult to contact externally. Molybdenum disulfide ($MoS_2$) presents an attractive candidate surface for the manufacturing of these atomic dangling bond wires, and as a substrate for a number of other atomic structures with interesting electronic properties. Controlled extraction of single sulfur atoms by a STM tip has been experimentally demonstrated.[2] In this project, we investigate from first principles the mechanisms governing the conductance properties of an atomic wire created by sulfur vacancies on a $MoS_2$ substrate. Furthermore, we investigate the physics behind a multilevel atomic junction formed by reinsertion of S atoms into an existing S vacancy wire.

## Approach

State of the art density functional theory[5] was used to obtain electronic band structures and accurate density of states for the periodic $MoS_2$ structure and all the modified surfaces. The conductance properties of the systems were studied using the elastic-scattering quantum chemistry (ESQC) method,[6] which describes using quantum chemistry techniques the scattering of electron waves by a defect embedded in a periodic system. In addition, extended Hückel molecular orbital descriptions were employed to gain a more qualitative understanding of the electronic picture of the system. The familiar language of molecular orbital theory and band structure theory provide a convenient framework to discuss the results.

## Results

### Sulfur Vacancy Atomic Wire

It has been theoretically demonstrated that a finite atomic line of S vacancies created on a planar $MoS_2$ substrate can function as a pseudo-ballistic wire for electron transport.[4] This wire can be formed by extracting S atoms from the $MoS_2$ surface.

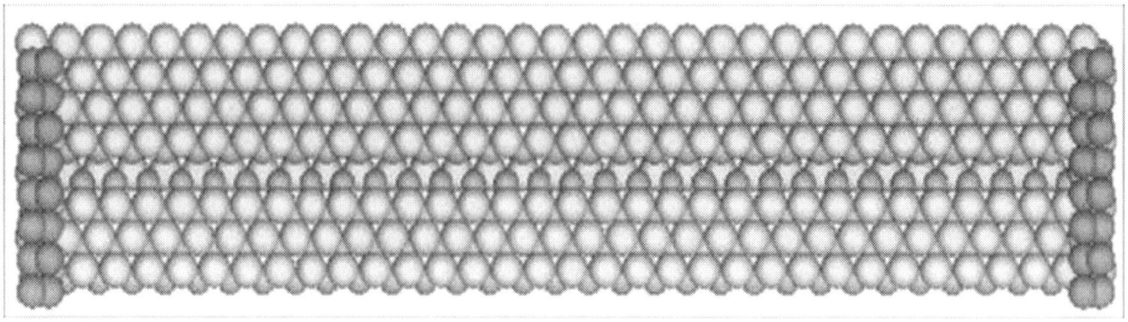

**Figure 1.** Model of a 10 nm atomic wire between 2 Au nano electrodes, created by extracting S atoms from a $MoS_2$ substrate

Removing surface S atoms introduces electronic states within the surface $MoS_2$ electronic band gap. Each S vacancy introduces a symmetric molecular orbital of mainly Mo $d_z^2$ character near the top of the valence band and a degenerate pair of molecular orbitals of mainly Mo $d_{xz}$ and $d_{yz}$ character near the center of the band gap. Mixing of the degenerate orbitals on neighboring vacancies causes the gradual formation of two overlapping energy bands near the Fermi level of the electrodes which create pseudo-ballistic electron transport channels.

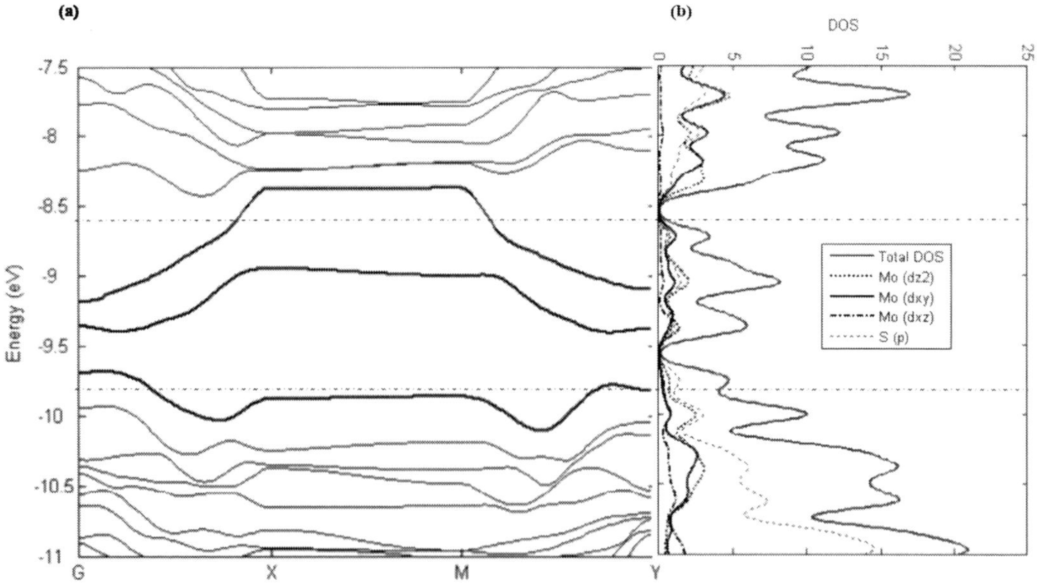


**Figure 2:** a) Band Structure for an infinite wire of S vacancies. Three bands (bold) characteristic of the wire are located within the original $MoS_2$ energy gap (dashed line). The upper two bands are responsible for the conductance. b) Density of States for the wire decomposed on atomic orbitals of different symmetry.

The conductance of the wire is independent of its length (Figure 3), but the overall channel transparency is not reaching unity, since electron waves are scattered between the Au nano electrodes and the wire of S vacancies. The conductance of an Au electrode – atomic wire – Au electrode junction can be enhanced by optimizing the Au electrode-wire distance and by increasing the lateral Au electrode-wire overlap by sliding the electrode over the wire.

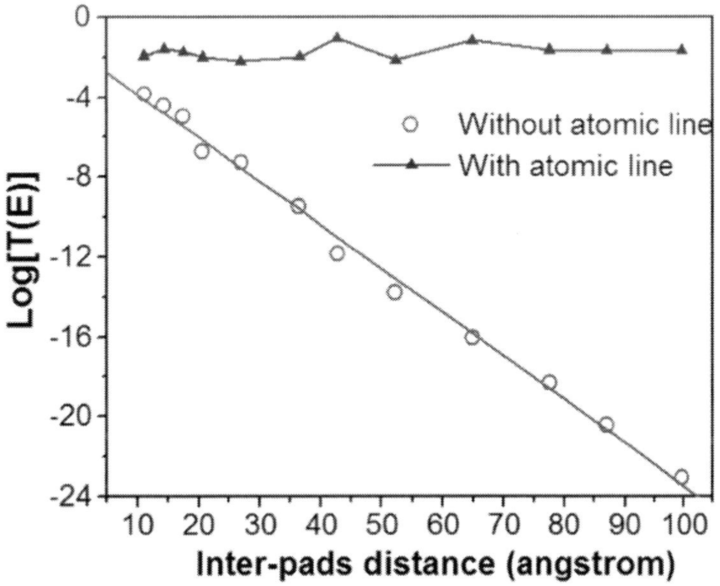


**Figure 3:** Variation of the transmission coefficient at the Fermi level as a function of the distance between the two Au nano-pads for a surface $MoS_2$ junction (o) and for an atomic wire of S vacancies fabricated on the $MoS_2$ surface (▲).

## Atomic Junction

Taking the concept of the atomic wire one step further, we studied the electronic properties of atomic junctions of various lengths that can be created by reintroducing S atoms in the wire (Figure 5). The introduced S atoms create an effective barrier and disconnect the two semi-infinite vacancy wires. Such a junction is fundamentally different from break junctions because the tunneling barrier created by the S atoms possesses a detailed electronic structure which contributes to current transmission. The transmission spectrum of the junction created by 2 S atoms (Figure 4) is a convolution of a tunneling current between the atomic wires, tunneling through sulfur states introduced by the sulfur atoms constituting the junction and interferences between electron waves travelling on the atomic wire and those tunneling through the S junction. The main features of the transmission spectra can however be illustrated with a tight binding model. The atomic junction discussed above provides a platform for connecting a molecule to two atomic wires and measure the conductance of the molecule (Figure 5).

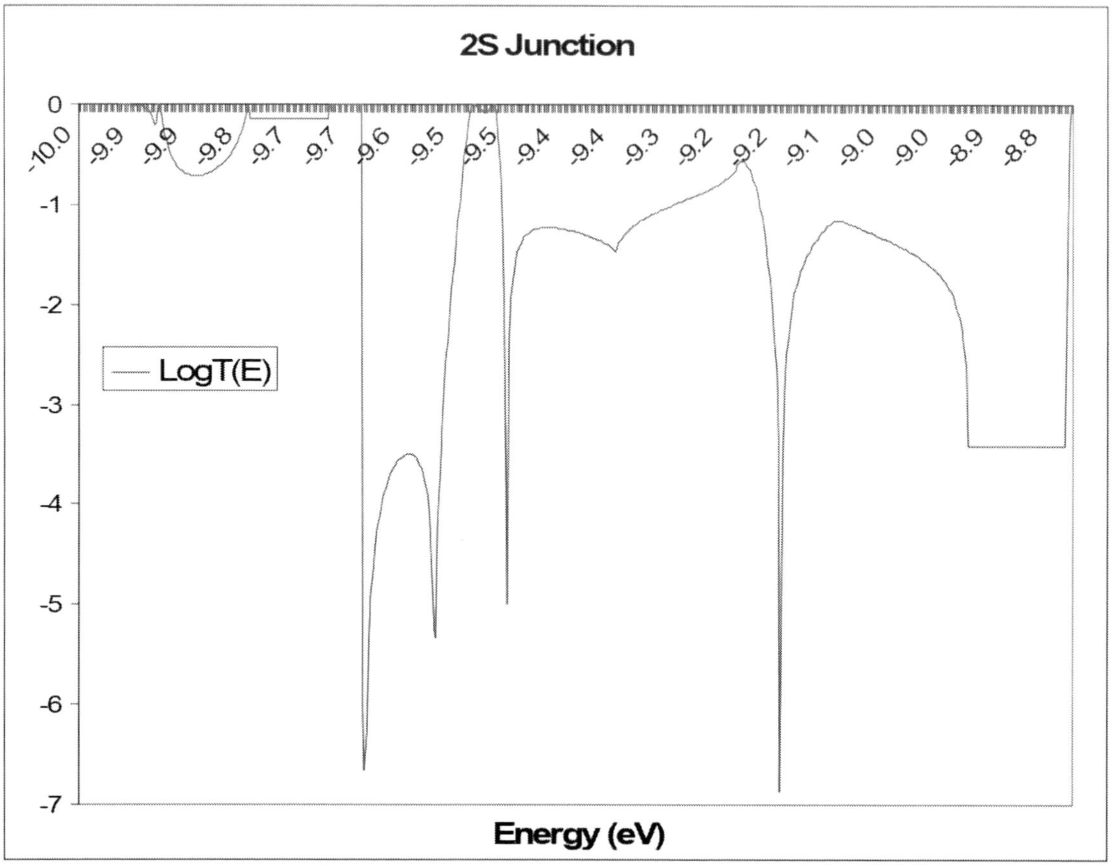


**Figure 4:** Transmission spectra for an atomic junction created by introducing 2 S atoms in an atomic wire. Sharp interference patterns are observed at well-defined energies.

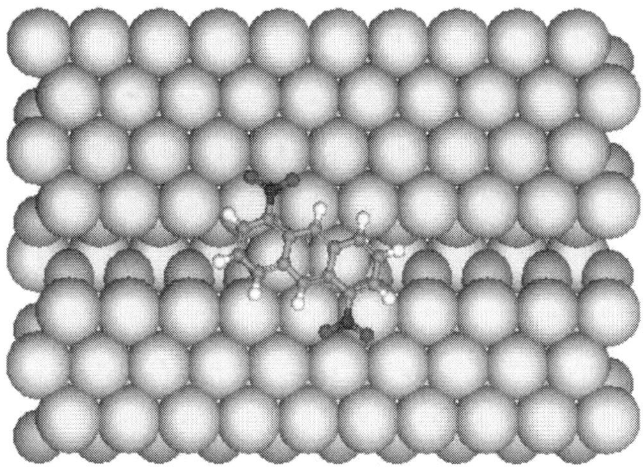

**Figure 5:** Model of an atomic junction created by insertion of 2 S atoms in an atomic wire with a single molecule logic gate inserted.

## Conclusion

In this study it is shown that a pseudo-ballistic atomic wire is created on a $MoS_2$ surface by extraction of sulfur atoms. Dangling bonds on the surface Mo atoms introduce states in the vicinity of the metal electrodes' Fermi level, which are responsible for the wire's conductance. Furthermore, it is shown that a structured atomic junction of controllable length can be constructed by reintroducing sulfur atoms into the wire's existing vacancies. The atomic wire and the junction offer a framework to study the conductance properties of single molecules.

## References


1 T.-C. Shen, C. Wang, G. C. Abeln, J. R. Tucker, J. W. Lyding, Ph. Avouris, and R. E. Walkup, Science 268, 1590 (1995)

2 Hosaka, S. et al. J. Vac. Sci Technol. B 13, 2813 (1995)

3 J. Yang, J. Deng, N. Chandrasekhar, and C. Joachim, J. Vac. Sci. Technol. B 25, 1694 (2007)

4 K.S. Yong, D.M.Otalvaro, I. Duchemim, M. Saeys, C. Joachim, Phys. Rev. B 77, 1 (2008)

5 J. P. Perdew, K. Burke, and M. Ernzerhof, Phys. Rev. Lett. 77, 3865 (1996).

6 P. Sautet and C. Joachim, Chem. Phys. Lett. 153, 511 (1988)

# 428c Atomic-Scale Analysis of the Effects of Thermal Treatment and Loading Conditions on the Mechanical Properties of Ultra-Low-Dielectric-Constant Mesoporous Amorphous Silica Films

**M. Rauf Gungor**[1], James J. Watkins[2], and Dimitrios Maroudas[1]. (1) Department of Chemical Engineering, University of Massachusetts, 686 N Pleasant Street, Amherst, MA 01003, (2) Department of Polymer Science and Engineering, University of Massachusetts, 120 Governors Dr., Amherst, MA 01003

In present generations of high-performance microelectronic devices, the constantly increasing density and larger-scale integration require closer spacing between interconnect lines and lower dielectric constant (low-k) of the dielectric material in order to reduce the capacitive coupling between the interconnect lines. Ultra-low-k materials in future-generation microelectronic technologies must also satisfy certain integration requirements, such as a minimum mechanical strength, threshold of thermal and electrical breakdown, and suitability for lithographic processing during chip manufacturing. Mechanical strength is particularly important: the dielectric material also is a structural component in the microchip that is required to ensure the structural integrity of the device under the thermomechanical loading conditions characteristic of semiconductor manufacturing processes, chip packaging, and device service. Among new materials for ultra-low-k applications, porous amorphous silicas have significant economical advantages due to their compatibility with current semiconductor manufacturing technologies.

In this presentation, we report results of molecular-dynamics (MD) simulations aiming at a fundamental understanding of the nano-scale mechanisms that control the mechanical behavior of mesoporous amorphous silica film structures and predicting the response of such structures to various mechanical loading conditions. The MD simulations employ a realistic classical potential that includes two-body and three-body interatomic interactions. The normal-density amorphous silica structures are prepared through MD starting from a crystalline beta-cristobalite solid structure and following a thermal processing sequence that includes melting, rapid quenching, and a thermal annealing schedule. We have generated "regular" mesoporous structures through the introduction of a regular array of spherical pores with nanometer-scale diameter by removal of atoms from the normal-density amorphous silica matrix and subsequent thermal annealing at the temperature of interest to ensure proper structural relaxation.

We present a systematic analysis of the mechanical response of these regular mesoporous amorphous silica structures under applied strains within the elastic limit near room temperature based on isostrain MD simulations using large-size computational supercells. The elastic moduli of the mesoporous structures are computed and their structural stability is analyzed under tensile and compressive strains as a function of density and pore diameter. We also analyze the anelastic characteristics of the mechanical behavior of these mesoporous structures based on nanosecond-scale MD simulations of dynamic deformation (i.e., constant strain rate) experiments under both tensile and compressive strains over a range of strain rates from $10^8$ to $10^{10}$ $s^{-1}$. Furthermore, we analyze the effects of thermal treatment near the glassy transition temperature of the mesoporous amorphous structures on their structural stability and mechanical strength under compressive strain.

## 428d Optimization of New Ultralow-K Materials for Vlsi Multilevel Interconnection

**Xuan Li** and James Economy. Department of Materials Science & Engineering, University of Illinois at Urbana-Champaign, 1304 W Green Street, Urbana, IL 61801

The demand for increased signal transmission speed and device density for the next generation of multilevel integrated circuits has placed stringent demands on materials performance. Currently, integration of the ultra low-k materials in dual Damascene requires chemical mechanical polishing (CMP) to planarize the copper. Unfortunately, none of the commercially proposed dielectric candidates display the desired mechanical and thermal properties for successful CMP. A new polydiacetylene thermosetting polymer (PTEB-DEB) which displays a low dielectric constant (low-k) of 2.7 was recently developed. This novel material appears to offer the only avenue for designing an ultra low k dielectric (1.85k), which can still display the desired modulus (7.7Gpa) and hardness (2.0Gpa) sufficient to withstand the process of CMP.

In this talk, we will present recent additional studies to further characterize the thermal properties of spin-on PDEB-TEB ultra-thin film. These include the coefficient of thermal expansion(CTE), biaxial thermal stress, and thermal conductivity. Thus the CTE in the perpendicular direction is 2.0*10-5K-1 and in planar direction 8.0*10-6K-1. The low CTE provides a better match to the Si substrate which also to minimize interfacial stress and greatly enhance the reliability of the microprocessors. Initial experiments with oxygen plasma etching suggest a high probability of success for achieving vertical profiles.

# Characterization and Processing/Property Relationships of Semiconducting PT and PPV Nanotubes by Template Wetting Nanofabrication

*Steven Bearden, Joseph Cannon, and Scott A. Gold*
*Institute for Micromanufacturing, Louisiana Tech University, 911 Hergot Ave., PO Box 10137, Ruston, LA 71272-0043*

**Abstract**

Nanotubules are fabricated from widely studied semiconducting poly(3-hexylthiophene) (P3HT) and a soluble poly(*p*-phenylene vinylene) (PPV) derivative using template wetting procedures. Ultraviolet-visible (UV-Vis) spectroscopy reveals a significant reduction in the bandgap in the non-crystalline PPV derivative when confined in the nanoporous structure. Analysis of space-charge-limited (SCL) currents demonstrated increases in low-field hole mobility by a factor of $10^2$. Each of these results are indicative of enhanced molecular ordering in the nanostructure, and neither result was observed in the semi-crystalline P3HT.

**Introduction**

Poor hole and electron mobilities inherent in disordered systems of conjugated, semiconducting polymers have been shown to limit efficiencies and performance in electronic devices such as polymer light-emitting diodes.[1] In particular, interchain hopping of carriers in the disordered systems found in traditional thin-film devices has been shown as the rate limiting step in carrier transport.[2] Exceptionally high hole mobilities have been observed in systems of self-ordering polymers like P3HT which produce strong interchain π-π orbital overlapping.[3] Thus, improving electronic transport in these materials has been the focus of much research, with specific interest on enhancement of order.

Several techniques including electrospinning,[4] template synthesis,[5] and template wetting[6] have all been demonstrated as viable means of producing nanoscale fibers or tubules from a wide variety of polymers, often with increased molecular order.[7-11] The template wetting approach has proven to be one of the simplest of these methods for the creation of such structures, making use of wetting phenomena to create uniform coatings of low surface energy materials, such as polymer solutions, upon high surface energy pore surfaces such as porous alumina. Steinhart *et. al.* later demonstrated curvature-directed crystallization of poly(vinylidene difluoride) by template wetting which produced a brush structure along the inner wall.[9] Such a structure should improve intermolecular π-π overlap, increasing carrier transport. Here, we examine poly[2-methoxy-5-(2-ethylhexyloxy)-1,4-phenylenevinylene] (MEH-PPV) and P3HT nanotubes fabricated using solution-based template wetting nanofabrication. Analysis of spectroscopic properties and space-charge-limited (SCL) currents are used to characterize the nanostructures.

## Experimental

***Preparation.*** MEH-PPV and regioregular P3HT were obtained from Sigma-Aldrich and used as received. Solutions of each polymer were created by dissolving in chloroform at a concentration of 2 mg/ml. Highly porous alumina membranes used as templates were 60 μm thick with pore diameters of 100 nm and 200 nm. Templates were wetted with polymer solutions by applying a droplet to the top surface of the membranes, and allowing the solvent to evaporate. Residual polymer on the template surface was subsequently removed by etching in a helium plasma. For spectroscopic analysis, templates were used directly after etching, and thin films were created by spincasting the polymer solutions onto quartz slides. Hole-only electronic devices were created by deposition of Au contacts on either side of the nanoporous membrane.

***Characterization.*** UV-Vis spectroscopy was performed using a Shimadzu UV 1650PC system in order to determine optical absorbance spectra and approximate band gaps of the material in various forms. Absorbance spectra were obtained for energies ranging from 1.5 to 3.5 eV. Current density – voltage (*J-V*) measurements were performed using a Keithley 236 source-measure unit. Devices were initially tested by repeatedly sweeping the applied bias between -100 V and 100 V in 1 V steps every 250 ms. Typically, electric fields applied to polymer thin-film devices are on the order of $10^5$ ~ $10^7$ V/cm$^2$. Here, we examine devices at significantly lower fields on the order of $10^3$ ~ $10^4$ V/cm$^2$ in order to accurately observe the SCL currents which dominate in this field regime.[1]

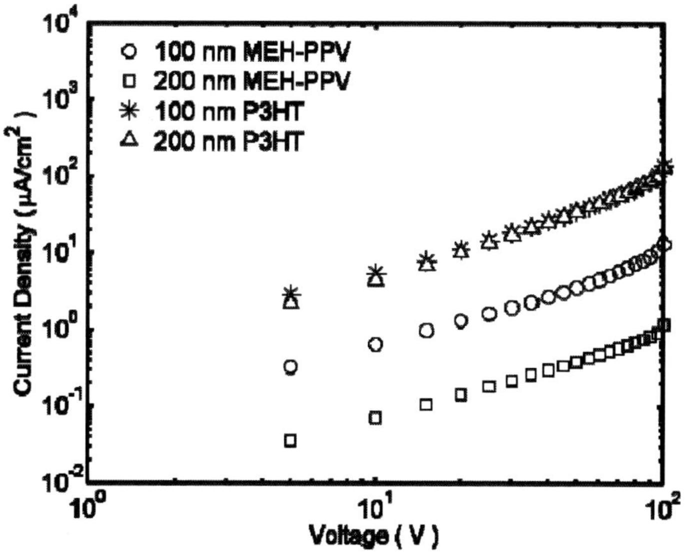


*Figure 1.* Current-voltage characteristics for MEH-PPV and P3HT. P3HT curves are indistinguishable.

## Results and Discussion

Electrical conduction was observed to be quadratic in all samples, indicative of SCL conduction. Fitting *J-V* data to the SCL model,

$$J = \frac{9}{8}\varepsilon_r \varepsilon_0 \mu_p \frac{V^2}{L^3}$$

where $V$ is the applied bias, $L$ is the device length, $\varepsilon_0\varepsilon_r$ is the permittivity of the material, and $\mu_p$ is the hole mobility.[12] Using this model, hole mobilities for 100 nm and 200 nm MEH-PPV nanotubes were calculated to be $10^{-4}$ cm²/V·s and $10^{-5}$ cm²/V·s, respectively. The mobility commonly noted in thin films of MEH-PPV is $10^{-6}$ cm²/V·s,[13] demonstrating improved mobility through progressive confinement. P3HT devices of both pore diameters produced mobilities of $10^{-2}$ cm²/V·s, in good agreement with values reported elsewhere for thin films.[3] These results are illustrated in Figure 1. The progressive increase in mobility with decreasing pore diameter in MEH-PPV suggests an enhancement of molecular ordering in the non-crystalline polymer, when compared with the thin film, given that similar increases in mobility were not observed in the semicrystalline P3HT.

These observations are further evidenced by UV-Vis spectroscopic analysis. Marked reduction in the bandgap of MEH-PPV was observed in the nanotube samples when compared with similar measurements on thin-films. The bandgap reduction was estimated approximately 0.6 eV in MEH-PPV, and the entire observed absorbance peak was correspondingly shifted. Little or no variation in the bandgap was observed in P3HT, suggesting, again suggesting little difference in the molecular ordering of P3HT nanotubes when compared to the thin-film. This short analysis suggests a more ordered structure in the template than is present in the film.

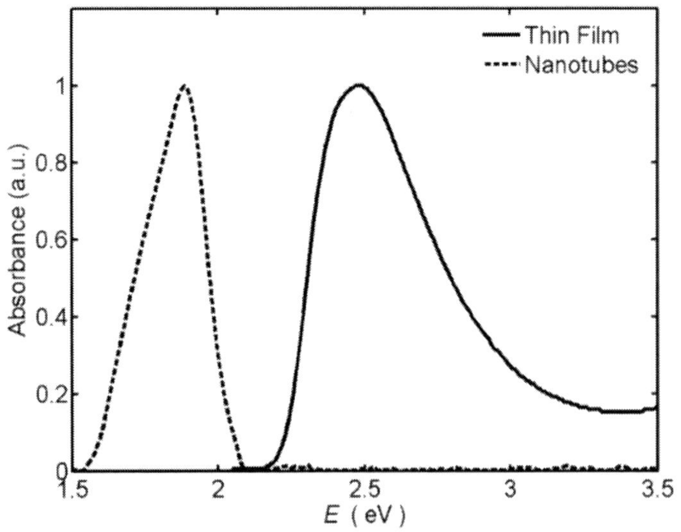


**Figure 2.** UV-Vis spectroscopy demonstrates significant red shifting of the absorbance in MEH-PPV nanotubes, and a corresponding reduction of the bandgap.

## Conclusions

We conclude that template wetting of conjugated polymers can significantly alter their electronic properties. The comparison of observations in materials known to be semicrystalline (P3HT) and amorphous (MEH-PPV) suggests that enhanced ordering of amorphous polymers is possible in template wetted nanostructures, suggesting improved interchain transport in the material. The observation that mobility in MEH-PPV is increased as pore size is reduced further indicates that these improvements are confinement induced.

## References


(1) Blom, P. W. M.; de Jong, M. J. M.; Vleggaar, J. J. M. *Applied Physics Letters* **1996**, *68*, 3308.

(2) Sirringhaus, H.; Wilson, R. J.; Friend, R. H.; Inbasekaran, M.; Wu, W.; Woo, E. P.; Grell, M.; Bradley, D. D. C. *Applied Physics Letters* **2000**, *77*, 406-408.

(3) Sirringhaus, H.; Brown, P. J.; Friend, R. H.; Nielsen, M. M.; Bechgaard, K.; Langeveld-Voss, B. M. W.; Spiering, A. J. H.; Janssen, R. A. J.; Meijer, E. W.; Herwig, P.; de Leeuw, D. M. *Nature* **1999**, *401*, 685-688.

(4) Reneker, D. H.; Chun, I. *Nanotechnology* **1996**, *7*, 216-223.

(5) Martin, C. R. *Science* **1994**, *266*, 1961-1966.

(6) Steinhart, M.; Wendorff, J. H.; Greiner, A.; Wehrspohn, R. B.; Nielsch, K.; Schilling, J.; Choi, J.; Gosele, U. *Science* **2002**, *296*, 1997.

(7) Liang, W.; Martin, C. R. *J. Am. Chem. Soc.* **1990**, *112*, 9666-9668.

(8) Parthasarathy, R. V.; Martin, C. R. *Chem. Mater.* **1994**, *6*, 1627-1632.

(9) Steinhart, M.; Senz, S.; Wehrspohn, R. B.; Gosele, U.; Wendorff, J. H. *Macromolecules* **2003**, *36*, 3646-3651.

(10) Li, D.; Babel, A.; Jenekhe, S. A.; Xia, Y. *Advanced Materials* **2004**, *16*, 2062-2066.

(11) Babel, A.; Li, D.; Xia, Y.; Jenekhe, S. A. *Macromolecules* **2005**, *38*, 4705-4711.

(12) Shockley, W.; Prim, R. C. *Physical Review* **1953**, *90*, 753.

(13) Campbell, I. H.; Smith, D. L.; Neef, C. J.; Ferraris, J. P. *Applied Physics Letters* **1999**, *74*, 2809.

## 428f Ultra-High Strength Due to Dislocation Depletion In Small-Volume Structures of Face-Centered Cubic Metals

**Kedarnath Kolluri**, M. Rauf Gungor, and Dimitrios Maroudas. Department of Chemical Engineering, University of Massachusetts, 686 N Pleasant Street, Amherst, MA 01003

Metallic small-volume structures, such as nanometer-scale-thick metallic films, are increasingly used in modern technologies, from microelectronics to various areas of nanofabrication. The mechanical behavior of small-volume structures of metals is very different from that of their bulk counterparts. Improvement of the reliability, functionality, and performance of nano-scale devices requires a fundamental understanding of the atomistic mechanisms that govern the response of strained small-volume structures in order to establish rigorous links between their structural evolution and their mechanical behavior.

In this presentation, we report an atomic-scale analysis of the mechanisms of dislocation starvation in small-volume structures of face-centered cubic (FCC) metals, focusing on free-standing ultra-thin copper films that are subjected to biaxial tensile strain. Our study is based on large-scale molecular-dynamics simulations of dynamic deformation experiments (i.e., at constant strain rate and temperature), using an embedded-atom-method parameterization to describe the interatomic interactions. The initial dislocation density of the thin films is on the order of $10^{16}$ $m^{-2}$. Our analysis of the films' mechanical response to the applied tensile strain reveals three stages of deformation. During the first stage, most of the dislocations in the material are unpinned; these dislocations glide under the application of biaxial strain in such a direction that they unzip the stacking faults they bound and the dislocation networks that they are a part of. With the stacking fault area reduced, there are fewer barriers to dislocation glide in the thin films. Consequently, the remaining dislocations glide faster and farther. During the second stage, gliding dislocations interact with the stacking faults. These interactions lead to dislocation dissociation and aid in dislocation annihilation. The obstacle stacking faults are unzipped almost always during these interactions. We have identified three classes of dislocation-stacking fault interactions, where the stacking faults act as barriers to dislocation glide and as sources for dislocation cross-slip. As the dislocation density decreases, the thin film's strength increases significantly and the mechanical behavior of the material is observed to be closer to that of a perfectly elastic solid. During the third stage of film deformation, continued application of strain leads to increase in the film's stress and, eventually, to nucleation of new dislocations in the thin film. Dislocation nucleation and depletion in the thin film continue in cycles until the failure of the film.

## 428g Surface Effects on Oxygen Defects In TiO2 for Nanoelectronic Materials

**Alice G. Hollister**, Chemical and Biomolecular Engineering, University of Illinois, Urbana - Champaign, 105 Roger Adams Lab, 600 S. Matthews, Urbana, IL 61801 and Edmund G. Seebauer, Department of Chemical & Biomolecular Engineering, University of Illinois at Urbana-Champaign, 600 South Mathews Avenue, Urbana, IL 61820.

The technologically useful properties of a solid often depend upon the types and concentrations of the defects it contains, particularly in semiconductors. Native point defects influence the effectiveness of nanoelectronic materials for devices in applications that include high voltage electronics, light emitting diodes, solid-state electrolyte sensors, and solar power generators. Recent work in our laboratory has shown that semiconductor surfaces can serve as especially efficient pathways for the generation and annihilation of point defects. At the nanoscale, such surface pathways become particularly important due to the large ratio of surface area to volume. A significant degree of control over the surface pathways has been demonstrated via the variation of surface bonding state: for example, many dangling bonds versus only a few.

We have employed measurements of isotopic self-diffusion to indirectly monitor the concentrations and behavior of native point defects in metal oxides such as titania that find use as nano-electronic materials. In particular, experiments employing secondary ion mass spectroscopy have yielded concentration profiles of the defect-mediated self-diffusion of isotopic oxygen into titania. Surface chemical bonding state greatly influences the observed diffusion rates, which increase when the surface is held essentially atomically clean in high vacuum. Such surfaces appear to promote the formation of a highly mobile defect within titania, although several distinct mechanisms are possible. We have developed three potential models for the mechanism. Simulations based on continuum equations for each of the three defect mechanisms have permitted the application of formal model discrimination methods to identify the most likely mechanism.

# 430a Supramolecular Templating of Nanoporous Catalysts

**Jackie Y. Ying**, Institute of Bioengineering and Nanotechnology, 31 Biopolis Way, The Nanos, #04-01, Singapore, 138669, Singapore

Nanostructured materials are of interest for a variety of applications. This talk describes the synthesis of novel nanoporous materials with tailored oxidation states, coordination chemistry and electronic structure. We have found that sol-gel processing can be combined with supramolecular templating agents in deriving well-defined porous structures of a variety of metal oxides. Through ligand interaction between metal alkoxide precursors and surfactant head groups, we have successfully derived a family of stable mesoporous and microporous transition metal oxide molecular sieves (termed TMS) with hexagonally-packed cylindrical pore structures. The compositional flexibility and pore size tailoring of the TMS system open new possibilities for catalytic applications beyond the silicate-based zeolitic materials or mesoporous MCM-41 structures. Besides surfactant templates, we have also utilized triblock copolymers for the synthesis of mesoporous tungstated metal oxides with nanocrystalline colloidal building blocks for solid acid catalysis. Lastly, the combination of triblock copolymer and oil has been introduced to derive siliceous mesocellular foams with open, ultralarge mesopores. These unique materials have been used to immobilize various organometallic, organic and enzyme catalysts for a broad range of reactions. These heterogenized catalysts provide for excellent activity, enantioselectivity and recyclability.

Web Page: www.ibn.a-star.edu.sg

## 430b Peptide Assembled Nanoparticles with Enzyme-Like Activity

**Joseph M. Slocik**, Air Force Research Laboratory, Nanostructured and Biological Materials Branch, Materials and Manufacturing Directorate, Wright-Patterson-AFB, OH 45433

There is significant interest in using enzymes for a variety of applications such as in biotransformations and as fuel cell components. In general, enzymes offer excellent activity and substrate specificity; but unfortunately, require cofactors, buffered solutions, denature at high temperatures, and suffer from instability and short shelf-lifes. Nevertheless, the ability to harness this extraordinary level of activity has inspired efforts in genetic engineering of enzymes and production of inorganic enzyme mimics. In the following, we have mimicked the activity of the nitrate reductase enzyme by using peptide assembled CdS-Pt nanoparticles. Notably, these nanoparticles exhibit a 23-fold increase in activity as compared to nitrate reductase, enhanced thermal stability, and increased activity at 75 C.

# 430c Development of Functional Nano-Biocatalysts through Novel Nanostructured Materials

**Jinwoo Lee**, Department of Chemical Engineering and School of Environmental Science & Engineering, Pohang University of Science and Engineering, xxx xxx xxx, Pohang, xxx, South Korea

Nanostructured materials have drawn much attention due to their interesting physical property and high surface area. We have developed multifunctional nano-biocatalyst using various designed nanostrucutred materials. Hierarchically ordered mesocellular mesoporous silica materials (HMMS) were synthesized using a single structure-directing agent. To improve the retention of enzymes in HMMS, we adsorbed enzymes, and then employed crosslinking using glutaraldehyde (GA). The resulting crosslinked enzyme aggregates (CLEAs) show an impressive stability with extremely high enzyme loadings. Magnetically separable and highly active enzyme system was developed using nanostrucutred magnetic materials. Magnetically switchable bioelectrocatalytic system was constructed using magnetic mesocellular carbon foam (Mag-MCF-C) with ultra-large mesopores.

## 430d Lipase Nanogel for Nonaqueous Biocatalysis

**Ge Jun**, Lu Diannan, Wang Jun, and **Liu Zheng**. Chemical Enigeering, Tsinghua University, Beijing, China

Lipase from Candida rugosa was prepared as nanogels with diameters around 30 nm by in situ polymerization using acrylamide as monomers. The lipase nanogel preserves over 90% of the initial activity while showed an essentially identical catalytic behavior to its native counterpart. The lipase nanogel remains its initial activity after 10 hour incubation at 50 °C while its native counterpart lost 50% activity within 50 min. The encapsulation also let to a significant enhancement in the tolerance to anhydrous dimethylsulfoxide (DMSO), dimethylformamide (DMF) and methanol.

Also included in the presentation was the assembly of monomer around enzyme as the essential step in the aqueous in situ synthesis of the lipase nanogel, as illustrated by molecular dynamic simulation and fluorescence resonance energy transfer spectrum using lipase/acrylamide as a model system. Application of the lipase was exemplified by the production of biodiesel, the esterification of ibuprofen and the synthesis of polysaccharide esters. It is concluded from above simulation and experiments that the nanogel provides a robust enzyme module for a full display of the biological power of enzymes.

# 430e Immobilization of Candida Antarctica Lipase B on Fumed Silica Nanoparticles

**Juan C. Cruz**, Peter H. Pfromm, and Mary E. Rezac. Chemical Engineering, Kansas State University, Manhattan, KS 66506

The development of highly active enzyme-based biocatalysts for synthetic applications in organic solvents represents a challenge for biomolecular researchers and engineers. Enzymatic catalysis in non-aqueous media can be useful when substrates or products have limited or poor solubility in water or if water is a reaction product. Improved thermal stability of enzymes in non-aqueous media and some advantageous effects on enantioselectivity have also been reported. It is however imperative to find inexpensive and simple immobilization techniques since enzymes are generally not soluble in solvents that do not denature them. The immobilized enzyme preparations needed for use in solvents should be easy to prepare, based on widely available materials, and should make excellent use of the catalytic competency of the enzyme. Our strategy comprises the immobilization of enzymes on fumed silica nanoparticles by lyophilization. The catalytic activity of Candida antarctica Lipase B (CALB) in hexane for a model esterification is comparable with that of the commercial preparation Novozym 435®. The immobilization is basically achieved in two sequential stages: (i) physical adsorption of the enzyme from aqueous suspension on fumed silica nanoparticles, and (ii) lyophilization. The kinetics of enzyme adsorption as a function of the fumed silica charge and pH of the aqueous phase before lyophilization will be discussed. The catalytic activity was assessed from initial reaction rate experiments and the physical enzyme distribution on the support was investigated by confocal microscopy imaging. Additionally, we explored the thermal and storage stability of our preparations. In light of our previous work with subtilisin Carlsberg our results with CALB are somewhat surprising since a counterintuitive loss of activity at very high fumed silica concentrations was detected. We hypothesize based on our data and recent models put forth by the research group of Professor Gross (Polytechnic University) that this apparent optimum enzyme-to-support ratio may be due to the unique protein-protein and surface-protein interactions during the adsorption of CALB on fumed silica.

# 430f Modeling the Effect of a Crowding Agent on the Activity of Enzyme-Au Bioconjugates

**Francisco G. Vital-Lopez**[1], Jacqueline Keighron[2], Christine D. Keating[2], Costas D. Maranas[1], and Antonios Armaou[1]. (1) Department of Chemical Engineering, The Pennsylvania State University, University Park, PA 16802, (2) Department of Chemistry, The Pennsylvania State University, University Park, PA 16802

Enzymes conjugated with nanoparticles (e.g., Au) are increasingly important in diverse areas including biosensing, nanocircuitry design and biocatalysis. Research on synthesis and characterization of bioconjugates has been prolific in recent years, especially for single enzyme-nanoparticle conjugates.

The construction of bioconjugates of two or more types of enzymes requires the consideration of additional factors such as the relative location of each type of enzyme (i.e., binding all types of enzymes in the same nanoparticle (multicomponent conjugates) or different types of enzymes on separate nanoparticles (single component conjugates)). In fact, experimental results show that the use of even the same crowding agent (PEG) has a differing effect on the sequential activity of two enzymes carrying out consecutive reactions (i.e., malate dehydrogenase and citrate synthase) in two different bioconjugate constructs. In one case (both enzymes bound on the same nanoparticle) the sequential activity increases almost linearly with PEG concentration whereas in the other case (each enzyme type is located on separate nanoparticles) the sequential activity remains almost constant.

In this work, we present a mathematical model to elucidate the mechanism behind these experimental observations. The model is based on a reaction-diffusion mechanism and an experimentally determined description of the effect of PEG on the kinetics of the considered enzymes. We observe a quantitatively good agreement between the proposed model and experimental results. The proposed modeling framework can be readily extended to systems involving a relatively large number of enzymes and it can offer a rational guide for designing synthetic metabolic pathways using enzyme-nanoparticle bioconjugates.

## 431a Self-Interaction Nanoparticle Spectroscopy: A Nanoparticle-Based Protein Interaction Assay

Aditya Bengali, Moumita Bhattacharya, and **Peter M. Tessier**. Dept of Chemical & Biological Engineering, Rensselaer Polytechnic Institute, Troy, NY 12180

Weak protein interactions have broad relevance to the fields of biology and biotechnology. For example, these interactions govern the assembly of proteins into crystals for structural analysis, as well as both the aggregation of therapeutic proteins in vitro and disease-related proteins in vivo. Despite the importance of these interactions, they are notoriously difficult to characterize, especially in a high-throughput manner. We are developing a nanoparticle-based platform, namely self-interaction nanoparticle spectroscopy (SINS), to detect protein self-interactions in a parallel manner (Tessier et al., J Am Chem Soc, 2008). In this presentation we will discuss different approaches to modifying gold nanoparticles for efficient, stable immobilization of proteins with diverse physical properties. Moreover, we will demonstrate that protein self-interactions can be characterized using SINS in good agreement with measurements of osmotic second virial coefficients. Finally, we will discuss future directions for integrating SINS into the design and selection of aggregation-resistant protein therapeutics.

Web Page: www.rpi.edu/~tessip

# Silver-copper alloy nanoparticles for metal enhanced luminescence

*Sanchari Chowdhury* [1], *Venkat R. Bhethanabotla* [1], *Rajan Sen* [2]

Department of Chemical and Biomedical Engineering [1], Department of Civil and Environmental Engineering [2]

University of South Florida, Tampa, Fl 33620-5350

## Abstract

Silver-copper alloy nanoparticles were explored for metal enhanced luminescence (MEL) applications in this study. Surface plasmon resonance (SPR) wavelengths of these Ag-Cu nanoparticles were tuned in the visible and near infrared regions by changing annealing temperature. We observed strong emission enhancement of luminophores (141.48 $\pm$ 19.20 times for Alexa Fluor 488 and 23.91 $\pm$ 12.37 times for Alexa Fluor 594) when the SPR spectrum of Ag-Cu nanoparticles was tuned to result in maximum spectral overlap of the emission and absorption spectra of the luminophores with the SPR spectrum of the nanoparticles. This finding opens new avenues for utilization of metal alloy nanoparticles in MEL-based applications.

## Introduction

Luminescence based measurements and devices are currently widely used methods in different fields such as biology, chemistry, materials science and medicine. Strong luminescence intensity is one of the most important desired properties for luminophores for these applications. It is possible to design and synthesis luminophores with desired spectral properties. The absorption and emission peaks of any luminophores can be predicted by analogy of known luminophores. But it is difficult to design luminophores with desired luminescence intensity. At the vicinity of conducting metallic nanoparticles such as those of silver and gold, the emission intensity of luminophores is known to be significantly influenced[1-4]. Using nanoparticles platform, it can be possible to increase the quantum yield of weakly luminescent probes by increasing it's emission efficiency by modifying radiative decay rate or by coupling the emission with surface plasmon resonance. It can also improve the intensity of luminophores that already have near unit quantum yield by enhancing their absorption by increasing the local field. Distance between the fluorophore molecules and nanostructured platform is also an important parameter in fluorescence enhancement[5]. Surface plasmon resonance (SPR) wavelength, one of the most important properties of nanostructures, dictates the choice of materials to be used for luminescence enhancement. Tam *et al.*[6] found that the enhancement is optimal when the plasmon resonance of the nanoparticles is tuned to the emission wavelength of the low quantum yield luminophores. Recently, some theoretical and experimental studies have suggested that luminescence enhancement is largest when emission wavelength is slightly red-shifted from the plasmon resonance[7,8]. One can expect that by tuning the position of the SPR peak of the nanoparticles over a wide range of wavelengths, metal enhanced luminescence (MEL) can be extended for a wide range of luminophores. Metal alloys offer additional degrees of freedom for tuning their optical properties by altering atomic composition and atomic arrangement. This motivates us to study alloy nanostructured platforms for MEL. Due to their interesting optical properties, we have chosen to study silver-copper alloy nanoparticles[9] for MEL application.

This work is focused on the use of alloy nanoparticles in the MEL. We demonstrate that SPR spectra of alloy nanoparticles can easily be tuned by manipulating only one experimental condition to result in maximum spectral overlap of the emission and absorption spectra of the luminophores with the SPR spectrum of the nanoparticles. We observed enhanced fluorescence emission from two thiol-reactive dyes, Alexa Fluor 594 and Alexa Fluor 488 (obtained from Molecular Probes, Invitrogen, Portland, OR), at the proximity of Ag-Cu alloy nanoparticles.

## Experimental procedure

In this study, Ag or Ag-Cu nanoparticles were deposited on 22×22 mm glass cover slips (Fisher finest cover glass, thickness approximately 140 microns) by using DC magnetron sputtering (Plasma Sciences CRC-100 Sputter Tool). Before the depositions, the cover slips were cleaned by air plasma (Harrick PDC-32G) for 10 minutes at 6.8 watts power setting. During deposition, the background pressure was of the order of $10^{-6}$ Torr, the Ar pressure was 5 mTorr and the current was 25 mA. An Ag target was utilized to deposit the Ag nanoparticles and a Cu foil attached on the Ag target was utilized for the Ag-Cu nanoparticle deposition. Varying the ratio of the surface area of Ag to Cu exposed for sputtering allowed for changing the composition of the Ag-Cu alloy nanoparticles. Surface morphology of the nanostructures was observed and characterized by transmission electron microscopy (FEI Tecnai F20 S-Twin TEM). An electrical furnace (Lindberg, Blue M) was used for annealing of the Ag-Cu nanoparticles. Annealing temperature ranged from 298 K to 523 K and the annealing time was 5 minutes. Annealing was done in vacuum (30 inch Hg vacuum) to minimize oxidation of the nanoparticles. An UV-vis spectrometer (JASCO, V-530) was used for measuring the light absorption spectra attributed to the SPR of these nanoparticles.

Mouse Immunoglobulin G (IgG), labeled with luminophores Alexa Fluor 488 and Alexa Fluor 594 was coated on samples following known methods[2]. The Leica DMI 4000b inverted fluorescence microscope equipped with Leica DFC340 FX CCD camera was utilized for MEL measurements. We obtained fluorescence intensities for each sample by analyzing a 1.64 mm × 2.19 mm image-section of each substrate.

## Result and Discussion

Transmission electron microscopy of the Ag-Cu alloy nanoparticles indicated the average size to be 7.03 nm ± 1.97 nm (derived from a population of 100 particles) (Figure 1). The absorption spectra attributed to SPR of Ag-Cu nanoparticles show a single peak in the visible range. With increasing copper percentage, this SPR peak shifts to higher wavelengths and becomes broader (data are not shown here). This result confirms that the nanoparticles are bimetallic form of silver and copper and not a mixture of silver nanoparticles and copper nanoparticles[10]. The SPR peak of Ag-Cu alloy nanoparticles blue shifts upon increasing the annealing temperature from 298 K to 523 K (Figure 2 ). With increase in annealing temperature, Cu atoms surface-segregate, thereby increasing the concentration of Ag in the nanoparticle core. As a result, the conductivity of the nanoparticle increases[9,11] and the SPR peak shifts to lower wavelengths, as well as becomes narrower.

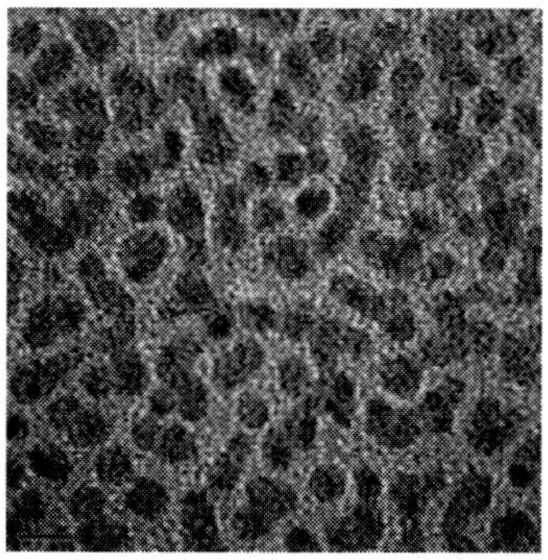

Figure 1: High resolution transmission electron micrograph of Ag-Cu alloy nanoparticles.

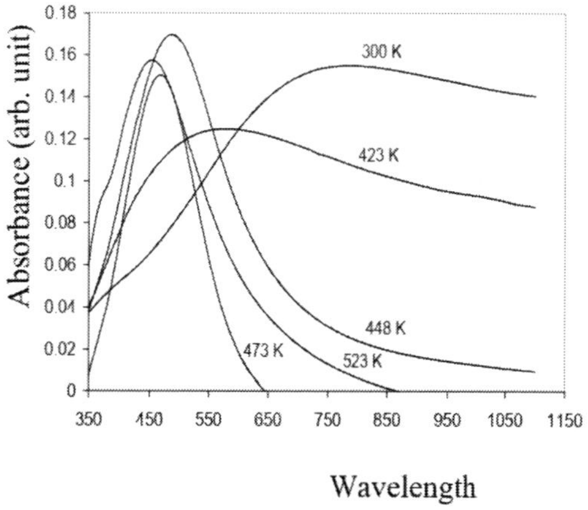


Figure 2. Absorption spectra of Annealed Ag-Cu nanoparticles

Luminescence of both Alexa Fluor 594 and Alexa Fluor 488 was observed to be enhanced significantly at the vicinity of these Ag-Cu nanoparticles. Enhancement ratio was calculated by comparing luminescence intensity of the sample with the luminescence intensity of the luminophore coated on a glass cover slip. As shown in Figure 3, the SPR spectrum of the 448 K annealed Ag-Cu nanoparticles nicely overlaps both the excitation and emission spectra of Alexa Fluor 488. This annealed Ag-Cu nanoparticle platform results in very strong enhancement ($141.48 \pm 19.20$ times) of luminescence of Alexa Fluor 488. The Ag-Cu nanoparticles at 298 K, which show less spectral overlap, also result in large enhancement ($100.87 \pm 10.21$ times). The lowest enhancement ($50.22 \pm 10.80$) was observed at the proximity of pure Ag nanoparticles (deposited at the same conditions as Ag-Cu nanoparticles), where the spectral overlap is least.

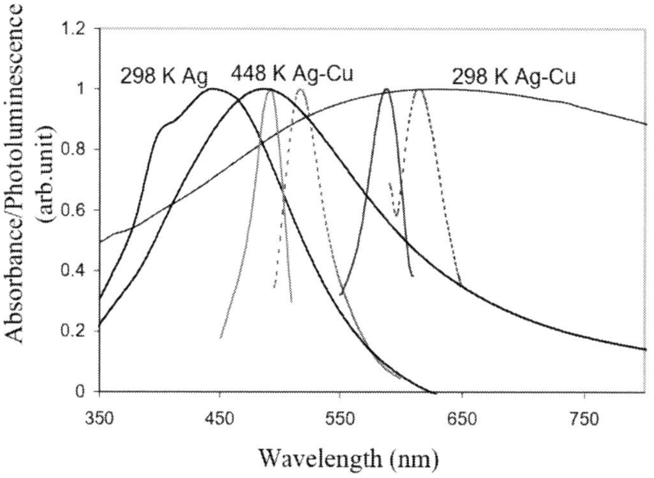


**Figure 3.** SPR spectrum of Ag-Cu nanoparticles and Ag nanoparticles used for MEL experiments and absorption and emission spectrum of Alexa Fluor 594 (----- emission spectrum, ‾‾‾‾ excitation spectrum) and Alexa Fluor 488 (----- emission spectrum, ____ excitation spectrum). All spectra were normalized for comparison.

The effect of spectral overlap on luminescence enhancement is also pronounced for Alexafluor 594. We found $23.91 \pm 12.37$ times enhancement of emission from Alexa Fluor 594 at the proximity of room temperature grown Ag-Cu nanoparticles. On the other hand, both pure Ag nanoparticles and the 448 K annealed Ag-Cu nanoparticles grown at similar conditions result in lower enhancement ($9.74 \pm 0.9$ times for 448 K annealed Ag-Cu nanoparticles and $6.89 \pm 3.06$

times for Ag particles) because of less spectral overlap. Luminescence enhancement is found to be a significant function of spectral overlap in this study.

## Conclusion

SPR wavelengths of these Ag-Cu nanoparticles were tuned in the visible and near infrared region by changing only one experimental condition, annealing temperature. We observed strong emission enhancement of luminophores ($141.48 \pm 19.20$ times for Alexa Fluor 488 and $23.91 \pm 12.37$ times for Alexa Fluor 594) at the vicinity of Ag-Cu nanoparticles when SPR spectrum was tuned to produce maximum spectral overlap. This study suggests that as SPR spectrum of Ag-Cu alloy nanoparticles can easily be tailored, this platform can be effectively used to enhance luminescence of different luminophores. This finding opens new avenue for the utilization of metal alloy nanoparticles in MEL applications.

## Acknowledgement

This study was funded by a grant from the National Science Foundation (No. CMS-409401). This funding is gratefully acknowledged.

## References

(1) Sokolov, K.; Chumanov, G.; Cotton, T. M. *Anal. Chem.* **1998**, *70*, 3898.

(2) Matveeva, E.; Gryczynski, Z.; Malicka, J.; Gryczynski, I.; Lakowicz, J. R. *Anal. Biochem.* **2004**, *334*, 303.

(3) Lakowicz, J. R.; Shen, Y.; D'Auria, S.; Malicka, J.; Fang, J.; Gryczynski, Z.; Gryczynski, I. *Anal. Biochem.* **2002**, *301*, 261.

(4) Kümmerlen, J.; Leitner, A.; A Brunner, H.; A Aussenegg, F. R.; A Wokaun, A. *Mol. Phys.* **1993**, *80*, 1031

(5) Ray, K.; Badugu, R.; Lakowicz, J. R. *Langmuir* **2006**, *22*, 8374.

(6) Tam, F.; Goodrich, G. P.; Johnson, B. R.; Halas, N. J. *Nano Lett.* **2007**, *7*, 496.

(7) Chen, Y.; Munechika, K.; Ginger, D. S. *Nano Lett.* **2007**, *7*, 690.

(8) Thomas, M.; Greffet, J.-J.; Carminati, R. *Appl. Phys. Lett.* **2004**, *85*, 3863.

(9) Hirai, M.; Kumar, A. *J. Appl. Phys.* **2006**, *100*, 0143091.

(10) Zhang, J.; Liu, H.; Wang, Z.; Ming, N. *J. Solid State Chem.* **2007**, *180*, 1291.

(11) Hirai, M.; Kumar, A. *Journal of Electronic Materials* **2007**, *36*, 1574.

## 431c Molecular Simulation Studies on Trapping Ions by a Nanoscale Paul Trap

**Xiongce Zhao** and Predrag S. Krstic. Center for Nanophase Materials Sciences, Oak Ridge National Laboratory, Oak Ridge, TN 37831

We found by molecular dynamics simulations that a low energy ion can be trapped effectively in a nanoscale Paul trap in both vacuum and aqueous environment when appropriate AC/DC electric fields are applied to the system. Using the negatively charged chlorine ion as an example, we show that the trapped ion oscillates around the center of the nanotrap with the amplitude dependent on the parameters of the system and applied voltages. Successful trapping of the ion within nanoseconds requires electric bias of GHz frequency, in the range of hundreds of mV. The oscillations are damped in the aqueous environment, but polarization of water molecules requires application of higher voltage biases to reach improved stability of the trapping. Application of a supplemental DC driving field along the trap axis can effectively drive the ion off the trap center and out of the trap, opening a possibility of studying DNA and other charged molecules using embedded probes while achieving a full control of their translocation and localization in the trap.

## 431d Probing Nanoscopic Origins of Adsorbate-Driven Liquid Crystal Ordering Transitions Using Localized Surface Plasmon Resonance of Gold Nanodots

**Gary M. Koenig Jr.** and Nicholas L. Abbott. Department of Chemical and Biological Engineering, University of Wisconsin-Madison, 1415 Engineering Dr., Madison, WI 53706-1691

This presentation will report that localized surface plasmon resonances (LSPRs) of gold nanodots immersed under liquid crystals (LCs) can be used to characterize adsorbate-induced ordering transitions of LCs. The adsorbate-induced nanoscopic changes in ordering of the LCs, as measured by LSPR, will be shown to give rise to macroscopic ordering transitions of the LCs that can be observed by polarized light microscopy. The results to be described in this presentation are significant because they demonstrate that (i) LCs can be used to enhance the sensitivity of LSPR-based detection of binding events, and (ii) that LSPR measurements using gold nanodots provide a means to characterize the nanoscopic origins of macroscopic, adsorbate-induced LC ordering transitions.

# 431e DNA Hybridization Detection Using Zinc Selenide Nanocrystals as Active Sensors

**Jun Wang**[1], Tracy Heckler[1], Bing C. Mei[1], Pedro Lei[2], Stelios T. Andreadis[2], and T. J. Mountziaris[1]. (1) Chemical Engineering, University of Massachusetts, Amherst, MA 01003, (2) Chemical and Biological Engineering, SUNY at Buffalo, Buffalo, NY 14260

Fluorescent labeling of biological molecules is used widely for analytical purposes in biotechnology and bioengineering. It typically involves the use of an organic dye linked to a moiety that selectively bonds a particular biological molecule, allowing the detection of the latter by the fluorescence of the dye. Typical organic fluorescent dyes have narrow excitation spectra and broad emission spectra that exhibit red tailing.

Semiconductor nanocrystals or quantum dots have emerged as a new class of fluorescent markers with distinct advantages over the traditional organic dyes [1-4]. Their attractive properties include a narrow, symmetric, and strong emission that is size-tunable, continuous excitation by any wavelength smaller than the emission wavelength, resistance to photobleaching, as well as excellent optical and chemical stability that allows their use in lengthy experiments, both in vitro and in vivo. The ability to synthesize different populations of quantum dots with narrow emission spectra permits multiplexing, a property that is very important for simultaneous detection of several analytes, that would be very tedious and expensive if done sequentially.

In addition to their use as multi-color "passive" tags for biomolecules, quantum dots have significant potential as "active" sensors, because their optical spectra change when the detection molecules attached to their surface hydridize with complementary ones in solution. The focus of this work is the development of new biological detection and DNA analysis schemes by using the changes in the optical spectra of ZnSe quantum dots.

We have synthesized highly-fluorescent ZnSe quantum dots, modified their surface with functional molecules to render them water-dispersible, and conjugated them with oligonucleotides. Hybridization of complementary oligonucleotides conjugated with ZnSe quantum dots results in a significant increase in the fluorescence intensity of the quantum dots and a measurable red shift in the emission wavelength.

A series of experiments was performed in which DNA hybridization was studied using ZnSe quantum dots conjugated with oligonucleotides as active sensors and free oligonucleotides in solution as targets. An increase in the emission intensity of the quantum dots was observed upon hybridization of the tagged oligonucleotides with free oligonucleotides. A dependence of the detected fluorescence emission intensity on the length and location of the hybridized part was detected, indicating that spectral changes can be used to identify the DNA stuctures attached to the quantum dots. Ongoing experiments are focusing on determining the sensitivity of this techique for detecting DNA mutations. Applications of these phenomena in the development of novel DNA analysis strategies will be discussed.

References

1. M. Bruchez, M. Moronne, P. Gin, S. Weiss and A.P. Alivisatos, Science 281, 2013 (1998).

2. W.C.W. Chan and S. Nie, Science 281, 2016 (1998).

3. M.-Y. Han, X. Gao, J. Z. Su, and S. Nie, Nature Biotechnology 19, 631 (2001).

4. W. C. W. Chan, D. J. Maxwell, X. Gao, R. E. Bailey, M. Han and S. Nie, Current Opinion in Biotechnology, 13, 40 (2002).

## 461a Award Submission: Engineering Mucus Penetrating Particles for Transmucosal Delivery

**Samuel K. Lai**[1], Ying-Ying Wang[2], Ming Yang[2], Amanda Pace[1], Benjamin C. Tang[1], Richard Cone[3], and Justin Hanes[1]. (1) Department of Chemical and Biomolecular Engineering, The Johns Hopkins University, 221 Maryland Hall, 3400 N. Charles Street, Baltimore, MD 21218, (2) Department of Biomedical Engineering, The Johns Hopkins University, 221 Maryland Hall, 3400 N. Charles Street, Baltimore, MD 21218, (3) Biophysics, Johns Hopkins University, 221 Maryland Hall, 3400 N. Charles St, Baltimore, MD 21218

Highly viscoelastic human mucus layers protect the lungs, GI tract, reproductive tract, eye and more from infectious pathogens and foreign particulates. The human mucus barrier was recently reported as impermeable to polymer nanoparticles as small as 59 nm. Particles that cannot move through the outer layers of mucus are cleared from the mucosal tissue within seconds to at most a few hours. This reality has strongly limited development of long-lasting controlled release systems to mucus-covered tissues, and likely has played a major role in thwarting gene therapy of mucosal tissues as well.

We hypothesized that adhesion to mucus was a critical rate-limiting barrier to nanoparticle transport through mucus layers. We sought to mimic the hydrophilic and net-neutral surface properties of viruses capable of rapidly moving through human mucus. Our initial search for a candidate material that could endow these surface properties on synthetic particles led us to poly(ethylene-glycol), or PEG. Paradoxically, PEG had a considerable history of use as a muco-adhesive.

Here, we describe our recent discovery that coating synthetic nanoparticles with high densities of low molecular weight PEG allows particles with sizes of at least 500 nm in diameter to rapidly transport through undiluted human mucus nearly as fast as they move through pure water. In contrast, high molecular weight PEG coating makes them even more adhesive to mucus than without coatings. We show that PEG density is especially critical as particle size diminishes from 500- to 100-nm. We also show that the spacings within the human mucus mesh are much larger than previously appreciated, which provides a significant opportunity for controlled drug delivery using large nanoparticles. The already large native mesh spacing can be further manipulated to allow micron-sized particles to penetrate mucus at speeds approaching water. We have further show that we can engineer mucus penetrating particles, using a variety of biomaterials, to rapidly penetrate various other human mucus secretions. Animal studies using MPP for gene therapy of the lungs as well as localized and sustained cancer therapy in mucosal tissues are underway.

# 461b Award Session: Single-Walled Carbon Nanotube Optical Biosensors for Multi-Modal and Single Molecule Detection of Genotoxic Agents

**Daniel A. Heller**[1], Hong Jin, and Michael S. Strano[2]. (1) Chemical Engineering, Massachusetts Institute of Technology, 77 Massachusetts Avenue, 66-565, Cambridge, MA 02139, (2) 66-566 Department of Chemical Engineering, Massachusetts Institute of Technology, 77 Massachusetts Avenue, Cambridge, MA 02139

Reactive oxygen species (ROS) and active alkylating chemotherapeutics are important bioanalytes but difficult to measure selectively, in vivo, or in real time. A new nanoscale optical sensor platform under development can now detect as well as identify these analytes in situ and in single-molecule quantities.

Carbon nanotube-DNA complexes, formed by individually encapsulating nanotubes with oligonucleotides, respond in real time to genotoxins via multiple optical modes, giving each analyte a distinct optical signature. Single-walled carbon nanotubes are environmentally sensitive and emit tissue-transparent near-infrared fluorescence. By their encapsulation in short strands of synthetic DNA, we introduce a selective handle for changes in their emission, which responds to the polarity of the solvent medium. The complexes consequently respond to DNA-damaging oxygen species or alkylating agents, by red-shifts in emission energy as well as optical quenching, giving multiple detection modes which result in analyte specificity and up to single-molecule sensitivity.

The nanotube-DNA complexes exhibit uptake into mammalian cells via endocytosis without cytotoxic effects and emit from within live tissues. Upon introduction of alkylating agents and oxidative species, the complexes transduce analyte activity information live and in situ. This platform is being developed as a research and diagnostic tool for free radical biology, drug discovery, and pharmacodynamics.

# 461c Award Submission: Targeting Colon Cancer Cells Using Pegylated Liposomes Modified with a Fibronectin-Mimetic Peptide

**Ashish Garg**, Alison W. Tisdale, and Efrosini Kokkoli. Department of Chemical Engineering and Materials Science, University of Minnesota, 421 Washington Ave. SE, Minneapolis, MN 55455

The ability to target cancer cells using an appropriate drug delivery system can significantly reduce the associated side effects from cancer therapies and can help in improving overall quality of life post cancer survival. Integrin $\alpha_5\beta_1$ is expressed on several types of cancer cells, including colon cancer, and plays an important role in tumor growth and metastasis. Thus, the ability to target the integrin $\alpha_5\beta_1$ using an appropriate drug delivery nano-vector can significantly help in inhibiting tumor growth and reducing tumor metastasis. In this study we have designed functionalized stealth liposomes (liposomes covered with polyethylene glycol (PEG)) that specifically target the integrin $\alpha_5\beta_1$. The PEG provides a steric barrier allowing the liposomes to circulate in the blood for longer duration and the functionalizing moiety, the PR_b peptide specifically recognizes and binds to integrin $\alpha_5\beta_1$ expressing cells. PR_b is a novel peptide sequence, designed in our lab, that mimics the cell adhesion domain of fibronectin, and includes four building blocks, RGDSP (the primary recognition site for $\alpha_5\beta_1$), PHSRN (the synergy site for $\alpha_5\beta_1$), a $(SG)_5$ linker, and a KSS spacer. In this study, we demonstrate that by optimizing the amount of PEG and PR_b on the liposomal interface we can engineer nano-vectors that bind to CT26.WT, HCT116, and RKO colon cancer cells in a specific manner and are internalized through $\alpha_5\beta_1$-mediated endocytosis. Stealth liposomes functionalized with an RGD containing peptide bind to colon cancer cells and internalize, but they have much lesser efficiency than PR_b-targeted stealth liposomes, and more importantly they are not as specific since many integrins bind to RGD. PR_b-targeted stealth liposomes are as cytotoxic as free 5-Fluorouracil (5-FU) and exert the highest cytotoxicity on CT26.WT cells compared to RGD-targeted stealth liposomes and non-targeted stealth liposomes. In order to further increase the efficacy of the system we have designed peptide-functionalized stealth liposomes that are pH-sensitive and exhibit triggered release under mild acidic conditions present in endocytotic vesicles. The proposed targeted delivery system has the great potential to deliver a therapeutic load directly to colon cancer cells, in an efficient and specific manner.

## 461d Award Submission: Morphology and Size-Controlled Self-Assembling of Enzyme-Responsive Peptides

**Weiping Wang** and Ying Chau. Department of Chemical Engineering, The Hong Kong University of Science and Technology, Clear Water Bay, Kowloon, Hong Kong

We have derived a novel reverse microemulsion process to impose temporal and spatial control of self-assembly. The ability of this process to transform morphology and size has been demonstrated using an enzyme-responsive dipeptide, Fmoc-Phe-Tyr-phosphate. In the presence of phosphatase, these molecules self-assemble into nanofibers of about 28nm in aqueous solution. By controlling the self-assembly process with an enzymatic trigger and confining it within nanoscale reverse micelles, the same molecules are fabricated into nanoparticles of around 6.5nm. We further investigate the morphology and size control of peptide assemblies by changing the process parameters, including the ratio of various reagents and incubation time. The resulting nanostructures are characterized by transmission electron microscopy and dynamic light scattering. The possible process for peptide nanoparticle formation inside reverse micelle is proposed according to the results. This method is applicable to other stimuli-responsive self-assembling blocks.

## 492a Design Considerations for CdTe Nanotetrapods as Electronic Devices

**Stephanie L. Teich-McGoldrick**, University of Michigan, Dept. of Chemical Engineering, 2300 Hayward Street, Ann Arbor, MI 48109-2136, Mathieu Bellanger, INSA (Institut National des Sciences Appliquées de Toulouse), 135 Avenue de Rangueil, Toulouse Cedex 4, France, Matthieu Caussanel, Laboratoire ELIAUS (Ancien LP2A), Université de Perpignan Via Domitia, 52 av. Paul Alduy, Perpignan Cedex, France, Leonidas Tsetseris, Department of Physics, Aristotle University of Thessaloniki, GR-54124, Thessaloniki, Greece, Sokrates Pantelides, Department of Physics & Astronomy, Vanderbilt University, Stevenson Center 6301, P.O. Box 1807 Sta B 37235, Nashville, TN 37235, Sharon C. Glotzer, Chemical Engineering and Materials Science and Engineering, University of Michigan, 2300 Hayward Street, Ann Arbor, MI 48109, and Ron. D. Schrimpf, Department of Electrical Engineering and Computer Science, Vanderbilt University, Box 1608, Station B, Nashville, TN 37235.

Multiple simulation techniques are used to study the feasibility of using CdTe nanotetrapods as circuit elements. Technology Computer-Aided Design tools are used to simulate the electrical behavior for both metal-semiconductor field-effect transistors and junction field-effect transistors. Our results show that by varying the doping concentrations and material composition, CdTe nanotetrapods have the potential to be useful circuit elements. Monte Carlo simulations provide insight into how control over inter-particle and particle-substrate interactions can lead to the directed assembly of ordered arrays of electrically gated nanotetrapods.

## 492b Synthesis of Boron Nanotube and Magnisium Boride Nanostructure

**Fang Fang**, Eswaramoorthi Iyyamperumal, Mathieu Pinault, Dragos Ciuparu, Codruta Zoican, and Lisa Pfefferle. Department of Chemical Engineering, Yale University, 9 Hillhouse Ave, New Haven, CT 06520

Synthesis of pure boron single-wall nanotubes by reaction of $BCl_3$ with $H_2$ over 1%Mg-MCM-41 and Mg-Ni co-incorporated MCM-41 at room pressure was previously reported by our group. Laboratory controllable parameters including catalyst nature, reaction pressure and temperature, boron precursors, are observed to significantly affect the boron nanostructures yield. Different reaction condition will result in various boron nanostructures, such as nanotubes, nanowires, and nanocapsules. A mechanism involving the physical templating of boron nanotubes is proposed. The boron nanotubes with diameter within 5nm are achieved from MCM-41 template with uniform diameter cylindrical pores. Boron nanowires (along with some nanopipes) with uniform 20nm diameter are achieved from Mg incorporated mesoporous alumina which has comparatively larger pores and less uniformity in porous structure. To further utilize the physical templating idea, surface modified single-wall carbon nanotubes were tested as the catalyst support for the synthesis of boron nanotubes by introducing active Mg such as Reike-Mg on the single-wall carbon nanotubes surface.

By doping Mg on to the as-synthesized boron nanotubes at high temperature in a sealed inert environment, magnesium boride nanotubes are obtained, as well as other nanostructures. TEM, EELS and Raman are used to confirm the existence of boron nanotubes and magnesium boride nanotubes. SEM with EDX is utilized to characterized larger nanostructure. NEXAFS of boron K-edge suggests the local coordination of boron nanostructure differs from the bulk phase boron compound. DC magnetization tests suggest the superconductive transition temperature of the different boron nanostructures range from about 35K, which is the same as bulk phase $MgB_2$ transition temperature, to roughly 100K. Magnesium boride nanotubes shows higher transition temperature compare to nanowire structures.

# 492c Synthesis of Short Chain Thiol Capped Au Nanoparticles and Their Stabilization Inside Functionalized Polymers for Organic Memory Applications

**Raju Kumar Gupta**, Chem and Biomol Engg, NUS, 4 Engineering Drive 4, Singapore, Singapore, MP Srinivasan, ChBE, NUS, 4 Engineering Drive 4, Singapore, Singapore, and Jianyong Ouyang, Mse, NUS, 4 Engineering Drive 4, Singapore, Singapore.

Interest in organic electronics is motivated by the wide variety of functionality available in organic materials, which may enable high-performance devices, coupled with the traditional strength of potentially low fabrication costs. In particular, use of nano-scale materials can lead to devices with superior performance. In order to overcome phase separation issues in polymeric-nanoparticle composite systems for memory device applications, we are investigating the employment of covalent binding between the nanoparticles and functionalized polymers.

In this work, we have synthesized short chain thiol (4-aminothiophenol) capped gold nanoparticles and studied the change in their morphology by changing the process conditions as well as the dispersion medium. Finally, we have stabilized these amine functionalized gold nanoparticles in solution which are prone to agglomerate due to amine-amine hydrogen bonding. X-ray photoelectron spectroscopy (XPS) confirms the absence of unreacted thiol. Transmission electron microscope (TEM) and UV-Vis spectroscopy have been used to study particle morphology.

## 492d Synthesis of Thiol Capped Gold Nanoparticles and Their Immobilization on a Substrate

**Raju Kumar Gupta**, Chem and Biomol Engg, NUS, 4 Engineering Drive 4, Singapore, Singapore and MP Srinivasan, ChBE, NUS, 4 Engineering Drive 4, Singapore, Singapore.

Ligand protected gold nanoparticles are being used in many fields of science from physics, chemistry to medicine. Ligand shell serves a critical role, providing the particles with a long series of properties as stability against coalescence, solubility in many solvents and assembly properties. Two dimensional assemblies of monodisperse metal particles with controlled size in the nanometer range present considerable interest in various applications of nanophysics, magnetic properties of ultrahigh density media storage and nanoelectronic devices.

Recently magnetism has been observed in gold NPs capped with dodecanethiol, in spite of the diamagnetic behavior of the bulk gold. For these free NP's, the magnetic moments can fluctuate because of nanoparticle rotation, hence prohibiting their application where permanent magnetic moments are required. Furthermore, the continuous movement of the nanoparticles prevents their use in creating patterns of stable magnetic fields. These problems can be overcome by creating an ensemble of nanoparticles.

In this work, we have synthesized 16-meracaptohexadecanoic acid capped gold nanoparticles by one-phase cold synthesis method to get uniform sized thiol stabilized Au particles without leaving hydrogen bound unreacted thiols. Later we report on the immobilization of these nanoparticles on Si/Au substrate through covalent binding to enhance their stability in magnetic field. The covalent bonding of nanoparticles is confirmed using the spectroscopic techniques such as X-ray photoelectron spectroscopy (XPS) and UV-Vis spectroscopy. The surface morphology of nanoparticles was studied by using Field emission scanning electron microscopy (FESEM).

## 492e 'Spring-Like' and Photomechanical Junctions Between Nanoparticles: An Avenue to Power Molecular-Machines by Compression Energy

Kabeer Jasuja and **Vikas Berry**. Chemical Engineering, Kansas State University, 1011 Durland Hall, Kansas State University, Manhattan, KS 66506

Integrating molecular elasticity and dynamics into nanoelectronics has potential to enable the development of next-generation nanoelectromechanical systems like energy storage devices, molecular timers and actuators, which could be integrated to build self-sustaining molecular machines. The major challenges to building such a system are (a) providing mechanism for application of confined forces acting on or generated by the molecular-junction, (b) fabrication of strong chemically-bonded molecular-junctions, which will not fail upon mechanical deformation and (c) having a nano-scale mechanically mobile system to achieve unrestrained mechanics. Such molecular mechanics cannot be achieved in a device construct where molecular junctions are incorporated between 'rigid' electrodes. We are studying the operation of a "molecular-spring" nanodevice, where controllable and confined forces are applied on molecular junctions and which can store compression and stretching energy. The system is built with covalently/electrostatically crosslinked polyelectrolyte (carbon-chains or photo-active) (cPE) molecules sandwiched between gold nanoparticles (GNP), where cPE molecular-junctions are reversibly compressed and stretched by applying electrically and centrifugally induced forces respectively, while GNPs play a dual role (a) of movable connectors to apply forces and (b) of nanoelectrodes to measure molecular deformation via change in electron tunneling conductivity. These 'molecular-spring' junctions can thus be compressed and stretched, and can subsequently apply forces on the nanoparticles to bring them back to their native state. We are studying the dynamics of these junctions. Also, we are investigating a light-actuated molecular junction (azo-group) between metal-nanoparticles. Here, the inherent mechanical actuation of azo-group is used to controllably manipulate the attached nanoparticles. We are investigating the dynamics of this system for reversible light-induced-actuation of nanoparticles. The ability to store compression energy in a molecular-device-architecture and to manipulate them by actuating junctions has the potential to power future molecular devices by stored molecular-energy and controlling the properties of nanocomponent based devices.

# 492f Polymer Collars in All-Copper Chip-to-Substrate Connections

Tyler Osborn[1], Charles Lightsey[1], and **Paul A. Kohl**[2]. (1) Chemical and Biomolecular Engineering, Georgia Institute of Technology, 311 Ferst Dr., Atlanta,, GA 30332, (2) Georgia Tech, 311 Ferst Drive N.W., Atlanta, GA 30328

A copper-to-copper bonding process was developed for an all-copper, chip-to-substrate interconnect technology. Copper pillars were fabricated by electroplating inside polymer molds. The chip-to-substrate, all-copper connections were formed by joining the two pillars with an electroless copper plating and low temperature anneal process. Excellent bond strength of the electrolessly joined pillars was achieved with a 180°C anneal. The bond strength of the copper pillar interconnects exceeded 165 MPa. The process was also characterized to have capabilities to compensate for planar and height mismatches between chip and substrate. Planar misalignment greater than the pillar diameter was overcome and successful bonded between pillars was achieved. A height mismatch of 65 μm or less between the pillars was compensated for in the joining process. Successful bonding between silicon and FR-4 has also been acheived.

The structures were modeled to assess the design space available for the compliant, all-copper chip-to-copper connections. The criteria includes thermo-mechanical compliance and acceptable electrical performance. The results show that there is an available design space (height, diameter) for the all-copper interconnects that satisfy mechanical compliance and maintain low-level electrical parasitics (inductance and capacitance).

# 492g Ultraviolet-Wavelength Franz-Keldysh Oscillations in Single-Walled Carbon Nanotubes

**Moon-Ho Ham**[1], Byung-Seon Kong[1], Woo-Jae Kim[1], Hee-Tae Jung[2], and Michael S. Strano[3]. (1) Department of Chemical Engineering, Massachusetts Institute of Technology, 77 Massachusetts Avenue, Cambridge, MA 02139, (2) Department of Chemical and Biomolecular Engineering, Korea Advanced Institute of Science and Technology, Daejeon, South Korea, (3) 66-566 Department of Chemical Engineering, Massachusetts Institute of Technology, 77 Massachusetts Avenue, Cambridge, MA 02139

We report large electroabsorption susceptibilities in the ultraviolet region for single-walled carbon nanotubes (SWNTs) supported on fused quartz substrates that are approximately 103 larger than the highest values reported to date for any system. The SWNT films were prepared on fused quartz substratres by a vacuum filtration method using a diluted suspension of HiPco SWNTs in H2O with 1 wt % sodium dodecyl sulfate. From electroabsorption spectra, the characteristic oscillatory behavior in the SWNT films was observed in the ultraviolet region, which decayed exponentially to the low energy side. It is easily described using a convolution of two sets of Airy functions in photon energy, attributing the effect to Franz-Keldysh oscillations [1]. The one-dimensional SWNT wave functions tunnel into the classically forbidden bandgap of the quartz substrate, creating a confined interfacial state demonstrating unusually strong Franz-Keldysh modulation. To address the origin of the unusually large susceptibility, the SWNT films with different ratios of metal to semiconductor content were prepared using a newly developed separation method [2]. By systematically varying the metallic and semiconducting SWNT composition of a series of films, the confinement energy is shown to correlate with the mean bandgap of semiconductors in the film. It is confirmed that the unusually large susceptibilities arise from a sub-percolated network of metallic SWNT that enhance the electric field by increasing the field gradient and decreasing the distance between conductive junctions.

References

[1] H. Shen and M. Dutta, J. Appl. Phys. 78, 2151 (1995).

[2] W.J. Kim, N. Nair, C.Y. Lee, and M.S. Strano, J. Phys. Chem. C in press.

## 493a Directed Assembly of Functional Nanoscale Biomolecular Catalytic Architectures

**Jonathan S. Dordick**, Chemical and Biological Engineering, Rensselaer Polytechnic Institute, 110 Eighth Street, Troy, NY 12180

Nature is unparalleled in its structural and functional diversity. Living organisms make fantastic materials under myriad conditions with properties we cannot emulate today using conventional approaches. In many cases, nature has provided us with a blueprint to design and assemble both natural and synthetic building blocks to create a new generation of functional, organized, and responsive materials. We have taken cues from nature to design materials with unique structural and functional properties, along with new process technologies with the ability to produce a wide range of biomimetic structures. Specifically, we have focused on the generation of nanostructures that are functionalized with and in some cases constructed from biological molecules, complete with tailored selectivities and biocatalytic activities. For example, these nanostructures have been exploited in the generation of biocatalytically functional polymeric films, coatings, and paints.

## 493b Micellar Immobilization for Bioelectrocatalysis

**Shelley D. Minteer**, Chemistry, Saint Louis University, 3501 Laclede Ave., St. Louis, MO 63103

Micelles have been used for decades to improve the lifetime, thermal stability, pH tolerance, and organic solvent tolerance of enzymes. This paper will detail the use of micellar polymers (specifically hydrophobically modified Nafion and hydrophobically modified chitosan) for oxidoreductase enzyme immobilization and stabilization. These oxidoreductase enzymes and their bioelectrocatalysis are of interest in biosensors and biofuel cells. This paper will detail the development and characterization of the micellar polymers, along with the use of these micellar polymers for enzyme immobilization and stabilization at carbon electrode surfaces. Experimental detail will be shown for the use of these enzyme modified electrodes for both sensor applications and biofuel cell applications.

## 493c Nanoscale Catalysts In Cell-Like Microcapsules

**Guanghui Ma**, Institute of Process Engineering, National Key Laboratory of Biochemical Engineering, Chinese Academy of Sciences, Beijing, 100080, China

A cell-like microcapsules containing nanoscale catalysts were constructed by combining nano-technology and micro-encapsulation method. The enzyme was first immobilized on a nano-particle, which was further encapsulated in a cell-like microcapsule. The cell-like microcapsule possess large internal cavity and porous wall. The nano-porous wall allowed the substrate and the product to pass through the wall freely. Nano-particles contributed to maintain the enzyme stability, and could move freely in the inside of the microcapsule. Furthermore, the high mechanical strength of microcapsules would allow it to be used in chromatographic column, as well as in a reactor, compared with conventional gel microcapsules. We have demonstrated that it could be a promising strategy to achieve more practical bio-transformations

# 493d Development of Cell Surface Engineered Whole Cell Biocatalysts for Biomass Conversion

**Akihiko Kondo**, Department of Chemical Science and Engineering, Graduate School of Engineering, Kobe University, Kobe, Japan

Because of the exhaustion of fossil fuels and environmental problems such as global warming and acid rain, the utilization of biomass as a resource of renewable, environmentally-friendly energy such like a bio-fuel, has attracted much attention in recent years. In particular, bioethanol produced from biomass, including fast-growing plant species and waste paper, has been suggested as a promising alternative fuel. Due to overcome to this global problem, we attempted the construction of

directed fermentation system from biomass to bioethanol by using cell surface engineered yeast. Several anchored proteins (such like a GPI-anchor, Flo1p [1], and cohesin) were used for displaying of biomass-catalyzed enzymes (amylase, and cellulase). In the case of glucoamylase-displaying yeast cell, glucose was maintained at a very low concentration and, a high ethanol-production rate was achieved. In addition, the glucose concentration in culture medium was usually maintained at a low level, because the recombinant yeast cells metabolize the glucose as soon as glucose was released from soluble starch by the glucoamylase displayed on the cell surface [2,3]. In the

fermentation of the sulfuric acid hydrolysate of wood chips, xylose and cellooligosaccharides were completely fermented after 36 h by the recombinant yeast having a fusion gene of A. aculeatus BGL1 and 3'-half

of ?-agglutinin for its cell-surface display. As a result, about 30 g/l ethanol was produced from 73 g/l total sugar added at the beginning [4-5]. These results demonstrate that the fermentation of the raw starch and lignocellulose hydrolysate could be performed efficiently by the

recombinant Saccharomyces strain with abilities for biomass degradation.

References

1. A. Kondo et al.: Appl. Microbiol. Biotechnol., Vol. 64, 28-40 (2004)

2. H. Shigechi et al.: Appl. Environ. Micobiol., Vol. 70, 5037-5040 (2004)

3. Y. Fujita et al.: Appl. Environ. Micobiol., Vol. 70, 1207-1212 (2004)

4. S. Katahira et al.: Appl. Environ. Micobiol., Vol. 70, 5407-5414 (2004)

5. S. Katahira, et al.: Enzyme Microbial Technol. in press (2008)

# 493e A Nanostructured Molecular Assembly in Ionic Liquids as a Novel Medium for Biocatalysis Reactions

**Masahiro Goto**, Kyushu University, 744 Motooka, Fukuoka, 819-0395, Japan

The insolubility of enzymes in most ionic liquids has been overcome by the formation of molecular assembly in a hydrophobic ionic liquid stabilized by a layer of anionic surfactant sodium bis(2-ethyl-1-hexyl) sulfosuccinate (AOT) and the catalytic activity one of the enzymes studied (lipase PS) became higher than in reverse micelles of AOT in isooctane.

The technological utility of enzymes can be enhanced greatly by their use in ionic liquids rather than in conventional organic solvents or in their natural aqueous reaction media due to their unique physicochemical properties. However, one important drawback regarding the use of enzymes in ILs is the fact that they are not soluble in most ILs. In our previous study, we reported the formation of nanometer-sized water domains in a hydrophobic IL [C8mim][Tf2N] stabilized by the layer of anionic surfactant sodium bis(2-ethyl-1-hexyl) sulfosuccinate (AOT) in the presence of 1-hexanol as a cosurfactant (Chem. Phys. Chem., 2008). The present study reports that various enzymes can be solubilized in such nano-scale water domains. Our study also addresses the question of whether such dissolved enzymes show catalytic activity. It was found that the catalytic activity one of the enzymes studied (lipase PS) became higher than in microemulsions of AOT in isooctane. To the best of our knowledge, this is the first report of an enzymatic reaction being conducted in a water-in-IL microemulsion. This work may open up new possibilities for studying enzymes in ionic liquids, which are receiving increasing attention as 'green' solvents for organic synthesis and catalytic processes.

## 493f Fast on–Line Digestion System for High-Throughput Proteomics

**Daniel López-Ferrer**, Pacific Northwest National Laboratory, xxx xxx xxx, Richland, WA 99352

Within the last decade, the field of proteomics has benefited biological research as well as, the pharmaceutical, food and biotech industries. The analysis pipeline is based in a bottom up approach, where proteins are digested into peptides using trypsin as enzyme, and the resulting peptides are then analyzed by liquid chromatography (LC)-mass spectrometry (MS). However, the enzymatic digestion is one of the most time-consuming steps in the analysis workflow as it is typically carried out manually. We present an efficient on-line digestion system that reduces the number of sample manipulation steps for high throughput proteomics. By incorporating a pressurized sample loop into a liquid chromatography-based separation system, both sample and enzyme (e.g., trypsin) can be simultaneously introduced to produce a complete, an ultra-fast digestion. Both standard proteins and a complex Shewanella oneidensis global protein extract were digested and analyzed using the automated on-line pressurized digestion system coupled to a mass spectrometer. The system denatured, digested, and separated product peptides in a manner of minutes, making it amenable to high-throughput applications. The on-line digestion system offers a powerful approach for high-throughput screening of proteins that could prove valuable in biochemical research; for example, quick screening of protein-based drugs.

## 494a Studying Single Molecule Receptor-Ligand Interactions on Self-Assembled Monolayer Platforms

Xiaojuan Zhang and **Vamsi K. Yadavalli**. Chemical and Life Science Engineering, Virginia Commonwealth University, 601 W Main Street, Richmond, VA 23284

Molecular recognition between specific receptors and ligands is a key step in diverse biological processes and forms the basis of many biosensors. These may include enzyme-substrate, antigen-antibody, lectin-carbohydrate binding. Force spectroscopy using an atomic force microscope (AFM) is a valuable tool for studying and detecting such interactions and molecular biorecognition at a single molecule level. The common practice is to apply a small concentration of protein to a surface and to use statistical techniques to infer the expected characteristics of the interaction between single molecule pairs. Whether the surface-bound proteins are indeed single and isolated remains unclear and often undesirable non-specific protein/surface interactions obscures informative features of the interaction. In this study, mixed self-assembled monolayers (SAMs) consisting of N-hydroxysuccinimide (NHS) and oligoethylene glycol (OEG)-terminated thiols on ultraflat gold surfaces were used to covalently immobilize proteins via lysine residues. By the optimization of attachment sites via lysine-NHS linkages on a protein-resistant layer of the OEG SAM, it is possible to isolate single proteins for study in a controlled fashion. Single protein distribution on the surface is clearly demonstrated by AFM imaging. The OEG-thiol also significantly reduces non-specific tip-surface interactions between the AFM cantilever and surface. The utility of this self-assembled monolayer platform is demonstrated in studies of the interactions of carbohydrate binding proteins (lectins) with sugars. The specific interaction profiles and rupture forces for lectin-carbohydrate binding under physiological conditions will be shown. This platform provides an attractive strategy to fabricate single molecule sensors and label free detection devices.

# 494b Single Conducting Polymer Nanowire Protein Biosensor

**Mangesh Ashok Bangar**[1], Dhammanand Shirale[1], Carlos Hangarter[1], Wilfred Chen[1], Nosang V. Myung[1], and Ashok Mulchandani[2]. (1) Chemical and Environmental Engineering, University of California, Riverside, Bourns Hall A242, University of California, Riverside, Riverside, CA 92521, (2) Department of Chemical and Environmental Engineering, University of California, Riverside, CA 92521

Advances in electronic detection based on nanowires (NWs) and nanotubes (NTs) has revolutionized our ability to provide label-free and real-time, yet sensitive and selective detection of a wide range of chemical and biological species using the NW or NT as the gate of a field effect transistor (FET). The selectivity of the nanosensors can be further enhanced by modification with specific bioreceptors.

Conducting polymers such as polypyrrole (Ppy) because of their electronic conductivity, environmental stability, easy and controlled processing by electrochemical polymerization and biocompatibility have emerged as promising materials in the development of biosensors. Here we report a cost-effective fabrication of a single Ppy nanowire based conductometric biosensor for label-free, sensitive, selective and rapid detection of proteins.

Two hundred nanometer diameter Ppy nanowires were fabricated using well established template method using alumina template and suspended in water after dissolution of template. Using electrophoretic alignment, single nanowire connections were obtained between 3 $\mu$m gap pair of prefabricated gold electrodes. Nanowire on the electrodes was subsequently secured on the electrodes by maskless gold electrodeposition. This also ensured good electrical contact between nanowire and the contact electrodes.

To demonstrate the utility of these devices as a biosensor, covalent surface functionalization of these nanowires was studied and an optimum strategy of functionalization was adopted. Ppy nanowires were functionalized with antibodies against carcinoembryonic antigen (CEA) and label-free detection and quantification of CEA achieved.

# Glucose Biosensor Based on Poly(N-methylaniline)/Chitosan Composite

*Ayşe Gül YAVUZ[1], Ayşegül UYGUN[1,*], Venkat R. BHETHANABOTLA[2,*]*

[1]Arts and Science Faculty, Department of Chemistry, Suleyman Demirel University, 32260 Isparta,TURKEY

[2]Sensors Research Laboratory, Department of Chemical and Biomedical Engineering, University of South Florida, Tampa, Florida, USA 33620-5350

## Abstract

Amperomertic glucose sensors were prepared by the electrochemical synthesis of new poly(N-methylaniline)/Chitosan(PNMANI)/Ch composites in 0.10 M $H_2SO_4$ acid medium. Glucose oxidase (GOD) was immobilized on to the surface of these composites by physical adsorption method. The composite was characterized by FTIR and UV-visible spectrophotometric methods and detailed response studies were performed with respect to glucose concentration and pH. The PNMANI/Ch/GOD electrode was utilized for glucose sensing for concentrations in the range of 0.06- 1.831 mM by the amperometric method. The electrode was found to have a response time of 70 s and the Michaelis-Menten constant was calculated as 1.71 mM.

## Introduction

The biopolymer chitosan is a type of matrix for enzyme immobilization with such attractive properties as excellent film-forming ability, high permeability for water, good adhesion, nontoxicity, and biocompatibility [1, 2]. However, its poor electrical conductivity results in a poor response time and a high operational voltage limits its applicability in devices. Composites have been prepared by incorporating a rigid conducting polymer (such as PANI) into a flexible matrix (such as chitosan) combining the good processability of the matrix and the electrical conductivity of the conducting polymer [3]. Additionally, conducting polymers have the ability to efficiently transfer the electric charges produced by biochemical reactions to electronic circuits[4].

Biosensors prepared by using a conducting polymer as a support material have a fast response time and a high storage and operational stability [5]. Polyaniline is extensively researched for its electrical, optical, chemical and electrochemical properties due to its simple synthesis method, stability in air, and potential range of applications [6]. It is the best known semi-flexible rod-like conducting polymer system with chemical and structural flexibility surrounding its amine nitrogen linkages for binding with biological materials [7]. However, the major disadvantages of PANI are its insolubility in common organic solvents and its infusibility. There are some possible methods for preparing soluble PANIs. One of these methods is to substitute one or more hydrogens by an alkyl, an alkoxy, an aryl hydroxyl, or an amino group, or halogen in an aniline nucleus [8,9]. As known from the literature both chitosan and polyanilines are sensitive to different enzymes such as lactate [10], horseradish [11], glucose oxidase [12].

The main goal of this work is to electrochemically synthesize poly(N-methylaniline)/Chitosan (PNMANI/Ch) composites in 0.10 M $H_2SO_4$ acid medium and to investigate their biosensing application to glucose sensing. The objective is to improve the response time and sensitivity of chitosan *via* the PNMANI/Ch composite. Glucose oxidase was selected as a model enzyme to construct an amperometric glucose biosensor, to explore for the first time, applicability of this composite polymer as a matrix for enzyme immobilization. For biosensor application, glucose oxidase (GOD) was immobilized on the surface of the composite by the physical adsorption method. A biosensor for glucose was developed based on the electrochemical oxidation of $H_2O_2$. Influence of pH on the activity of the PNMANI/Ch-GOD electrode was also studied.

## Experimental Procedure

All electrochemical experiments were carried out with the conventional three-electrode cell, using the modified Pt electrode as a working electrode, a saturated calomel electrode (SCE) as a reference electrode, and a platinum wire as an auxiliary electrode. The working electrode was a platinum electrode with 1 $mm^2$ of surface area. 4μL of 0.01g chitosan/ml solution in 2 wt% acetic acid was dropped onto the Pt and dryed at room temperature. Then, 0.2 ml NMANI was added to 10 ml of 0.10 M $H_2SO_4$ acid. The potential was cycled between 0 and 1.05 V versus saturated calomel reference electrode at a scan rate of 50 mV/s for 30 cycles. The amount of

PNMANI/Ch composite was increased for characterization using a platinum electrode with bigger surface area. The composite was washed with 2 wt% acetic acid and deionized water until it became colorless. The final composite was dried at 50 °C for 24 hours. Electrochemical measurements were carried out with a Gamry PCI4/750 potentiostat

**Result and Discussion**

Figure 1 shows the cyclic voltammogram recorded between 0.00 and 1.05 V, with a scan rate 50 mVs$^{-1}$ during the synthesis of PNMANI using a modified Pt electrode with chitosan, as a working electrode. The obtained electrode was ready for use in sensing after the final wash with water and denoted as PNMANI/Ch/GOD. The cyclic voltammetric growth of PNMANI shows only one well-defined anodic peak at 0.38 V [13].

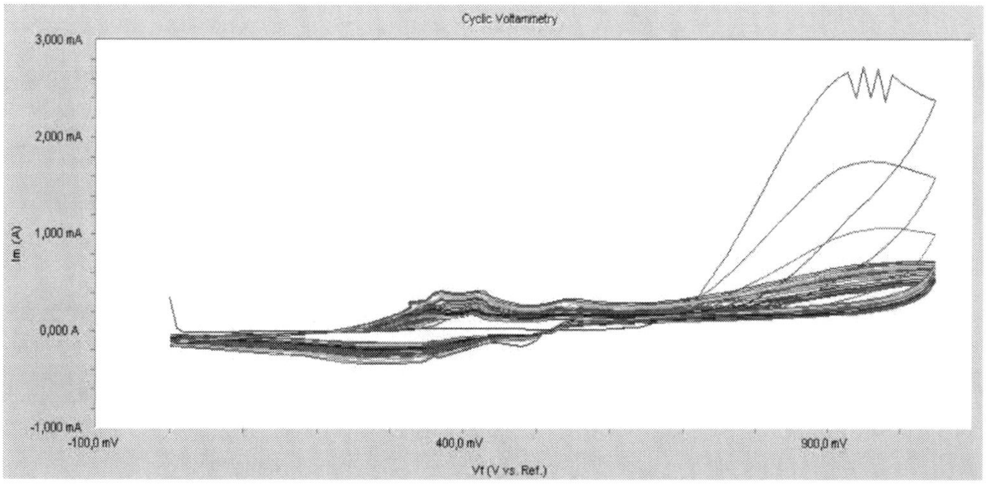

Figure 1 Electrochemical polymerization cycles of PNMANI/Ch

Electrochemically synthesized films of PNMANI/Ch composite were prepared in 0.10 M $H_2SO_4$ for immobilizing glucose oxidase (GOD). The enzyme electrode was prepared by immersing the PNMANI/Ch composite film into 5mg/ml glucose oxidase in 0.10 M phosphate buffer solution for 20 minutes. Then, the enzyme electrode was washed with 0.10 M phosphate buffer after which, it was ready to use. Free enzyme electrodes and chitosan enzyme electrodes (without any conducting polymer) were also prepared for comparison of glucose sensing capabilities.

The effect of pH on the behavior of the enzyme electrode was studied with 0.10 M phosphate buffer solution and 5 mg/ml GOD at the steady state potential at 0.6 V as a function of the pH. The electrochemical response is quite good at pH ranging from 7.0-8.0 and the maximum potential was observed at pH 7.4 in phosphate buffer.

The effect of the glucose concentration was investigated on the response of chitosan, PNMANI/Ch and free enzyme electrodes. The response current increases somewhat linearly with glucose concentration from 0.06 mM to 0.60 mM for the chitosan/GOD, from 0.06 mM to 1.54 mM for free enzyme and from 0.06 mM to 1.83 mM in the case of PNMANI/Ch/GOD. The PNMANI/Ch/GOD composite biosensor has a bigger linear range and higher sensitivity (larger slope) than free enzyme and chitosan/GOD biosensors (Figure 2).

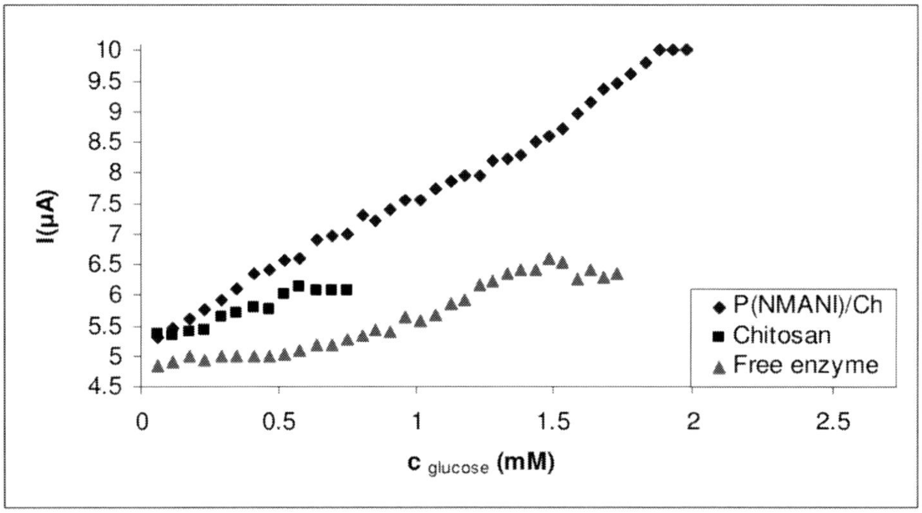


Figure 2. Changes in the response of free enzyme, chitosan/GOD, PNMANI/Ch/GOD enzyme electrode with glucose concentration (at 0.6 V in 0.10 M phosphate buffer solution (pH 7.4), 5 mg/ml GOD, 25 °C )

According to the lineweaver-Burk form of the Michaelis– Menten equation, the relation between the reciprocal of the response current and the reciprocal of glucose concentration is linear. The apparent Michaelis–Menten constant $Km_{app}$ (slope of this line) of the biosensor based on the PNMANI/Ch/GOD electrode is fitted to be 1.707 mM glucose.

## Conclusions

Upon immobilizing glucose oxidase phy physical adsorption methods, an electrode for glucose sensing, the PNMANI/Ch/GOD electrode, was prepared and tested for its performance in amperometric sensing of glucose. The PNMANI/Ch/GOD composite electrode shows wider linear range, higher sensitivity, and smaller response time, when compared to the Ch/GOD or free GOD electrodes due to the increased conductivity of the composite material. This glucose sensor material is expected to have superior biocompatibility, which can be revealed by further studies.


## Acknowledgements

The authors gratefully acknowledge financial support from the Scientific and Technological Research Council of Turkey and the Süleyman Demirel University Fund (Project number: 1506-D-07). This work is also partially funded by NSF grant ECCS-0801929, and a USF-FMM grant, which support is gratefully acknowledged.


## References


(1) Luo, X.L.; Xu, J.J.; Du, Y.; Chen, H.Y., Analytical Biochemistry, 2004, 334, 284–289.

(2) Du, Y.; Luo, X.L.; Xu, J.J.; Chen, H.Y., Bioelectrochemistry, 2007, 70(2), 342-347.

(3) Kim, S.J.; Shin, S.R.; Spinks, G.M.; Kim, I.Y.; Kim, S.I., Journal of Applied Polymer Science, 2005, 96(3), 867-873

(4) Tahir, Z.M.; Alocilja, E.C.; Grooms, D.L.,Biosensors and Bioelectronics, 2005, 20, 1690-1695.

(5) Mu S.; Xue, H., 1996. Sensors and Actuators B: Chemical 31(3), 155-16.

(6) W.J. Winokur, in Handbook of conducting polymers T. A. Skotheim, R. L. Elsenbaumer, J. R. Reynolds, Eds. (Marcel Dekker, Inc,), 1998, 707-726.

(7) Genies, E.M.; Boyle, A.; Lapkowski, M.; Tsintavis, C., Synthetic Metals, 1990,36(2), 139-182.

(8) Cataldo, F.; Paolo, M.,European Polymer Journal, 2002, 38, 1791-1803

(9) Gök, A.; Sarı, B.; Talu, M., Synthetic Metals, 2004, 142, 41-48.

(10) Tsai Y.C.; Chen S.Y.; Liaw H.W., Sensors and Actuators B, 2007, 125(2), 474–481.

11.Jin, Z.; Su, Y.; Duan,Y., Synthetic Metals, 2001, 122( 2), 237-242.
12. Ismail, Y.A.; Shin, S.R.; Shin, K.M.; Yoon, S.G.; Shon, K.; Kim, S.I.; Kim ,S.J., Sensors and Actuators B, 2008,129(2), 834-840.
13.Sivakumar, R.; Saraswathi R., Synthetic Metals, 2003, 138(3), 381-390.

# 494d Antibody Functionalization of Nanostructured Diatom Biosilica for Selective Detection of Biomolecules by Enhanced Photoluminescence

**Debra K. Gale**[1], Timothy Gutu[2], Jun Jiao[2], Chih-hung Chang[1], and **Gregory L. Rorrer**[1]. (1) Oregon State University, Department of Chemical Engineering, Corvallis, OR 97331, (2) Department of Physics, Portland State University, Portland, OR 97207

Diatoms are single-celled algae that make silica shells or "frustules" with intricate features patterned at the nano- and microscales. We report that antibody-functionalized diatom biosilica frustules may serve as a biosensor platform to selectively report antibody-antigen immunocomplex formation through enhanced photoluminescence. Biosilica frustules of 10 micron diameter isolated from cultured cells of centric diatom Cyclotella sp. were mounted on glass and then covalently functionalized with the model antibody Rabbit Immunoglobulin G (IgG) to yield a uniform, nanotextured antibody surface that binds selectively to its complimentary antigen, Goat anti-Rabbit IgG. Diatom frustules possess an intrinsic capacity to emit blue light when excited with a UV laser light source, a property called photoluminescence, which is enabled by the nanostructure of the diatom biosilica. Binding of the antibody-functionalized diatom frustule with its complimentary antigen selectively intensified the intrinsic photoluminescence intensity of the diatom biosilica through a nucleophilic enhancement mechanism, whereas challenging the antibody-functionalized diatom frustule with its non-complimentary antigen (Goat anti-Human IgG) did not change the photoluminescence intensity. It was proposed that the nucleophilic immunocomplex increased the PL intensity by donating electrons to non-radiative defect sites on the photoluminescent diatom biosilica, thereby decreasing non-radiative electron decay and increasing the radiative emission, resulting in label-free detection of the complimentary antigen. Consequently, antibody-functionalized diatom frustules hold considerable promise for the standoff detection of target biomolecules.

# 494e Solution-Phase ELISA Enabled by Magnetic Nanoparticles and Vc-Containing Chitosan Nanocapsules

**Lizeng Gao**, Jiamin Wu, and Di Gao. Chemical and Petroleum Engineering, University of Pittsburgh, 1249 Benedum Hall, 3700 O' Hara street, Pittsburgh, PA 15261

We combine the use of magnetic particles and Vitamin C-containing chitosan capsules to develop a liquid-phase ELISA technology, where targeted analytes are sandwiched between magnetic particles (intrinsic activity of horseradish peroxidase) and the Vc-containing chitosan capsules (inhibitor), and binding of analytes to antibodies takes place in liquid phase instead of on the walls of the plates in conventional ELISA. Such an approach greatly shortens the diffusion path of analytes in order to reach antibodies and therefore, accelerates the detection speed of the ELISA technology. The technology has great potential for a wide range of applications in medicine, biotechnology, and environmental monitoring industries.

# 494f Single-Walled Carbon Nanotube Optical Biosensors for Multi-Modal and Single Molecule Detection of Genotoxic Agents

**Daniel A. Heller**[1], Hong Jin, and Michael S. Strano[2]. (1) Chemical Engineering, Massachusetts Institute of Technology, 77 Massachusetts Avenue, 66-565, Cambridge, MA 02139, (2) 66-566 Department of Chemical Engineering, Massachusetts Institute of Technology, 77 Massachusetts Avenue, Cambridge, MA 02139

Reactive oxygen species (ROS) and active alkylating chemotherapeutics are important bioanalytes but difficult to measure selectively, in vivo, or in real time. A new nanoscale optical sensor platform under development can now detect as well as identify these analytes in situ and in single-molecule quantities.

Carbon nanotube-DNA complexes, formed by individually encapsulating nanotubes with oligonucleotides, respond in real time to genotoxins via multiple optical modes, giving each analyte a distinct optical signature. Single-walled carbon nanotubes are environmentally sensitive and emit tissue-transparent near-infrared fluorescence. By their encapsulation in short strands of synthetic DNA, we introduce a selective handle for changes in their emission, which responds to the polarity of the solvent medium. The complexes consequently respond to DNA-damaging oxygen species or alkylating agents, by red-shifts in emission energy as well as optical quenching, giving multiple detection modes which result in analyte specificity and up to single-molecule sensitivity.

The nanotube-DNA complexes exhibit uptake into mammalian cells via endocytosis without cytotoxic effects and emit from within live tissues. Upon introduction of alkylating agents and oxidative species, the complexes transduce analyte activity information live and in situ. This platform is being developed as a research and diagnostic tool for free radical biology, drug discovery, and pharmacodynamics.

## 548a Concentration-Specific Hydrogen Sensing Behaviour In Monosized Pd Nanoparticle-Based Thin Films

**Einar Kruis**[1], Manika Knanuja[2], Shubhra Kala[2], and Bodh Raj Mehta[2]. (1) Faculty of Engineering, University Duisburg-Essen, Bismarckstr 81, Duisburg, Germany, (2) Physics, IIT Delhi, New delhi, India

A unique concentration-specific H2 sensing behaviour with a 'pulsed' response at larger H2 concentrations and 'saturated' response at lower concentrations has been observed in monosized and monocrystalline Pd nanoparticle-based films. The threshold concentration required for transition from 'saturated' to 'pulsed' response is very sensitive to measurement temperature and lies in the 2 to 10 % concentration range, this would allow to fabricate a simple hydrogen gas alarm system triggered by a decrease in resistance.

In the present study, H2 sensing response of monosized, monocrystalline and spherical Pd nanoparticle-based thin films with a controllable size in 15-25 nm range with a narrow size distribution ($\sigma g < 1.10$) has been investigated. The sensitivity and response of the electronic and geometric effects during H2 sensing have been determined as a function of nanoparticle size, H2 concentration and measurement temperature. It has been shown that the sensing behavior is highly concentration- specific and threshold concentration defining the change over from 'saturated' response to 'pulsed' response is a strong function of the measurement temperature. The present study provides a 'proof of concept' for developing a sensor capable of detecting H2 concentration levels along with fast response of EE and high sensitivity of GE. Its characteristic change in mechanisms can be tuned by choosing a suitable operating temperature so that the reaching the H2 explosion limit results in a sudden decrease in resistance.

## 548b Volatile Organic Compound Sensor Based on Conducting Polymer-Metal Nanoparticle Composites

**Sreeram Vaddiraju**, Department of Chemical Engineering, Massachusetts Institute of Technology, 25 Ames Street, Bldg. 66, Rm. 419, Cambridge, MA 02139 and Karen K. Gleason, Chemical Engineering, Massachusetts Institute of Technology, 25 Ames Street, Bldg. 66, Rm. 419, Cambridge, MA 02139.

Chemically selective, cheap and robust chemresistor platforms, based on novel metal nanoparticle-conducting polymer composites, for sensing volatile organic compounds were developed and presented. These composites were obtained by covalent attachment and assembly of metal nanoparticles on top of functionalized conducting polymer platforms. This integration strategy allowed for the uniform assembly of metal nanoparticles on top of the conducting polymer films, while preserving the active metal surface. The fabrication of these composites involved the following steps: (i) Use of oxidative Chemical Vapor Deposition (oCVD) for obtaining uniform films of poly(3,4-ethylenedioxythiophene-*co*-thiophene-3-acetic acid), P(EDOT-*co*-TAA), and (ii) covalently tethering metal nanoparticles on top of these conducting polymer films using conjugated linker molecules, such as 4-aminothiophenol.

Metal nanoparticle layers on top of these platforms served as the sensing elements in these platforms. Nobel metals exhibit a change in work function on exposure to gases and vapors. The magnitude of this change depends on the gas (or vapor) it is exposed to and its concentration. By assembling metal nanoparticles on top of conducting polymers, these changes in work function could be translated into changes in resistance of the composites and detected easily. Selectivity in sensing volatile organic compounds was accomplished by varying the material of the metal assembled on top of the P(EDOT-*co*-TAA) films. Selective sensing of acetone and toluene was accomplished by assembling palladium and nickel on top of the conducting polymer films, respectively. Concentrations of acetone and toluene as low as 171 ppm and 51 ppm, respectively, were detected using these platforms. The response of these composites to these volatile organic compounds was found to be instantaneous (few seconds), and recovery was also observed to be fast (20-40 seconds).

## 548c Development of Peptide and Protein Assays Based on Enzyme-Multiplied Assay Technique

**May L. Chiu**, Chemical and Biomolecular Engineering, University of California, Los Angeles, 420 Westwood Plaza, 7820 Boelter Hall, Los Angeles, CA 90095 and Harold G. Monbouquette, Department of Chemical and Biomolecular Engineering, University of California, Los Angeles, 420 Westwood Plaza, 5531 Boelter Hall, Los Angeles, CA 90095.

Enzyme-multiplied assay technique (EMAT) is a homogeneous method widely used in urine and blood testing due its simplicity, sensitivity, and fast response time. We are currently aiming to extend EMAT for protein and peptide detection. With this capability, EMAT can potentially be used in areas such as bioterrorism defense and infectious disease detection.

Conceptually, EMAT consists of an analyte-reporter enzyme conjugate and an anti-analyte antibody. When the antibody binds to the analyte-enzyme conjugate, enzyme activity is repressed due to physical blockage or conformational change of the active site. When introduced, free analyte competes for antibody binding sites and at least partially prevents repression, thus the reporter enzyme can more readily turn over substrate, and a signal related to the quantity of free analyte present is generated.

As a proof-of-the-concept study, a peptide assay based on EMAT was prepared by conjugating a peptide sequence from influenza hemagglutinin A (HA) to glucose-6-phosphate dehydrogenase (G6PDH) via a crosslinker. Matrix-assisted laser desorption/ionization mass spectrometry shows that an average of 14 peptides attach to each G6PDH. 70% repression is achieved upon antibody binding. Preliminary data shows that repression can be relieved with ≥100 nM of free HA. Lower detection limits can be achieved by optimizing the antibody to HA-G6PDH ratio. With a similar approach, assays that are useful for biotoxin detection can be developed.

# 548d Rapid Screening of DNA Due to Differences In the Viscoelastic Properties

**Ashish S. Yeri**[1], Lizeng Gao[2], and Di Gao[2]. (1) Department of Chemical and Petroleum Engineering, University of Pittsburgh, 1249 Benedum Hall, Department of Chemical and Petroleum Engineering, Pittsburgh, PA 15261, (2) Chemical and Petroleum Engineering, University of Pittsburgh, 1249 Benedum Hall, 3700 O' Hara street, Pittsburgh, PA 15261

As the role of disease causing genetic mutations are more clearly understood, the need for quick DNA screening is rapidly increasing. Here we show a rapid characterization technique which makes use of the single strand DNA's ability to form a particular configuration in aqueous solution which is directly related to its base sequence. When immobilized onto a substrate, the film formed will have viscoelastic properties which are characteristic of the base sequence in the ssDNA.

In this study we evaluate the differences in the viscoelastic properties between the P53 Wild type DNA molecule and the P53 mutant R type DNA molecule using a QCM-D. The QCM-D is a surface acoustic wave based sensor which can measure extremely low masses with very high precision. It makes use of a quartz crystal with a resonant frequency of around 5 Mhz (crystals with different resonant frequencies can also be used). For a rigid film depositing on the crystal, the frequency of resonance decreases linearly with the amount of mass deposited according to the Saurberey's equation and so, a highly sensitive measurement of frequency would enable us to calculate the amount of mass deposited in the order of ngs per square cm. By measuring the energy dissipation of the film at very short intervals, the instrument enables us to evaluate the viscoelastic properties of the film. For a film that is viscoelastic (not rigid with an appreciable energy dissipation), the Saurberey's equation becomes insufficient and viscoelastic modeling (usually the Voight Maxwell model) becomes necessary.

We have studied the viscoelastic properties of the film formed by the deposition of biotinylated P53 ssDNA, both wild type and the mutant type by immobilizing it via Avidin to a biotin-thiol surface. We have shown that the wild type and the mutant type ssDNA form films of different properties which are due to interactions between the bases in single strand DNA molecules in aqueous solution leading to the formation of particular configurations.

The value of the shear modulus was evaluated for the film formed by the wild type DNA and the film formed by the mutant DNA. The appreciable difference in the values suggests that this method of rapid screening has much promise.

## 548e Calibration of the Charge-Coupled Device Camera Based Microscopic Particle Tracking and Microrheology

**Pei-Hsun Wu** and Yiider Tseng. Chemical Engineering, University of Florida, Bldg 723 Room 118, Gainesville, FL 32611

Video-based particle-tracking technique has been widely applied as a tool to promote our understanding in both physical and biological systems. In biological system, this technique has been used to document the dynamics of a probe-labeled sub-cellular component, such as a nuclear body and a transmembrane protein. On the other hand, the mechanical properties of a complex fluid system can be probed by single- or two-particle microrheology, in which video-based particle-tracking is essential to validate the accuracy of the results from these measurements. In recent years, this technique becomes even more prominent since the physical and biological applications have been emerging together as ballistically intracellular nanorheology (BIN) to examine the mechanical changes of the cytoplasmic region of a mammalian cell upon the extracellular chemical and mechanical stimuli. However, while the applications of this technique are promising, the spatial errors generated from the imaging acquisition process could significantly hinder the progression and further block the extension of this technique. Therefore, in this report, we systematically address the how to experimentally refine the spatial errors from a video-based particle-tracking system to advance this technique.

## 549a Adjuvant Characterization for Virus-Like Particle-Based Influenza Vaccines

**Harvinder S. Gill**, Fu-Shi Quan, Sang-Moo Kang, and Richard W. Compans. Department of Microbiology and Immunology, Emory University, Atlanta, GA 30332

Influenza virus continues to pose serious pandemic risk largely due to antigenic drift. Current influenza vaccines predominantly provide strain specific protection. Our laboratory has engineered and fabricated novel virus-like particles that mimic native influenza viruses. Although we have demonstrated that multiple doses of influenza VLPs show protective immune response in mice, limited knowledge is available on the ability to enhance immune response using adjuvants through the intramuscular route to induce protection against a broader spectrum of influenza virus strains. It is also of interest to enable a single dose vaccination regimen as opposed to the current two dose influenza vaccination schedule.

In this study we investigated the effect of different water soluble adjuvants such as muramyl dipeptide, granulocyte macrophage colony-stimulating factor (GMCSF), CpG, poly (I:C) and gardiquimod on the immune response in mice. Adjuvants at a 1 μg dose were administered with influenza VLPs intramuscularly. Mouse serum antibodies were analyzed at weeks 2 and 4 post immunization with a booster dose to determine the effect of adjuvants. VLPs alone at a dose of 1 μg per mouse induced protection from lethal virus challenge. While all the different adjuvants tested showed increase in antibody levels, gardiquimod and CpG showed the highest increase.

In conclusion, we have fabricated virus-like particles that mimic native influenza viruses and produce protective immune response in mice. Gardiquimod and CpG were found to be potential adjuvants for VLPs. Effects of different doses of adjuvants on the immune response and on single dose-induced protective immunity by virus-like particle vaccine will also be discussed.

## 549b Biodegradable Polyanhydride Nanosphere Interactions with Antigen Presenting Cells

**Bret D. Ulery**[1], S. Avanti Sarkar[2], Maria Torres[1], Yashdeep Phanse[2], Jenny Wilson[2], Bryan Bellaire[2], Michael J. Wannemuehler[2], and Balaji Narasimhan[1]. (1) Chemical and Biological Engineering, Iowa State University, 2114 Sweeney Hall, Ames, IA 50011, (2) Veterinary Microbiology and Preventative Medicine, Iowa State University, 2180 Veterinary Medicine, Ames, IA 50011

Use of efficacious vaccines remains one of the most cost effective methods to prevent disease. New vaccine delivery vehicles are needed to improve effectiveness and patient compliance. Biodegradable polyanhydrides have shown great promise as single dose vaccine delivery vehicles that also function as superb adjuvants. Our current studies are designed to test polyanhydride nanosphere uptake and intracellular localization in antigen presenting cells since these are key determinants in antigen presentation. Copolymers of sebacic acid and 1,6-bis(p-carboxyphenoxy)hexane and copolymers of 1,8-bis(p-carboxyphenoxy)-3,6-dioxaoctane and 1,6-bis(p-carboxyphenoxy)hexane were tested as potential vaccine adjuvants. Polyanhydride nanospheres were fabricated using the nanosphere antisolvent precipitation technique. Murine bone marrow derived dendritic cells were incubated in vitro with nanospheres and analyzed for increases in cell surface marker expression by flow cytometry. Internalization and intracellular distribution of fluorescein-loaded nanospheres within human monocytic cells were visualized using laser scanning confocal microscopy. It was observed that internalization and particle trafficking were strongly dependent on specific copolymer chemistry and particle size. These studies provide evidence that polyanhydride nanospheres are promising candidates for vaccine delivery. In addition, the effect of size and chemistry can be exploited to design delivery systems that facilitate appropriate and efficacious immune responses.

## 549c High-Throughput and High-Content Screening of Antibody Responses from Single Cells

**J. Christopher Love**[1], Adebola Ogunniyi[1], Eliseo Papa[2], Eduardo Guillen[3], Jae H. Choi[3], and Craig M. Story[4]. (1) Chemical Engineering, Massachusetts Institute of Technology, 77 Massachusetts Ave., Bldg. 66-456, Cambridge, MA 02139, (2) Health Sciences & Technology, Massachusetts Institute of Technology, 77 Massachusetts Ave., Cambridge, MA 02139, (3) Whitehead Institute, 9 Cambridge Center, Cambridge, MA 02139, (4) Biology, Gordon College, 255 Grapevine Rd., Wenham, MA 01984

The development of new vaccines is limited in part by the challenges of profiling the diversity in antibody responses generated to infectious diseases or candidate vaccines to diseases. This talk will describe a soft lithographic technology that uses an array of subnanoliter containers to isolate large numbers of individual antibody-secreting cells and print protein microarrays of the corresponding antibodies from each cell. The microarrays can be interrogated to characterize the specificity, isotype, and affinity of each clone. It is also possible to identify antibodies reacting to particular epitopes such as phosphorylated tyrosines. The cells of interest identified in the screens can be retrieved for expansion or genetic sequencing.

## 549d Anionic Nano-Polymersomes for Enhanced Circulation Kinetics In Vivo

**David A. Christian**, Diana M. Bowen, Ryan J. Dunn, and Dennis E. Discher. Chemical and Biomolecular Engineering, University of Pennsylvania, Room 129 Towne Building, 220 South 33rd Street, Philadelphia, PA 19104-6393

Amphiphilic diblock copolymers are well known to form a variety of supramolecular morphologies – spherical micelles, worm-like micelles, and bilayer vesicles (polymersomes) – that have been studied extensively for drug delivery applications. Previous studies with these polymer-based carriers have shown that their circulation kinetics in vivo can be enhanced with the presence of polyethylene glycol (PEG) on their outer surface. This neutral PEG brush delays the opsonization by serum proteins that allows the body's immune system to recognize and clear the drug carrier from the body. While these carriers have shown promising circulation kinetics ($t1/2 \sim 18$ hours for 100 nm polymersomes), longer circulation times are certainly possible based on red blood cells with lifetimes of many weeks to months (110 days in humans).

Cell membranes have a glycocalyx with a net negative charge, which motivates studies of polymers other than PEG. The diblock copolymer poly(acrylic acid)-polybutadiene (PAA-PBD) carries a large negative charge, and assemblies of this charged copolymer in water have been shown to undergo morphological phase transitions based on aqueous solution pH and salt concentration. Blends of PAA-PBD and a polymersome-forming PEG-PBD are fully miscible in organic solvent and can be dried to form a homogeneous polymer film. By hydrating this film in aqueous conditions suited to polymersome formation, the surface charge of the resulting polymersomes has been shown to be linearly tunable with the blend fraction of PAA-PBD. By incorporating only a small percentage of PAA-PBD, polymersomes with a dense PEG brush can be made to have a surface charge similar to that of red blood cells. These polymersomes can be sized down to a diameter of 100 nm, and studies in blood plasma have demonstrated that these anionic nano-polymersomes are stable at physiological conditions for at least 24 hours. Ongoing experiments examine the in vivo stability and circulation.

# 549e A Trojan Horse Approach for the Non-Invasive Delivery of Nanotherapeutics to and through the Lungs

**Balaji Bharatwaj**[1], Libo Wu[1], Judith A. Whittum-Hudson[2], and Sandro R. P. da Rocha[1]. (1) Chemical Engineering and Materials Science, Wayne State University, Detroit, MI 48201, (2) Immunology and Microbiology - School of Medicine, Wayne State University, Detroit, MI 48201

Inhalation therapy has been long utilized as an efficient route for the local (lung) delivery of therapeutics. More recently, it has been recognized as an efficient pathway to the systemic circulation. It has improved patient compliance, fast drug uptake - in the order of minutes or even seconds, and improved drug bioavailability. Inhalation therapy can also be used for the local and systemic delivery of therapeutic biomolecules. Nanotechnology offers great promise in combination with inhalation therapies. Nanoparticles can be used as efficient carriers of nucleotides and other therapeutic biomolecules. Nanoparticles can also be designed so as to target specific cell organelles or for the controlled release of therapeutics. However, a suitable formulation is required in order to deliver very small particles to the lungs. The optimal aerosol size for lung deposition lies between ~0.4 – 5 $\mu$m, larger particles being deposited in the mouth/throat, while smaller particles may be exhaled. In this work we will discuss a novel propellant-based inhalation formulation that can be used to efficiently deliver nanoparticles to the deep lungs. We show that biodegradable and biocompatible core-shell particles with solute-laden nanoparticles may be effectively dispersed in hydrofluoroalkane propellants, generating aerosols with enhanced characteristics. The applicability of the technology is demonstrated in vitro by investigating the ability of the nanoparticles in the core-shell formulation to reach the inclusions of chlamydial-infected lung epithelial cells.

# 549f Selective Control of Mucus Mesh Size and Nanorheology for Prevention of Infectious Diseases

**Samuel K. Lai**[1], Ying-Ying Wang[2], Kaoru Hida[2], Richard Cone[3], Denis Wirtz[4], and Justin Hanes[1]. (1) Department of Chemical and Biomolecular Engineering, The Johns Hopkins University, 221 Maryland Hall, 3400 N. Charles Street, Baltimore, MD 21218, (2) Department of Biomedical Engineering, The Johns Hopkins University, 221 Maryland Hall, 3400 N. Charles Street, Baltimore, MD 21218, (3) Department of Biophysics, Johns Hopkins University, 221 Maryland Hall, 3400 N. Charles St, Baltimore, MD 21218, (4) Department of Chemical & Biomolecular Engineering, Johns Hopkins University, 3400 N. Charles St., Baltimore, MD 21218

Mucus covers the luminal surfaces of the respiratory, digestive, and reproductive tracts, as well as the surface of the eye and peritoneal surfaces of intra-abdominal organs. The ability of mucus to accomplish its critical functions as a lubricant and protective barrier against infection rests on its viscoelastic properties, which remain poorly understood at the length scales in which it functions as a selective barrier against pathogens and bacteria.

We engineered synthetic particles that readily penetrate human mucus, and used them as non-destructive nanoprobes to accurately measure the structural architecture of fresh undiluted human cervicovaginal mucus for the first time. We discovered that the average interfiber spacing of mucus is much larger than expected—large enough to allow nearly unhindered penetration of toxins and the largest viruses if they are not slowed by adhesive interactions. We then showed that mucus behaves as a viscoelastic solid barrier to objects 1 um and larger (phase angle ~12-30 degrees), but as a permissive viscoelastic liquid to objects with sizes ~500 nm and below (phase angle >60 degrees). At the 200 nm length scale, the local viscosity of human mucus is more than 3-orders of magnitude lower than its macro- (bulk) viscosity, highlighting the length scale-dependent barrier behavior of human mucus. Most viruses have evolved to possess sizes of ~100 nm or smaller and, therefore, are small enough to access the low-viscosity pathways in human mucus.

We have subsequently discovered that we can significantly tighten the mesh spacing of human mucus, and thereby vastly improve the barrier properties of mucus at smaller length scales, by application of a nonionic detergent, N9. Mucus treated with N9 behaves as a highly viscoelastic solid to 200 and 500 nm particles. This shift in nanoscale barrier behavior is caused by a partial dissociation of bundled mucin fibers, which shifts the viscoelastic solid-liquid transition of CVM to below 200 nm. In contrast to classical theories for crosslinked or entangled semiflexible polymer networks, N9 treatment did not alter the macroscopic viscous and elastic moduli of mucus, probably because N9 reduces adhesive (enthalpic) mucin interactions that counteract the increased entanglement (entropic) contributions to macro-rheology. This unexpected and highly evolved behavior of human mucus provides new clues for development of disruptive new technologies, from medical therapies to prevent infection to production of biomimetic materials with highly tunable behavior.

## 550a Liquid Crystalline Assemblies of Inorganic Nanorods

Shanthi Murali[1], Khristine Pizarro[1], Kyle Taylor[1], and **Virginia A. Davis**[2]. (1) Chemical Engineering, Auburn University, 212,Ross Hall, Auburn, AL 36849, (2) Department of Chemical Engineering, Auburn University, Auburn, AL 36849

The recent development of facile routes for the production of inorganic nanorods, nanowires, and nanowhiskers has resulted in an array of materials with intriguing optical, electronic, magnetic and structural properties. However, the production of many functional materials requires aligning nanorods on the micro- to macro- scales. The combination orientational self-assembly of anisotropic materials into liquid crystalline phases and shear alignment is a well established means to produce high performance polymeric materials such as bullet-proof vests and liquid crystalline displays. The general foundation for the self-assembly of anisotropic particles in solution dates back to Onsager (1949), understanding the liquid crystalline phase behavior of nanorod dispersions is a nascent field formed by the intersection of nanotechnology, liquid crystalline science and colloid science. We report on the assembly of silver and silicon nitride nanorods into liquid crystalline phases and aligned films. Optical microscopy, RAMAN spectroscopy, Differential Scanning Calorimetry and Rheological characterizations were used to characterize the dispersions and films. This work represents a step forward in developing a fundamental understanding about the impacts of nanorod concentration, nanorod aspect ratio, and nanorod-nanorod and nanorod-solvent interactions on liquid crystalline phase behavior.

# 550b Parallel Patterning of Single Inorganic Nanowires with Arbitrary Composition Using Optoelectronic Tweezers

**Peter Pauzauskie**[1], Arash Jamshidi[2], Aaron T. Ohta[2], Justin Valley[2], Peidong Yang[3], and Ming Wu[2]. (1) CMELS, Chemical Sciences Division, Lawrence Livermore National Laboratory, 7000 East Ave., Livermore, CA 94551, (2) Electrical Engineering, University of California, Berkeley, Cory Hall, UC Berkeley, Berkeley, CA 94720, (3) Department of Chemistry, University of California, Berkeley; Materials Sciences Division, Lawrence Berkeley National Laboratory, Berkeley, CA 94720

The synthesis of inorganic nanowires has advanced in the last decade to a point where a vast range of insulating, semiconducting, and metallic materials are available for use in heterogeneous, integrated optoelectronic devices at nanometer scales. However, a persistent challenge for the application of this family of materials has been the development of general strategies for the massively parallel manipulation of single nanowires with arbitrary composition. For instance, although laser tweezers are capable of addressing both insulating and semiconducting particles in three dimensions, extreme heating effects prevent the stable manipulation of metallic nanowires with highly focused lasers [1]. Optoelectronic tweezers (OET) have emerged in recent years as a powerful tool for the manipulation of micron scale particles, including living cells. When particles are much smaller than 1 micron the forces generated with OET typically are overcome by thermal (Brownian) fluctuations because DEP forces scale with material volume. In this presentation we demonstrate that individual semiconducting and metallic nanowires with diameters below 20 nm, are addressable with forces generated by optoelectronic tweezers (OET) [2]. Using 100,000x less optical power density than optical tweezers, OET is capable of transporting individual nanowires with speeds 4x larger than maximum speeds achieved by optical tweezers. A real-time array of silver nanowires is formed using photopatterned virtual-electrodes, demonstrating the potential for massively parallel assemblies. Digital video microscopy is used to quantify the dipole interaction between adjacent silver nanowires. The use of a photocurable polymer gel is presented as a means of fixing patterns of individual nanowires in place, opening up opportunities for novel device architectures.

References:

[1] Pauzauskie, P. J.; Radenovic, A.; Trepagnier, E.; Shroff, H.; Yang, P. D.; Liphardt, J. Nature Materials 2006, 5, 97-101.

[2] Jamshidi, A.; Pauzauskie, P. J.; Schuck, P. J.; Ohta, A. T.; Chiou, P. Y.; Chou, J.; Yang, P. D.; Wu, M. C. Nature Photonics 2008, 2, 85-89.

## 550c Gas Phase Production of Metal Oxide Nanowires and Nanoparticles Using a Microwave Plasma Reactor

**Vivekanand Kumar**[1], Jeong H. Kim[2], and Mahendra K. Sunkara[1]. (1) Department of Chemical Engineering, University of Louisville, 106 Ernst Hall, Department of Chemical Engineering,, University of Louisville, Louisville, KY 40292, (2) Chemical Engineering, University of Louisville, 106 Ernst Hall, Department of Chemical Engineering,, University of Louisville, Louisville, KY 40292

Nanowires have attracted great interest because of their promise in a variety of applications such as electronics, optoelectronics, nano electro-mechanical devices, catalysis and several other fields. Hence, the synthesis of bulk quantities of nanowires with controlled composition, and high crystallinity is important. However, bulk production of nanowires has been challenging using the techniques known so far.

A novel reactor and method for bulk production of high purity metal oxide nanowires using a high throughput micro wave plasma reactor has been developed. The metal oxide nanowires are produced directly in the gas phase without the need of any catalyst, substrate or template. Each run for processing metal powders into metal oxide nanowires took less than a second. So, this process is capable of making a kilogram per day. We have produced nanowires of tin oxide, zinc oxide, titanium oxide, alumina and copper-zinc alloy so far. Nanoparticles of Alumina and Titania were also produced at the temperatures above the melting points of their corresponding metals. Pure nanowires are purified by a dispersion in 1-methoxy 2-proponal followed by horn sonication and gravity sedimentation.A simplified analysis followed by experiments suggest that the metal powders melt in the plasma primarily with the heat generated from chemical reactions, i.e., radical recombination and oxidation reactions on particle surface.

## 550d Length and Tortuosity Control of Te Nanowires

**G. Daniel Lilly** and Nicholas A. Kotov. Chemical Engineering, University of Michigan, 2300 Hayward, HH Dow Building #3074, Ann Arbor, MI 49109

One-dimensional semiconductor nanowires (NWs) have many unique optical and electronic properties that make them desirable for use in new sensing, electronic, and photonic devices. For practical NW application to these uses, synthesis techniques must be manipulated to allow large numbers of uniform NWs to be produced. Our paper is motivated by the desire to improve the control of NW growth, and to better understand the mechanism of NW formation. We discuss the formation of Te NWs from CdTe nanoparticles (NPs) using a modification of current techniques involving stabilizer depleted CdTe NPs in a growth solution of DMSO and deoxygenated water. The length and tortuosity of these Te NWs are controlled by the addition of CdSe NPs to the growth solution.

## 550e Fabrication of High Aspect Ratio Gold-Polypyrrole Bilayer Nanobelts as Nanoscale Actuators

**Nicha Chartuprayoon**[1], Youngwoo Rheem[1], Bong Young Yoo[2], and Nosang V. Myung[1]. (1) Chemical and Environmental Engineering, University of California, Riverside, Bourns Hall A242, 900 University Ave., Riverside, CA 92521, (2) Department of Materials and Chemical Engineering, Hanyang University, Ansan, South Korea

One dimensional conducting polymer nanostructures exhibit unique physical and chemical properties for many nanotechnology applications and devices such as nanoscale actuators for the NEMS (nanoelectromechanical system). These promising materials can be potentially used for artificial muscle and robotic system. Polypyrrole, PPy, is one of the conducting polymers that can undergo reversible oxidation and reduction reaction, resulting in dilation and contraction in the polymer volume, respectively. Many polypyrrole based macro- and microscale actuators have been demonstrated; however, actuation behavior on PPy-based nanoscale actuators has not been realized. In this research, single gold-polypyrrole bilayer nanobelts will be fabricated via newly developed Lithographically Patterned Nanowire Electrodeposition (LPNE) method. Electrochemically controlled reversible bending of gold-polypyrrole bilayer nanobelts will be carried out and demonstrated.

# 550f Molecularly Guided Assembly of Silver into 1 Dimensional Structures

**Miguel Castro**[1], **Roberto Irizarry**[2], and Madeline Leon[1]. (1) Chemical Imaging Center, Department of Chemistry, University of Puerto Rico, Car. 108 km 2.0, Mayaguez, PR 00680, (2) Electronic Technologies, DuPont, 14 TW Alexander Drive, Research Triangle Park, Raleigh, NC 27709

I am having problems with the abstract. Will submit shortly after receiving confirmation.

## 588a The Nanolab: An Interdisciplinary Undergraduate Science and Engineering Laboratory at the University of Texas at Austin

**Brian A. Korgel**, University of Texas at Austin, Department of Chemical Engineering, Austin, TX 78712, John G. Ekerdt, Chemical Engineering, University of Texas at Austin, University of Texas at Austin, Austin, TX 78712, and Andrew Heitsch, The University of Texas at Austin, 1 University Station C0400, Austin, TX 78712.

At The University of Texas at Austin, we have developed an upper-division undergraduate laboratory, called the NANOLAB, which is designed introduce engineering and natural sciences students to nanoscale science and engineering (NSE) concepts. For the NANOLAB, we have developed four new 6-hour experiments: 1) "Fabrication of gold nanoparticles using self-assembled templating", 2) "Optical and redox properties of colloidal semiconducting quantum dots", 3) "Acid-doped polyaniline nanofiber sensor for vapor detection", and 4)"Gold nanorod synthesis and optical properties". The NANOLAB experiments are integrated with existing undergraduate laboratories, currently in five different departments—Chemistry/Biochemistry, Physics, Chemical Engineering, Biomedical Engineering and Mechanical Engineering—across both the Colleges of Engineering and Natural Sciences. The NANOLAB is not a stand-alone course, but rather is a laboratory "station" that interfaces with existing laboratory courses in the participating departments. For example, a student enrolled in physical chemistry laboratory performs one of the NANOLAB experiments as a required experiment in that course. A chemical engineering student in the unit operations laboratory does likewise. In the NANOLAB, students also work in a multidisciplinary team of two natural sciences and two engineering majors to promote the interdisciplinary communication that is fundamental to NSE. Approximately 960 students per year participate in the NANOLAB; thus, providing broad exposure to a large student population at a public university with a hands-on active NSE learning experience. This presentation will provide a snapshot of the program and a summary of successes and failures and lessons learned thus far.

## 588b The Ethics of "Nanoethics"

**Michael Bennett**, Social Sciences, Michigan Technological University, 1400 Townsend Dr, Houghton, MI 49931

Text Not Available.

## 588c Nue: Active Learning Experiences for Undergraduates

**Miguel Castro**, Edmy Ferrer, Madeline Leon, and Eunice Mercado. Chemical Imaging Center, Department of Chemistry, University of Puerto Rico, Car. 108 km 2.0, Mayaguez, PR 00680

Active learning is used to train undergraduates in nanotechnology.

# Nanotech Innovations: Nanotechnology Enterprise at Michigan Technological University

**John A. Jaszczak**, Department of Physics, Michigan Technological University, 1400 Townsend Dr, Houghton, MI 49931, **Mary Raber**, Enterprise Program, Michigan Technological University, 1400 Townsend Dr, Houghton, MI 49931, **Michael Bennett**[1], Department of Social Sciences, Michigan Technological University, 1400 Townsend Dr, Houghton, MI 49931, **Nasser Alaraje**, School of Technology, Michigan Technological University, 1400 Townsend Dr, Houghton, MI 49931 and **Paul L. Bergstrom**, Department of Electrical and Computer Engineering, Michigan Technological University, 1400 Townsend Dr, Houghton, MI 49931

An innovative, goal-oriented, and active-learning educational experience called the Michigan Tech Enterprise Program (http://www.enterprise.mtu.edu) was initiated a decade ago in order to provide a vehicle for undergraduate engineering students to experience the development and operation of a commercial enterprise and to actively address industry's needs for engineers with strong technical, communication, interpersonal, leadership, entrepreneurship and business skills. This highly successful program now supports 27 enterprises with over 700 students- more than 10% of the total undergraduate student enrollment at the university. Within this framework, a new nanotechnology-related enterprise called Nanotech Innovations Enterprise (NIE) was launched in January 2008, with the help of an National Science Foundation Nanotechnology Undergraduate Education in Engineering grant (Fig. 1). The mission of NIE is to engage students in a comprehensive hands-on entrepreneurial educational experience, including management, market research, technical research, product

Fig. 1. Screen shot of the Nanotech Innovations Enterprise website's front page at http://nano.mtu.edu/ni/

development, service, outreach, and education related to nanoscale science, engineering and technology (NST). The primary goals of NIE are to:

1. Integrate nanotechnology into engineering and science education at Michigan Tech via a continual core of dedicated students (the NIE members).
    a. Develop a sustainable business model, branding, and student leadership team;
    b. Develop, build and market an affordable STM system and accessories for distribution to participating high school teachers;
    c. Research, develop and market other selected nanotechnology products, tools, and services, particularly those related to sensing technologies;
    d. Significantly impact ~180 undergraduate engineering students over the first 2-years through NIE involvement, explorations, and new modules and courses;
    e. Maintain a balanced emphasis on the foundational science, exciting applications, and societal implications of NST, building on the successful nanotechnology education framework already established at Michigan Tech[2,3];
    f. Develop curricular modules and courses in
        i. scanning probe microscopy (AFM and STM),
        ii. nanotechnology commercialization,
        iii. nanotechnology business plan development,
        iv. intellectual property ethics.
2. Enhance and broaden the undergraduate educational experience of a wide variety of Michigan Tech students, including non-engineering students.
3. Contribute to the advancement of NST education, including societal implications, for up to 12 high school teachers and 500 students through workshops and on- and off-campus presentations.

These goals are in addition to the overall goals of the Michigan Tech Enterprise Program[4,5]:

1. Create an environment for students to facilitate the transition from their undergraduate program to the professional workforce;
2. Provide opportunities for students and faculty to develop leadership and entrepreneurial skills in a learning setting that closely resembles an industrial or professional environment;
3. Give the students ownership of a portion of their academic program that connects strongly to career goals;
4. Give students a taste of the rewards and accountability associated with creating new products and working with paying clients; and
5. Utilize the students' fundamental background in science and engineering in the context of a problem when non-technical issues, such as cost or societal impacts, are of equal importance.

Students choosing to participate in NIE come from a wide variety of disciplines, including non-engineering disciplines. Currently, 16 students from the following engineering programs are enrolled: biomedical engineering (6), mechanical engineering (2), electrical and computer engineering (3), and materials science and engineering (1). In addition, 3 physics majors and 2 business majors are participating. Primary participation in the enterprise is through enrollment in a 1-credit enterprise project course each semester, often starting in the student's sophomore year. Seniors, who are expected to take more leadership roles, may sign up for two credits each semester. Students may also take 1-credit modules in enterprise-related courses such as leadership, teaming, communications, entrepreneurship, business planning, ethics in engineering design and implementation, product/process development, or technical courses such as industrial health and safety or the new course in practical scanning probe microscopy. Participation in the enterprise can count toward free electives, Senior Design projects, an Enterprise Certificate program, a university minor in Enterprise, or toward the research requirements of the university minor in Nanotechnology.

The Nanotech Innovations Enterprise is currently developing seven major projects in addition to developing structural and policy aspects of the enterprise. Structural aspects include the election of officers (President, Vice President for Public Relations, Vice President of Finance, and Vice President of Records), lab and office spaces, recruiting, web site development, etc. Draft policies, dealing with issues of election procedures, job descriptions, meetings, project documentation, and grading, were developed during the first semester of the project and are currently being refined as the team learns first-hand what may or may not work, and what issues are currently of pressing interest. Students may work on up to two projects unless they are an officer, in which case they can only participate in one project team. The enterprise meets weekly as a whole group for project updates, brief business, and a topical presentation. Topics for presentations that are planned for this semester include societal implications issues, communications and conflict resolution, technical themes, and several on business plan development. Projects each have a project manager and additional meeting times.

Several primary projects center around nanoscale science and technology education and outreach activities for high school students and teachers, as well as first- and second-year undergraduate students. All NIE members are expected to participate in education and outreach activities, especially early in their enterprise involvement. The Outreach team's primary function is to develop and coordinate these educational activities. A focal point of many of these activities is the demonstration of scanning probe microscopes, of the enterprise has an atomic force microscope (AFM) and a portable scanning tunneling microscope (STM), both of which are easyScan2 models manufactured by Nanosurf AG (Switzerland). In addition to being portable, these instruments are very robust, provide excellent usability and results, and are ideal educational purposes (Fig. 2). For first-year engineering students in the required Engineering Fundamentals class, a two-hour "exploration" introducing NST topics and providing live demonstrations of AFM, STM, scanning electron microscopy, and focused ion beam milling will be offered. Undergraduates will also gain exposure to NIE through participation in annual exhibitions such as the university's annual Enterprise Expo in which all enterprise and senior design projects exhibit their work. Outreach and educational activities for high school students are offered both on and off campus during the academic year as well as in the summer through the university's Summer Youth Programs and the annual *YES! Expo*[6] held at Ford Field in Detroit, Michigan. These activities will be further developed and expanded in cooperation with high school and middle school teachers through a series of teacher workshops being developed to introduce teachers to NST topics and to assist teachers in developing related curricular activities they can use in their classrooms.

An associated project is the development and building of a Lego® Mindstorms-based scanning probe imaging machine that is computer interfaced and that can

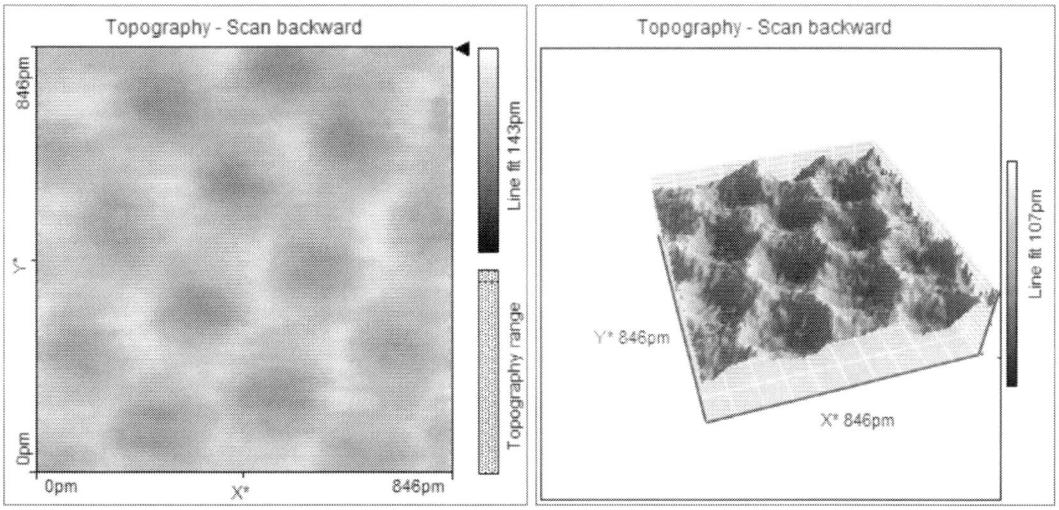


Fig. 2. Atomic-scale scanning tunneling microscope (STM) images of the surface of highly oriented pyrolitic graphite (HOPG) obtained in the Nanotech Innovations Enterprise laboratory using the Nanosurf ® easyScan2 STM (http://www.nanosurf.com).

demonstrate the fundamental principles of scanning probe microscopy but is more amenable to interactive classroom or outreach presentations. Based on a prototype design, middle school or high school students could possibly form teams to further explore, enhance and build their own designs. NST-related sample kits and supporting documentation, graphite crystals for graphene research and development, scanning tunneling microscope design and development. Perhaps the most ambitious project is the development of a simple STM that would cost under $1,000 but be sufficiently robust for use in high school laboratories. Yet another associated project team has goals of developing know how, student laboratories (high school and undergraduate), and devices to supplement and improve the operation of STMs for education.

Because of its exotic electronic and magnetic properties and its potential uses in new electronic devices, there has been tremendous recent interest in the material known as graphene[7,8]. This two-dimensional material- composed solely of $sp^2$-bonded carbon atoms one atomic layer thick- can be produced by mechanically separating the layers of carbon in graphite and has only recently been verified experimentally. Interest among scientists in obtaining high-quality graphite crystals for graphene research, and also for use as STM samples and AFM sample substrates, has led to a market for the sale of high-quality natural graphite crystals. NIE has established a graphite team to isolate, characterize, and market high quality natural graphite crystals through a subsidiary, Naturally Graphite, and associate web site graphitecrystals.com (Fig. 3).

Taking advantage of nanotechnology-related projects being conducted under the direction of faculty across numerous disciplines across campus, an exciting project involves the development of educational sample kits for scanning probe microscopy. Currently, a prototype kit is under development that will contain six different samples that can successfully be imaged by AFM or STM. The kit will also contain informational materials explaining how the samples can be successfully imaged, why the samples are particularly interesting, descriptions of how they are fabricated, and selected references for further reading. A supplemental CD will contain animations giving further illustration

Fig. 3. Screen shot of the *Naturally Graphite* web page at http://graphitecrystals.com, advertising high quality graphite crystals for research and industry. Crystals are most commonly sought for scanning probe microscopy applications or for the fabrication of graphene.

of the applications and fabrication of the sample in the kit. It is anticipated that the first prototype kit will contain samples of:

1. Natural graphite crystal with growth spirals (relevant to materials science, mineralogy, and geology);
2. Carbon nanotubes grown by chemical vapor deposition;
3. Photonic crystal devices;
4. Single electron transistor device;
5. Microporous silicon;
6. Collagen fibers for nerve regeneration research

The enterprise also plans to develop self-contained laboratory guides for use in high school or advanced undergraduate laboratories appropriate to each type of sample.

Interest in the nanotechnology enterprise by students has been very gratifying. Enrollment started with 10 students in the first semester and has grown to 12 students in the summer, and 16 now in its second academic semester. Inquiries by interested students are also continuing as current NTE members work to recruit new members. Starting a new enterprise has been both rewarding and challenging. Perhaps the most challenging issue from the standpoint of advising is learning to appropriately balance the degree of direct involvement with both projects and structural issues of the enterprise. Another challenge faced was the development of policies and procedures while at the same time all participants were just learning about the enterprise program and the nanotechnology-related projects. As might be expected, the enterprise has experienced firsthand some conflicts and communication shortcomings, which have served to emphasize the importance of these issues in developing new collaborative teams. On the other hand, it is very gratifying to see students from diverse backgrounds, majors, and personalities working together and so quickly realizing many of the goals of the enterprise. All necessary teams are formed, project goals have been identified and progress documented, and the Nanotech Innovations Enterprise web sites are now public. The first of the teacher workshops is scheduled for February 2009 and already there is more interest among teachers than we are likely to be able to accommodate. The graphite project team has also launched its website, begun marketing its products, and has already made several sales. The goal-oriented nature of the enterprise, the intrinsic "student ownership" of the enterprise, and the focus on an exciting emergent technology are highly engaging for the students. We are excited to see students learning here the skills that will help them to continue to "create the future" when they leave Michigan Tech.

We thank Sue Hill for help with web site design and development, Ruth Kramer and Brad Baltensperger for assistance with advising outreach and teacher workshop projects, and Jarek Drelich for developing and teaching the Practical Scanning Microscopy course. We gratefully acknowledge support from National Science Foundation grant ECE 0741490.

---

[1] Current Address: Science, Technology and Society, Vassar College, Poughkeepsie, NY 12604-0246

[2] J. A. Jaszczak and B. E. Seely, "Developing nano education at a technological university: Science, technology and societal implications of nanotechnology" in *Nanoscale Science and Engineering Education: Issues, Trends and Future Directions,* edited by A. E. Sweeney and S. Seal, American Scientific Publishers (2008).

[3] J. A. Jaszczak and B. E. Seely, "Planting Seeds: Introducing Nanotechnology Education into Engineering Curricula". In *Education in Nanoscience and Engineering*, edited by R. Carpenter, S. Seal, N. Healy, N. Shinn, W. Braue. (Mater. Res. Soc. Symp. Proc. **931E**, Warrendale, PA, 2006), 0931-KK01-08.

[4] M. R. Plichta and R. O. Warrington, "Active, Discovery-Based Learning Through the Enterprise Program in the College of Engineering at Michigan Technological University," Proceedings of the ASEE/SEFI/TUB Conference, October 2002, Berlin, Germany, CD-ROM.

[5] D. Stone, M. B. Raber, S. Sorby, and M. Plichta, "The Enterprise Program at Michigan Technological University," *International Journal of Engineering Education* **21**, 212 (2005).

[6] http://www.yes.mtu.edu

[7] A. Geim and P. Kim, "Carbon Wonderworld," *Scientific American* **298**(4), 90 (2008).

[8] A. K. Geim and A. H. MacDonald, "Graphene: Exploring carbon flatland," *Physics Today* **60**(8), 35 (2007).

## 588e Teaching Nanoscale Science and Engineering to Chemical Engineers at Georgia Tech

**Sankar Nair**, School of Chemical & Biomolecular Engineering, Georgia Institute of Technology, 311 Ferst Drive NW, Atlanta, GA 30332-0100

This presentation will discuss the incorporation of nanoscale science and engineering into the graduate and undergraduate chemical engineering curriculum at Georgia Tech. The talk will be centered around the development and teaching of the course "Chemical Engineering in Nanoscale Systems", which has been offered since 2004. The objectives of the course, desired outcomes, selection and scope of topics, and the implementation of the course will be discussed both in the context of graduate and undergraduate students. In addition to lecture materials, the involvement of students in the conceptual design of nanoscale systems will be illustrated through the use of term project examples. Some quantitative and qualitative data will be presented in regard to assessment of the course outcomes and impact on undergraduate and graduate students. The adaptation of the course materials towards developing nanotechnology curricula for K-12 institutions will be illustrated through an ongoing project. The role of this course in larger nanotechnology education efforts at Georgia Tech will be discussed, for example its incorporation in the NanoScience and Technology (NaST) course cluster. The talk will conclude with a discussion of potential future developments and issues, particularly relating to the incorporation of hands-on laboratory experiences for students.

## 588f Learnings from Teaching An Interdisciplinary Class on Nanotechnology

**David H. Gracias**, Department of Chemical and Biomolecular Engineering, Johns Hopkins University, 125 Maryland Hall, 3400 N Charles Street, Baltimore, MD 21218

For the last five years, I have introduced a new class entitled micro and nanotechnology that is cross listed in the undergraduate and graduate curriculum. A couple of years ago, I also tried to incorporate a laboratory component to the class. In this talk, I will describe my learnings from the class that is normally attended by 30-55 students. I will also discuss the need for funding for such classes especially those with a laboratory component.

# 588g Nanomaterials Education for City College of New York Science and Engineering Undergraduates

**Ilona Kretzschmar**, Chemical Engineering, The City College of the City University of New York, Steinmann Hall, 140 ST. and Convent Avenue, New York, NY 10031

Approximately 2 million engineers and scientists will be needed worldwide in the field of nanotechnology within the next decade.[1] The growing number of nanomaterials and – technology related articles in both the public sector and in the scientific literature exemplifies the expanding interest in the field and the need for a mature perspective amongst young scientist, engineers and the general public. Exposure of students at all levels in Science and Engineering as well as non-Science and non-Engineering majors is needed to prepare the US for interdisciplinary nature of nanotechnology.

A team of 13 professors from Chemistry, Physics, Biomedical, Chemical, Electrical, and Mechanical Engineering have created a coherent modular nanomaterials course and laboratory that reflect the interdisciplinary nature of nanomaterials science. The nanomaterials course has been taught twice and the laboratory once. The nanomaterials course and laboratory are offered through the chemical engineering department and have been able to draw students from different disciplines. The nanomaterials course content has been implemented in such a way that the modules can be used as lecture material in courses from each of the participating disciplines. Additionally, we have assembled a speaker forum of industrial partners to provide the outside experience part of the course. The laboratory consists of 5 experimental modules which are housed in the laboratories of the participating departments.

We will present an overview of the nanomaterials course and laboratory structure. We will describe the course development strategy the group has applied as well as present evaluation data for both the course and the laboratory. [1] Roco, M. C. "Converging Science and Technology at the Nanoscale: Opportunities for Education and Training" Nature Biotech. 2003, 21, 1247-1249.

## 588h A Resource Sharing Approach to Nanotechnology Education: The PA Nanotechnology Partnership

**Amy Brunner**, Jamie Houseknecht, William Mahoney, Terry Kuzma, Robert Ehrmann, and Stephen Fonash. Center for Nanotechnology Education and Utilization, The Pennsylvania State University, University Park, PA 16801

Bringing hands-on nanotechnology to 2-year and 4-year degree students often is hampered, especially in smaller institutions, by the need for proper safety facilities, equipment, faculty expertise, and technician support capabilities. To solve this problem in Pennsylvania and thereby bring in-depth nanotechnology education to every corner of the State, the Commonwealth supports the PA Nanotechnology Partnership, a nanotechnology education coalition of 27 two-year degree and four-year degree institutions across Pennsylvania. This Partnership offers nanotechnology education paths to PA students in Chemistry, Chemical Engineering, Chemical Engineering technology, and related areas. The common element of these pathways is the Partnership`s "Capstone Semester", which is required for all 2-yr. and 4-yr. degree nanotechnology degrees and concentrations available at Partner institutions. The 18 credit "Capstone Semester" is an emersion taught as a service to Partnership institutions by Penn State at its University Park campus. To "level the playing field" to allow all these students from 27 schools to have access to this education opportunity, the State of Pennsylvania makes up the difference between a student's Partner institution tuition and the actual cost of the "Capstone Semester". This also makes the program "seamless" for the student since he or she simply pays the tuition of the institution in which they are enrolled. Since many of the students live at home while attending their 2-yr. or 4-yr. institutions, the State also provides a room and board grant for the semester spent at Penn State. The NMT Partnership resource-sharing model, which brings micro- and nanotechnology expertise, equipment, and facilities to students in every corner of Pennsylvania, would not be possible without this State support. To date, over 500 students have completed the "Capstone Semester" and have taken jobs at over 70 Pennsylvania micro- and nanotechnology industries.

Web Page: www.cneu.psu.edu

## Incorporation of Nanotechnology into the ChE Curriculum at Oregon State University

*Milo D. Koretsky, Alexandre Yokochi, Sho Kimura, and Sarah Herzog*
*School of Chemical, Biological and Environmental Engineering*
*Oregon State University*

### Introduction

With the substantial investment into the development of nanotechnology infrastructure for the 21st century and beyond, there is a need to adapt engineering and science curricula to equip students with the skills and attributes needed to contribute effectively in manufacturing based processes that rely on nanotechnology.[1-3] The incorporation of nanotechnology into the undergraduate engineering curriculum represents both an opportunity and a challenge. On the one hand, nanotechnology can revitalize undergraduate programs by engaging students with interesting nanotechnology related concepts, examples, and experiments. On the other hand, due to its inherent interdisciplinary nature, programs will need to accommodate greater degrees of interdisciplinary teaching and research. Chemical and biological processes will play a significant role in the manufacturing operations. Chemical and biological engineers have the advantage of a solid background in chemical kinetics, reactor design, transport phenomena, thermodynamics and process control to undertake the challenges in the high volume manufacturing of nanotechnology-based products. Thus, these processes fall well within the purview of chemical and biological engineering undergraduate programs. However, at the same time, the products rely on principles based on other disciplines such as physics, mechanical engineering and electrical engineering. Thus research and development of new processes based on new products is inherently interdisciplinary in nature. The curricular challenge that needs to be addressed is how to design a program that reinforces the ChE undergraduate's core skills (Depth) in a way that can be applied towards manufacturing nanotechnology-based products while simultaneously providing the Breadth to interact effectively on the multidisciplinary teams which span the wide range of opportunities enabled by this emerging area.

It has been proposed that as the chemical engineering profession takes its next evolutionary step towards applying molecular scale engineering to a set of new and emerging technologies, the core undergraduate curriculum needs associated reform.[4] However, as topics from these emerging molecular-based technologies are incorporated, there is a legitimate concern of dilution of the core content due to staffing issues.[5] At OSU, the Chemical, Biological and Environmental Engineering programs have recently joined into a single administrative structure. This structure alleviates the staffing issue in two ways. First, a significant portion of the courses for all three programs are jointly taught. This set of eleven core courses covers fundamentals germane to all three disciplines (e.g., material and energy balances, transport processes, thermodynamics and process data analysis) while reducing the number of instructors needed. Second, the Option areas in chemical engineering are taken from topics that have core research faculty. In two of the Options, biological processes and environmental processes, students take elective classes from amongst those offered by the other programs. In this way, some of the key elements identified in the "New Frontiers in Chemical Engineering Education" workshops are integrated into the undergraduate curriculum while, simultaneously, holding students accountable for the same depth of learning which has served OSU ChE graduates for many years. Moreover, this integration is accomplished in a reasonable scope commensurate with the resources of the program.

This paper presents the curriculum developed to incorporate nanotechnology education in the College of Engineering (CoE) at Oregon State University OSU. The approach is twofold: (1) to develop a Nanotechnology Processes Option in the Chemical Engineering (ChE) Program and (2) to develop a survey course within the CoE that is broadly available to all engineering undergraduates. The curricular development fits in well with the growing research and commercialization activity of the Oregon Nanoscience and Microtechnologies Institute (ONAMI), and is consistent with the evolutionary vision developed by leading chemical engineering educators in the three-workshop series "New Frontiers in Chemical Engineering Education."[4]

**Nanotechnology Processes Option in Chemical Engineering**

To meet all the ABET engineering topics and advanced science requirements, ChE students are required to take five to six technical elective classes outside the ChE core. These courses may be taken in any area as long as they have the appropriate engineering or science content as prescribed by ABET and AIChE. However, taken in an *ad hoc* manner, students were getting little satisfaction or career enhancement. The ChE Department has established Options to aid students in selection of elective courses. This also helps to broaden and strengthen the undergraduate ChE curriculum, potentially attracting more students to the department. To be eligible for an Option, the student must fill out and present a Student Petition for Option Program in Chemical Engineering to the faculty "champion" for the desired area. The champion is a faculty member with expertise in the area of the Option. Additionally an Option must contain at least 21 credits. Three options have been available at OSU: (1) Biochemical Processes; (2) Environmental Processes and (3) Microelectronics Processes and Materials Science. These areas correspond to strengths in the OSU ChE program. A fourth option, the Nanotechnology Processes Option, has been developed. An outline of the curricular requirements is listed in Table 1. It contains six courses, five required courses and an elective, including two sophomore level courses. Four of the five required classes are laboratory-based and emphasize hands-on experience. The Nanotechnology Processes Option was approved at the university level in Fall 2006. In Fall 2008, there were 10 seniors and 12 juniors who have selected this option.

The Science, Engineering and Social Impact of Nanotechnology (ENGR 221) is a general engineering survey course that provides students from Chemical, Biological, Electrical, Environmental, Industrial, Manufacturing and Mechanical Engineering exposure to the field of nanotechnology; therefore, there is inherently a multidisciplinary approach. On the other hand, Material and Energy Balances in Nanotechnology (ChE 214) is a ChE specific laboratory-based course, emphasizing how the fundamental skills students have learned couple to nanotechnology. For ChE students, the approach is to provide students both a breadth of multidisciplinary experiences and a depth of specific technical applications within the discipline. Thus,

**TABLE 1.** Nanotechnology Processes Option

| Class# | Credits | Title |
|---|---|---|
| ENGR 221 | 3 | The Science, Engineering and Social Impact of Nanotechnology (lec) |
| ChE 214 | 4 | Material and Energy Balances in Nanotechnology (lec/lab) |
| ChE 416 | 3 | Chemical Engineering Lab III (lab)* |
| ChE 417 | 4 | Analytical Instrumentation in Chemical, Environmental and Biological Engineering (lec/lab) |
| ChE 444 | 4 | Thin Film Materials Processing (lec/lab) |
| | 3 | Elective |

* The capstone laboratory project will be in the area of nanotechnology

they are exposed to these complementary experiences early in their undergraduate studies. These sophomore level courses lead into three upper division courses already in place. This duality (Breadth plus Depth Pedagogy) is reinforced in senior laboratory (ChE 416), through which students synthesize both aspects in their capstone project, and potentially through their Honors College thesis.

**ENGR 221 - The Science, Engineering and Social Impact of Nanotechnology**

ENGR 221 is a general engineering survey course with the objective of ensuring all engineering students have access to a course offering basic understanding of the engineering field of nanotechnology. The course learning outcomes are presented in Figure 1 below. The concepts of nanotechnology have been divided in several sections, with each one spanning roughly one to two weeks. The course includes several features intended to promote active learning including hands-on activities and demonstrations and a capstone ethics project where students complete a risk assessment of the impact of nanotechnology on society. In addition to introducing technical knowledge surrounding the field of nanotechnology, this course focuses on synthesizing fundamental concepts in science and engineering within the context of nanotechnology.

ENGR 221 was delivered for the first time in winter 2007 with an enrollment of 31, and again in winter 2008 with an enrollment of 45. The success of the course was assessed in terms of the achievement of these learning objectives and the effectiveness of the different modes of delivery used during the course. Assessment methods for this course primarily relied on pre and post assessments of one kind or another. Overall course pre and post concept inventory assessments were administered, in addition to pre and post worksheets for two class activities. The other major methods of assessment of student learning were an end of term survey and an analysis of critical thinking of the final ethics paper.

The course was assessed in terms of both achievement of learning objectives and modes of delivery such as lecture, lab, or projects. The following assessment methods were used:

- Overall course pre and post concept inventory assessments
- End of the term survey
- Pre and post worksheets for a lecture
- Wise learning tool
- Pre and post worksheets for a hands-on laboratory activity
- Final ethics project analysis

The detailed assessment will be presented elsewhere.[6] One of the interesting results is from analysis of an end-of-term survey that asked students to discuss in more detail one concept that they applied, and how it related to nanotechnology. These responses were then rated as declarative, procedural, or schematic (conceptual) knowledge.[7] In their responses, 21 out of 36 students showed a conceptual understanding of the material they discussed. Students were able to take concepts introduced in other classes, build on them in the context of nanotechnology, and develop that knowledge into a strong, conceptual understanding of both the basic material and its relation to nanotechnology. Schematic learning is valuable due to its transferability. That so many students are displaying this type of learning is a great success for the course.

After successful completion of this course, students become able to:

1. Define nanotechnology.
2. Discuss how nanotechnology may impact society.
3. Identify products based on nanostructured materials.
4. Explain how the properties of nanostructured materials differ from their non-nanostructured (conventional) material counterparts.
5. Explain how these unique properties may adversely impact human health and the environment; define the concerns with nanotoxicity research and summarize the status in this area.
6. Explain the difference in approach of top-down and bottom-up manufacturing methods.
7. Describe major manufacturing methods used to produce nanostructured materials and devices and discuss issues in this area.
8. Identify some common methods used for nanomaterials characterization; describe the principles by which each method works and the type of information obtained.
9. Compare two prevalent ethical theories, utilitarianism and absolutism
10. Perform a risk assessment to determine the best direction for nanotechnology development.

**FIGURE 1.** Course Learning Objectives for ENGR 221

## ChE 214 - Material and Energy Balances in Nanotechnology

ChE 214 is a chemical engineering lab course intended to give students an immediate way to apply what they learned in Material and Energy Balances (the first strongly technical chemical engineering course students take) in a nanotechnology setting. The course learning objectives are presented in Figure 2. For most students, this course will directly follow ENGR 221, and students will already have a solid background in nanotechnology.

CHE 214 is a lab course, consisting of a two hour lecture period and a four hour lab period each week. Each lab is not an isolated occurrence, but instead builds on all previous labs. For example, the catalyst the students create in the first lab is used throughout the course to grow nanotubes. The lecture periods consist of one hour of new material followed by an hour-long quiz. Students are given weekly homework assignments intended to prepare them for lab.

After successful completion of this course, students become able to:

1. Quantitatively describe the rate of reaction through real-time measurements of changes in the mass of product carbon nanotubes.
2. Calculate molar and mass concentrations based on flow rates of mixture-gas components and correlate them to GC based concentrations.
3. Calculate the fractional conversion of limiting reactant based on the reactant inlet and outlet flow rates.
4. Calculate product yields based on the gas-flow rates and correlate them to mass-based product yields.
5. Use temperature measurements at the reactor inlet and outlet to explain heats of reaction in conjunction with endothermic and exothermic reaction concept.
6. Predict reactor outlet temperature and compare it to actual temperature measurements.

**FIGURE 2.** Course Learning Objectives for ChE 214

As a course specific to chemical engineering, the students in CHE 214 have exclusively been chemical engineering majors. There were 15 students enrolled in the course during spring 2007 and 12 in spring 2008. In 2008, the enrollment was evenly divided between sophomore and seniors. Again the success of the course was assessed in terms of the achievement of the learning objectives. Since the course consisted primarily of laboratory sessions, observations and survey of these sessions were the primary tools of assessment. In addition, a pre and post test was administered, an end of term survey and analysis of the final project (which covered most material introduced in the course).

The survey was intended to reveal the student's perception of what they were expected to learn and the concepts they employed in each lab. Again the responses to each question of each survey were categorized in terms of declarative, procedural, or schematic knowledge. The first conclusion from this analysis is that seniors are better able to think about the lab material schematically than sophomores. A second conclusion is that students are more able to respond schematically when asked directly about a concept than they are when asked about what they were intended to learn in the laboratory. In fact, when asked about what they learned in lab, students are much more likely to describe the physical system and its operation than the concepts behind why the system behaves as it does. This is especially true for sophomore students.

An unsolicited comment from a senior read as follows:

> Throughout my career as a chemical engineering student, I was always curious about the research and experimentation process of creating and testing a new product. It may not be a primary goal, however, CHE 214 does a great job of taking students through the entire laboratory process of research experimentation. I really appreciated starting from making a catalyst and following through to product characterization using scanning electron microscopy. I found this to be very beneficial since many of the techniques taught in CHE 214 can be applied to testing much more than carbon nanotube production.
>
> One of the most exciting aspects of this class was the opportunity to work on creating a product which is very close to the cutting edge of new technologies. It was exciting to be the only undergraduate students in the country to be creating carbon nanotubes as an instructional lesson on chemical engineering principles. It was very helpful to having a significant amount of time in which questions could be answered on the barriers in carbon nanotube and nanotechnology in general.

## References


1. S. J. Fonash, "Education and training of the nanotechnology workforce" *J. Nanoparticle Res.* **3**, 79, (2001).
2. M. C. Roco, "Broader societal issues of nanotechnology" *J. Nanoparticle Res.* **5**, 181, (2003).
3. M. C. Roco, "Science and Technology Integration for Increased Human Potential and Societal Outcomes" *Ann. N.Y. Acad. Sci.* **1013**, 1, (2004).
4. "Frontiers in Chemical Engineering Education," CCR/NSF Workshops, http://mit.edu/che-curriculum/index.html. Accessed 7/15/08.
5. Joseph Shaeiwitz and Richard Turton, "The Changing ChE Curriculum - How Much Change is Appropriate," *ASEE Annual Conference and Exposition Proceedings*, 2006.
6. Milo D. Koretsky, Alexandre Yokochi, Sho Kimura, and Sarah Herzog, "Elements of Student Learning in Nanotechnology," *Int. J. Eng. Edu., in prep.*
7. Richard J. Shavelson, Maria A. Ruiz-Primo, and Edward W. Wiley, "Windows of the Mind," **49**, 413 (2005).

# 603f Targeted Block Copolymers for the Prevention of Post-Surgical Adhesions

**John M. Medley**[1], Thomas D. Dziubla[1], Eugene Kaplan[2], and Eric J. Beane[1]. (1) Chemical and Materials Engineering, University of Kentucky, 177 F. Paul Anderson Tower, Lexington, KY 40506, (2) Urology & Pelvic Reconstructive Surgery, Center for Advanced Gynecologic Surgery and Pelvic Floor Medicine, 120 La Casa Via, Suite 209, Walnut Creek, CA 94598

Following abdominal surgery, the internal wound healing can result in fibrous tissue deposits that connect adjacent tissues, resulting in formations called post-surgical adhesions (PSAs)[1, 2]. While often asymptomatic, these adhesions can lead to serious problems including bowel obstructions, chronic pain, and infertility. In the United States, an estimated 440,000 adhesiolysis surgeries are conducted annually to correct issues arising from the formation of PSAs [3].

To date, no satisfactory method has been developed to prevent all PSA types. While improvements in surgical techniques have reduced instances of PSA for some procedures, the most successful methods to prevent adhesions have relied on physical barriers that separate damaged tissue surfaces[4]. Current barriers methods are limited by the fact that the barriers must be applied directly to the area of damage. This limitation precludes their application in laparoscopic surgeries and limits their effectiveness, as the damaged tissue cannot necessarily be identified or accessed by the physician during surgery.

We have previously demonstrated the ability of block copolymers to adsorb to model surfaces based on charge affinity and inhibit the non-specific adsorption of proteins from solution. Since this protein adsorption event is important in the biochemical cascade that leads to postsurgical adhesion formation, this approach shows great promise as a method to reduce PSA formation. Such an approach overcomes some of the limitations of current treatments; specifically, since the polymer can is applied as a solution and self-assembles at the site of injury, this technique is readily adaptable to laparoscopic procedures and can be employed to treat damaged inaccessible or unidentified tissue damage.

We have extended the power of this method by incorporating peptide domains which have specific binding affinity for damaged tissue protein markers. By controlling the polymer architecture and optimizing the targeting moieties, the function of the material can be controlled. Using a Quartz Crystal Microbalance with Dissipation (QCM-D), the adsorption to model surfaces has been investigated. This technique allows the simultaneous evaluation of binding kinetics, protein blockade, and adsorbate conformation.

**References**

1. Ellis, H., et al., Adhesion-related hospital readmissions after abdominal and pelvic surgery: a retrospective cohort study. Lancet, 1999. 353(9163): p. 1476-80.

2. Risberg, B., Adhesions: preventive strategies. Eur J Surg Suppl, 1997(577): p. 32-9.

3. Matthews, B.D., et al., Assessment of adhesion formation to intra-abdominal polypropylene mesh and polytetrafluoroethylene mesh. J Surg Res, 2003. 114(2): p. 126-32.

4. Boland, G.M. and R.J. Weigel, Formation and prevention of postoperative abdominal adhesions. J Surg Res, 2006. 132(1): p. 3-12.

## 605a Low Temperature Tetherless Grippers for Medicine

**Timothy Leong**[1], Christina Randall[2], Bryan Benson[1], and David H. Gracias[1]. (1) Department of Chemical and Biomolecular Engineering, Johns Hopkins University, 125 Maryland Hall, 3400 N Charles Street, Baltimore, MD 21218, (2) Department of Biomedical Engineering, Johns Hopkins University, Baltimore, MD 21218

We demonstrate the concept of mass-producible, thermally triggered functional tools in the form of microgrippers. The microgrippers close when exposed to a temperature above 40°C, and their response is triggered by changes in material properties engineered within. Thus, the microgrippers do not require any tubes, wiring, or other tethers and are truly mobile. The microgrippers can be set-free in large numbers or individually manipulated magnetically with ease, even through highly coiled tubes. We have utilized the microgrippers to retrieve beads from substrates and cells within narrow tubes; an in vitro biopsy was also demonstrated.

## 605b Engineering Mucus Penetrating Particles for Transmucosal Delivery

**Samuel K. Lai**[1], Ying-Ying Wang[2], Ming Yang[2], Amanda Pace[1], Benjamin C. Tang[1], Richard Cone[3], and Justin Hanes[1]. (1) Department of Chemical and Biomolecular Engineering, The Johns Hopkins University, 221 Maryland Hall, 3400 N. Charles Street, Baltimore, MD 21218, (2) Department of Biomedical Engineering, The Johns Hopkins University, 221 Maryland Hall, 3400 N. Charles Street, Baltimore, MD 21218, (3) Department of Biophysics, Johns Hopkins University, 221 Maryland Hall, 3400 N. Charles St, Baltimore, MD 21218

Highly viscoelastic human mucus layers protect the lungs, GI tract, reproductive tract, eye and more from infectious pathogens and foreign particulates. The human mucus barrier was recently reported as impermeable to polymer nanoparticles as small as 59 nm. Particles that cannot move through the outer layers of mucus are cleared from the mucosal tissue within seconds to at most a few hours. This reality has strongly limited development of long-lasting controlled release systems to mucus-covered tissues, and likely has played a major role in thwarting gene therapy of mucosal tissues as well.

We hypothesized that adhesion to mucus was a critical rate-limiting barrier to nanoparticle transport through mucus layers. We sought to mimic the hydrophilic and net-neutral surface properties of viruses capable of rapidly moving through human mucus. Our initial search for a candidate material that could endow these surface properties on synthetic particles led us to poly(ethylene-glycol), or PEG. Paradoxically, PEG had a considerable history of use as a muco-adhesive.

Here, we describe our recent discovery that coating synthetic nanoparticles with high densities of low molecular weight PEG allows particles with sizes of at least 500 nm in diameter to rapidly transport through undiluted human mucus nearly as fast as they move through pure water. In contrast, high molecular weight PEG coating makes them even more adhesive to mucus than without coatings. We show that PEG density is especially critical as particle size diminishes from 500- to 100-nm. We also show that the spacings within the human mucus mesh are much larger than previously appreciated, which provides a significant opportunity for controlled drug delivery using large nanoparticles. The already large native mesh spacing can be further manipulated to allow micron-sized particles to penetrate mucus at speeds approaching water. We have further show that we can engineer mucus penetrating particles, using a variety of biomaterials, to rapidly penetrate various other human mucus secretions. Animal studies using MPP for gene therapy of the lungs as well as localized and sustained cancer therapy in mucosal tissues are underway.

## 605c Targeting Colon Cancer Cells Using Pegylated Liposomes Modified with a Fibronectin-Mimetic Peptide

**Ashish Garg**, Alison W. Tisdale, and Efrosini Kokkoli. Department of Chemical Engineering and Materials Science, University of Minnesota, 421 Washington Ave. SE, Minneapolis, MN 55455

The ability to target cancer cells using an appropriate drug delivery system can significantly reduce the associated side effects from cancer therapies and can help in improving overall quality of life post cancer survival. Integrin $\alpha_5\beta_1$ is expressed on several types of cancer cells, including colon cancer, and plays an important role in tumor growth and metastasis. Thus, the ability to target the integrin $\alpha_5\beta_1$ using an appropriate drug delivery nano-vector can significantly help in inhibiting tumor growth and reducing tumor metastasis. In this study we have designed functionalized stealth liposomes (liposomes covered with polyethylene glycol (PEG)) that specifically target the integrin $\alpha_5\beta_1$. The PEG provides a steric barrier allowing the liposomes to circulate in the blood for longer duration and the functionalizing moiety, the PR_b peptide specifically recognizes and binds to integrin $\alpha_5\beta_1$ expressing cells. PR_b is a novel peptide sequence, designed in our lab, that mimics the cell adhesion domain of fibronectin, and includes four building blocks, RGDSP (the primary recognition site for $\alpha_5\beta_1$), PHSRN (the synergy site for $\alpha_5\beta_1$), a $(SG)_5$ linker, and a KSS spacer. In this study, we demonstrate that by optimizing the amount of PEG and PR_b on the liposomal interface we can engineer nano-vectors that bind to CT26.WT, HCT116, and RKO colon cancer cells in a specific manner and are internalized through $\alpha_5\beta_1$-mediated endocytosis. Stealth liposomes functionalized with an RGD containing peptide bind to colon cancer cells and internalize, but they have much lesser efficiency than PR_b-targeted stealth liposomes, and more importantly they are not as specific since many integrins bind to RGD. PR_b-targeted stealth liposomes are as cytotoxic as free 5-Fluorouracil (5-FU) and exert the highest cytotoxicity on CT26.WT cells compared to RGD-targeted stealth liposomes and non-targeted stealth liposomes. In order to further increase the efficacy of the system we have designed peptide-functionalized stealth liposomes that are pH-sensitive and exhibit triggered release under mild acidic conditions present in endocytotic vesicles. The proposed targeted delivery system has the great potential to deliver a therapeutic load directly to colon cancer cells, in an efficient and specific manner.

## 605d Designing Nanotherapeutics Via Prediction of Nanoparticle-Protein Binding Interactions

**Esther S. Jeng**, Chemical Engineering, Massachusetts Institute of Technology, Building 66-580, 77 Massachusetts Ave., Cambridge, MA 02139 and Michael S. Strano, 66-566 Department of Chemical Engineering, Massachusetts Institute of Technology, 77 Massachusetts Avenue, Cambridge, MA 02139.

While many biological processes can be controlled by small molecule inhibitors in the binding site of a specific protein, control of most biological processes requires disruption of protein-protein interactions. Nanoparticles provide a scaffold that allows for the design of specific functional groups on a surface for binding to a larger protein-protein interfacial site. Nanoparticles can also be further functionalized with reporter molecules for tracking and sensor purposes. Although many different nanoparticle materials have been functionalized and subsequent binding with proteins has been measured experimentally, these binding events have not been modeled to date. We present a model to predict binding energies of functionalized nanoparticles with biomolecules. A standalone nanoparticle functionalizer and builder were created to allow for binding calculations between target proteins and moieties of various charges, sizes, and configurations. This model can inform about potential candidates for binding systems or reject structures to direct subsequent synthesis of the nanoparticles.

## 605e Gold Nanorod Architectures for Targeted Delivery of Combination Treatments to Cancer Cells

Huang-Chiao Huang[1], Jared Burdick[2], Joseph Wang[3], and **Kaushal Rege**[1]. (1) Department of Chemical Engineering, Arizona State University, ECG 202, Tempe, AZ 85287-6006, (2) Department of Mechanical Engineering and The Biodesign Institute, Arizona State University, Tempe, AZ 85287-6006, (3) Departments of Chemical Engineering and Chemistry and The Biodesign Institute, Arizona State University, Tempe, AZ 85287-6006

The complexity associated with advanced cancer disease necessitates repeated administration of radio / chemotherapy primarily due to recurrent disease and sub-optimal drug delivery. Combination therapies have emerged as powerful alternatives to 'single-agent therapies' since they enhance cell death typically by overcoming multiple cell-death resistance mechanisms in cancer cells. We describe the design and generation of nanoscale architectures based on gold nanorods for delivering combination treatments selectively to cancer cells. Multisegmented gold nanorods with spatially defined regions for cancer cell targeting, combination treatment, and optical (near infrared, NIR) imaging were designed and generated using electrochemical synthesis techniques. These nanorods have significant advantages over commonly used nanoparticles due to the ability to dedicate specific regions at the nanoscale to biological function (i. e. targeting, cell ablation, and optical imaging). Related to this, we describe a newly developed method for the high throughput generation of gold nanorods using a templated electrodeposition approach with an eye towards increasing the yield of nanorods. The uptake, cytotoxic efficacy, selectivity, and mechanisms of cell death induced upon chemotherapeutic drug delivery using multisegmented gold nanorods were evaluated with chemoresistant human prostate cancer cell lines. It is anticipated that these nanorod architectures will have significant impact on the delivery of combination treatments for advanced cancer disease.

# 605f Nanoparticle-Aptamers: An Effective Growth Inhibitor for Human Cancer Cells

**Jong Hyun Choi**[1], Kok Hao Chen[2], Amanda Chaffee[2], and Michael S. Strano[3]. (1) Chemical Engineering, MIT, 77 Massachusetts Avenue, 66-580, Cambridge, MA 02139, (2) Chemical and Biomolecular Engineering, University of Illinois, Roger Adams Laboratory, 600 S. Mathews Avenue, Urbana, IL 61801, (3) 66-566 Department of Chemical Engineering, Massachusetts Institute of Technology, 77 Massachusetts Avenue, Cambridge, MA 02139

Semiconductor nanocrystals have unique optical properties due to quantum confinement effects, and a variety of promising approaches have been devised to interface the nanomaterials with biomolecules for bioimaging and therapeutic applications. We present a novel scheme of synthesizing fluorescent nanocrystal quantum dots (NQDs) using DNA aptamers for growth inhibition of human cancer cells. An anti-cancer aptamer DNA sequence is used as a capping ligand to directly synthesize highly fluorescent NQDs in aqueous solution. The DNA aptamer retains its unique secondary structure on the NQD surface, which is necessary for selective and specific binding to its target protein, nucleolin. The aptamer-functionalized NQDs are proposed as proliferation inhibitors, and their efficacy is examined with MCF-7 human breast cancer cells using hemacytometry and MTT assay. The NQDs show 3-4 times greater inhibition efficacy than DNA drugs alone. In vitro cytotoxicity of these NQDs is investigated with mouse fibroblast cells, and the NQDs are found to be non-toxic under the experimental conditions. The nanoparticle-based therapeutic schemes devised in this work should be valuable in developing a multifunctional drug delivery and imaging agent for biological systems.

J.H. Choi, K.H. Chen, and M.S. Strano; J. Am. Chem. Soc. 128, 15584 (2006)

## 605g Microfluidic Blood Cleansing Device for Sepsis Therapy

**Chong Wing Yung**, Surgery - Vascular Biology Program, Harvard Medical School & Children's Hospital Boston, Karp Family Research Lab, 300 Longwood Ave., 11.005D, Boston, MA 02115, Jason Fiering, Microsystems, Charles Stark Draper Laboratory, 555 Technology Square, Cambridge, MA 02139-3563, and Donald E. Ingber, Surgery & Pathology, Harvard Medical School & Children's Hospital Boston, Karp Family Research Lab, 300 Longwood Ave., 11.127, Boston, MA 02115.

Sepsis is a major killer of man because a high microbial load in blood overcomes even the most powerful existing antibiotic therapies, resulting in multi-systems failure. This is a particularly significant problem in premature infants and immunocompromised patients with fungal infections since antibiotic therapies are dose-limiting. Development of an extracorporeal blood cleansing device that can rapidly clear the blood of pathogens without removing normal blood components could help remedy this problem. We have developed a prototype microfluidic-micromagnetic blood cleansing device that rapidly clears relevant fungal pathogens from flowing human whole blood with high efficiency and throughput. This device is composed of an array of microfluidic channels fabricated from poly-dimethylsiloxane. Magnetic opsonins were created by coating 1-micrometer superparamagnetic microbeads with pathogen-specific antibodies. Using this device, we have selectively removed nearly 80% of living fungal pathogens from a continuously flowing stream of heparinized human whole blood (20 ml/hr/device) in a single pass. We are now working to further enhance throughput and adapt our system for continuous *in vivo* testing in a rabbit sepsis model. The future of our work is to develop this new therapeutic platform microtechnology into a highly useful point-of-care medical device for both civil and military applications. (Funded by CIMIT)

## 610a Nanoscale Engineering for Cytochrome P450 System

**Teruyuki Nagamune**, Department of Bioengineering, School of Engineering, The University of Tokyo, 7-3-1 Hongo, Bunkyo-ku, Tokyo, Japan

Cytochrome P450s are heme-containing monooxygenases involved in various physiological processes from archaea to mammals. They catalyze diverse types of reactions containing hydroxylations, epoxidations, dealkylations and heteroatom oxidations, which make P450s attractive industrial catalysts. P450s require electrons from NAD(P)H to convert molecular oxygen into reactive species. However, excess amounts of electron transfer partner proteins (more than 10 times) are necessary to obtain sufficient reaction rate and coupling efficiency and these properties are bottlenecks for practical applications of P450s.

Here, we show a site-specific branched fusion protein of P450 with its electron transfer proteins constructed using enzymatic cross-linking with a transglutaminase from Streptomyces mobaraensis (TGase). A branched fusion protein of P450 system, which was composed of one molecule each of P450 from Pseudomonas putida (P450cam) which catalyzes stereo-selective hydroxylation of d-camphor, putidaredoxin (Pdx) and putidaredoxin reductase (Pdr), showed higher catalytic activity (306 min-1) and coupling efficiency (99 %) than the reconstitution system due to the intramolecular electron transfer. Many substrates for P450s are highly hydrophobic and P450-catalyzed reactions are preferable to be conducted in nonaqueous media. The regeneration of NADH is also required due to its cost. We finally examined the hydroxylation of d-camphor by the fusion protein using NADH-regeneration driven by a hyper thermostable alcohol dehydrogenase in a reversed micella system. The hydroxylation of d-camphor and the regeneration of NADH were observed even in a reversed micellar system, while those were not observed using the reconstitution system.

This unique site-specific branched structure simply increased local concentration of proteins without serious loss of freedoms and activities of each protein. Therefore, enzymatic posttranslational protein manipulation can be a powerful alternative to conventional strategies for the creation of multi-component enzyme systems with novel proteinaceous architecture.

## 610b Structure and Transport in High-Rate Biocatalytic Electrodes

**Scott Calabrese Barton**, Ramanujam Kothandaraman, and Hao Wen. Chemical Engineering and Materials Science, Michigan State University, 2527 Engineering Building, East Lansing, MI 48824

Attainment of high power density in biocatalyzed fuel cells will require careful attention to electrode design. Enzyme electrocatalysts immobilized on high-surface-area porous supports can achieve high utilization and current density, as long as (1) the microscopic surface area is large, (2) the enzyme interacts with the surface in an way that is conducive to both activity and stability, and (3) transport of electrons and reactants is facilitated adequately.

This talk will review the many approaches to achieving these conditions that appear the literature, and will discuss our own efforts to create high-current density enzyme electrodes. We have designed a multi-scale carbon material that can be used to efficiently support and achieve electrical contact with enzymes. The material is produced by growing carbon nanotubes on carbon fibers using chemical vapor deposition (CVD). With this technique, an increase of more than two orders of magnitude in the surface area available for enzyme immobilization was obtained, resulting in a ten-fold increase in achievable current density using a glucose oxidase-catalyzed glucose electrode.

We have modelled this electrode as supported electrode using a macrohomogeneous approach that considers an active, high surface area film supported on cylindrical carbon fibers. Electrode morphology and biocatalytic film volume are known, and bi-bi ping-pong enzyme kinetics are assumed. We will discuss the use of such a model to study catalyst distribution and predict performance of the multiscale biocatlyzed electrode.

## 610c New Possibility: Biofuel Cells with Stabilized Enzyme Activity

**Su Ha**, School of Chemical Engineering and Bioengineering, Washington State University, PO Box 642710, Pullman, WA 99164-2710

We have developed miniature biofuel cells (BFCs) with dimensions as small as 12x12x9 mm by adopting the design of stackable proton exchange membrane (PEM) fuel cells. The enzymatic anodes were constructed by using stabilized glucose oxidase (GOx) in the form of crosslinked enzyme clusters (CECs) on the surface of carbon nanotubes (CNTs). The combination of stabilized GOx and unbuffered fuel solution resulted in stabilized performance of miniature BFCs under continuous operation for more than 16 hours. This unprecedentedly high operational stability of miniature BFCs opens up new possibilities for many BFC applications.

## 610d Structure, Function, and Stability of Enzymes Covalently Attached to Oxidized Carbon Nanotubes

**Prashanth Asuri**, Sandeep S. Karajanagi, Ravindra C. Pangule, Dhiral A. Shah, Shyam Sundar Bale, Ravi S. Kane, and Jonathan S. Dordick. Chemical and Biological Engineering, Rensselaer Polytechnic Institute, 4005 Center for Biotechnology, Troy, NY 12180

We report the activity, stability, and reusability of enzyme-carbon nanotube conjugates in aqueous solutions. A variety of functionally unrelated enzymes were covalently attached to oxidized single- and multi-walled carbon nanotubes (SWNTs and MWNTs). These conjugates were soluble in aqueous buffer, retained a high fraction of their native activity; and the intrinsic high length of the nanotubes led to facile filtration for reuse. The nanotube-enzyme conjugates were also more stable at elevated temperatures relative to their solution phase counterparts. Furthermore, the high surface area of SWNTs afforded high enzyme loadings, which allowed for a detailed characterization of the secondary and tertiary structure of the immobilized proteins by circular dichroism and fluorescence spectroscopies, respectively. These enzyme conjugates represent novel preparations that possess the virtues of both soluble and immobilized enzymes, thus providing a unique combination of useful attributes such as low diffusional resistance, high activity and stability, and reusability making them attractive choices for applications ranging from diagnostics and sensing to drug delivery.

# 610e Highly Enhanced Fluorescence Detection of Biomolecular Microarray on Large-Area Metal Nanopatterned Substrate

**Hee-Tae Jung**, Department of Chemical and Biomolecular Engineering, Korea Advanced Institute of Science and Technology, Daejeon, South Korea

The fluorescence enhancement can be significantly improved by the use of the periodicity of an ensemble of metal nanostructures, since highly ordered metal nanostructure is far greater than that of aperiodic array or randomly roughened metal surface. As yet, however, very few attempts have been made in the use of metal nanopattern as the MEF, largely due to the difficulty of generating perfect nanoarrays over large-area and placing and immobilizing the biomolecules at predefined positions afforded by the periodic regions of the templates.

Here, we developed a platform of the use of large-area of metal nanopattern (2.7 nm × 2.7 mm) using capillary lithography with high-throughput and subsequent layer-by-layer (LBL) deposition for fluorescence enhancement in a form of DNA microarray, in which their optical properties are adjusted by varying fabrication parameters such as a feature size, periodicity and LBL thickness, leading to a tailored nanoscale design of ultrabright fluorescence emission system with practical manner. The present developments potentially open the way for a variety of nano-biotechnology and optical device applications based on highly periodic patterns over large-areas.

References

1. Generation with capillary force lithography of various complex patterned structures from a single ellipsoidal dot prepattern, J.M. Jung, F. Stellacci, H.-T. Jung, Advanced Materials 19(24), 4392 (2007)

2. Gold-conjugated Protein Nanoarrays via Block Copolymer Lithography: From Fabrication to Biosensors Design, J.-M. Jung, K. Y. Kwon, T.-H. Ha, B. H. Chung, H.-T. Jung, Small 2, 1010 (2006)

3. Internal structure visualization and lithographic use of periodic toroidal holes in liquid crystals, DK Yoon, MC Choi, YH Kim, MW Kim, OD Lavrentovich, HT Jung, Nature Materials 6(11), 866 (2007)

4. Covalent Attachment and Hybridization of DNA Oligonucleotides on Patterned Single-Walled Carbon Nanotube Films, D.-H. Jung, B. H. Kim, Y. K. Ko, M. S. Jung, S.O. Jung, S. Y. Lee, and H.-T. Jung, Langmuir, 20(20), 8886-8891 (2004)

## 610f Protein Patterning through Microcontact Printing with Sortase

**Ranganath Parthasarathy**[1], Scott Retterer[2], Mitchel J Doktycz[2], and Eric T. Boder[3]. (1) Chemical and Biomolecular Engineering, University of Tennessee, 1512 Middle Drive, Knoxville, TN 37996-2200, (2) Cnms, Oak Ridge National Laboratory, PO Box 2008 MS 6123, Oak Ridge, TN 37831-6123, (3) Department of Chemical and Biomolecular Engineering, University of Tennessee, 1512 Middle Drive, 437 Dougherty Engineering Bldg., Knoxville, TN 37996-2200

Sequence-, orientation-, and site-specific ligation of proteins to surfaces presents considerable challenges. We have shown that the transpeptidase, Sortase A, from the Gram-positive bacterium Staphylococcus aureus is a versatile enzyme that can accomplish sequence-specific ligation of proteins to surfaces as well as to soluble species. The ligation requires a carboxy-terminal pentapeptide tag, ---LPXTG on one reactant (R1-LPXTG) and an amino-terminal (Gly)n , n≥2, on the other (GGG-R2). A protein tagged with -LPETG can thus be ligated to a surface with an amino-GGG resulting in orientation-specific presentation of the protein with a tether at its carboxy terminus.

Site-specific modification of surfaces with high spatial resolution has been accomplished using either dip-pen lithography or, more recently, with topographically etched stamps, wherein tethering an enzyme to the stamp greatly increases spatial resolution. We have used Sortase A to spatially pattern protein onto surfaces by contact printing. Sortase A was immobilized on a micro-patterned PDMS stamp and used to couple LPETG-tagged enhanced green fluorescent protein (eGFP) to Gly3-modified glass surfaces. The stamped protein is covalently coupled to the glass and demonstrates high fidelity replication of the spatial pattern of the PDMS stamp. Since one of the benefits of microcontact printing is the possibility of cross-patterning using multiple enzymes with unique target sites, we have also conducted preliminary work to characterize Sortase A from Corynebacterium diphtheriae. In vivo, this enzyme replaces the Glyn substrate of the S. aureus homolog with a lysine side chain contained within a conserved stretch of amino acids (the "pilin box"). Solution data suggest functional overlap between the two homologs in terms of recognition sites. The rate of transition state hydrolysis is faster with the C. diptheriae Sortase A relative to the S. aureus enzyme.

Sortase-mediated ligation has the advantages of using gentle conditions, requiring recognition sequences that are only 3-5 amino acids long, and specifically coupling proteins from a complex mixture; these features render the method versatile and compatible with the functions of many proteins. We anticipate that Sortase-catalyzed microcontact printing will be a powerful tool with which to modify surface physical properties on very small scales in addition to the obvious applications in protein array construction.

# 611a Supercritical Carbon Dioxide Processing of Nano-Clays and Polymer/ clay Nanocomposites

**Mihai Manitiu**[1], Robert Bellair[1], Steven E. Horsch[1], Esin Gulari[2], and Rangaramanujam M. Kannan[1]. (1) Chemical Engineering, Wayne State University, 5050 Anthony Wayne Drive, Detroit, MI 48202, (2) Clemson University, Clemson, SC

Effective dispersion of the fillers in a polymer matrix and improvement of polymer-clay interactions are two key challenges in the field of nanocomposites. We explore a novel processing method that utilizes the unique properties of supercritical carbon dioxide (sc$CO_2$) to disperse nano-clay and prepare a series of polymer/clay nanocomposites with enhanced properties. Significant dispersion was achieved using the sc$CO_2$ process with both Cloisite 93A and Cloisite 10A without the presence of an organic phase as evident by the absence of $d_{001}$ diffraction peak in WAXD. SEM revels that the individual tactoids in the sc$CO_2$ samples have delaminated from each other and lost their parallel registry. The expanded flexible structure of the sc$CO_2$ processed particles/tactoids exposes more of the available surface area allowing for more contact between the polymer and clay surface, and the platelets should be easier to disperse into a polymer matrix than the "as received" clay. Furthermore, multiple sc$CO_2$ processing of the nano-clays leads to an increase level of dispersion which opens up a larger surface area.

We have also prepared a series of polystyrene/clay nanocomposites using the sc$CO_2$ processing method to investigate the effects of sc$CO_2$ depressurization rate, solution blending, clay weight fraction, and pre-dispersion on clay dispersion, nanocomposites morphology and polymer-clay interactions. High molecular weight polystyrene is not processable at low enough temperatures that would prevent clay modifier degradation and a co-solvent is used to improve polymer mobility during processing. Although significant dispersion of the nano-clay is achieved when processes without polymer present, the WAXD of the high molecular weight polystyrene/clay nanocomposites processed in the presence of a co-solvent reveals a strong intercalation peak that is similar in all the nanocomposites regardless of the processing conditions. However, despite the lack of differentiation in WAXD between all the nanocomposites studied, TEM images reveal that there is a larger fraction of dispersed platelets present in the sc$CO_2$ sample compared to solution blended control sample which displays much larger tactoids and a lack of individual clay sheets. There is a three fold reduction in the number of tactoids in the sc$CO_2$ samples leading to an increase in the surface area available for polymer-clay interactions. Processing polymer and clay in sc$CO_2$ followed by quick depressurization leads to significant reinforcement of the polymer matrix compared to solution blending analog alone, implying that clay platelets are effectively being dispersed by the dramatic $CO_2$ expansion. Replacing "as received" clay with pre-dispersed clay increases the surface area available for polymer-clay interactions, resulting in a significant doubling of

G' at low frequencies over the standard $scCO_2$ processed sample. The $scCO_2$ samples show solid-like behavior at loadings as low at 2 wt%, and elastic modulus improvements as high as 3 orders of magnitude in the 10 wt% nanocomposites over the pure polymer.

[1] Manke, C. W.; Gulari, E.; Mielewski, D. F.; Lee, E. C. System and method of delaminating a layered silicate material by supercritical fluid treatment. October 2002.

[2] Mielewski, D. F.; Lee, E. C.; Manke, C. W.; Gulari, E. System and method of preparing a reinforced polymer by supercritical fluid treatment. June 2004.

[3] Horsch, S.; Serhatkulu, G.; Gulari, E.; Kannan, R. M. Polymer 2006, 47, (21), 7485.

Web Page: www.kannangroup.com

## 611b Exponential Growth of Lbl Films with Incorporated Inorganic Sheets

**Paul Podsiadlo**[1], Marc Michel[1], Jungwoo Lee[2], Eric Verploegen[3], Nadine Wong Shi Kam[1], Jaebeom Lee[4], Vincent Ball[5], Ying Qi[6], Anastasios J. Hart, Paula T. Hammond[7], and Nicholas Kotov[1]. (1) Chemical Engineering, University of Michigan, 2300 Hayward St., 3074 H. H. Dow Bldg., Ann Arbor, MI 48109-2136, (2) Biomedical Engineering, University of Michigan, 2300 Hayward St., 3074 H. H. Dow Bldg., Ann Arbor, MI 48109-2136, (3) Materials Science and Engineering, Massachusetts Institute of Technology, Cambridge, MA 02139, (4) University of Michigan, 2300 Hayward, HH Dow Building #3074, Ann Arbor, MI 49109, (5) Institut National de la Santé et de la Recherche Médicale, (6) Electron Microbeam Analysis Laboratory, University of Michigan, 2455 Hayward St., Ann Arbor, MI 48109, (7) Chemical Engineering, Massachusetts Institute of Technology, 77 Massachuhsetts Avenue, 66-525, Cambridge, MA 02139

The fastest growth pattern of layer-by-layer (LBL) assembled films is exponential LBL (e-LBL), which has both fundamental and practical importance. It is associated with "in-and-out" diffusion of flexible polymers, and thus was considered to be impossible for films containing clay sheets with strong barrier function, preventing diffusion. Here, we demonstrate that e-LBL for inorganic sheets is possible in a complex tri-component film of poly(ethyleneimine) (PEI), poly(acrylic acid) (PAA), and Na+-montmorillonite (MTM). This system displayed clear e-LBL patterns both in terms of initial accumulation of materials and unusually thick individual bilayers later in deposition process with film thicknesses reaching 200 ìm for films composed of 200 pairs of layers. Successful incorporation of MTM layers was observed by scanning electron microscopy and thermo-gravimetric analysis. Surprisingly, the growth rate was found to be nearly identical in films with and without clay layers, which suggests fast permeation/reptation of polyelectrolytes between the nanosheets during the "in-and-out" diffusion of polymer. Considering these findings, e-LBL growth property is expected for a wide array of available inorganic nanoscale components and have a potential to greatly expand the e-LBL field and LBL field altogether. The large thickness and rapid growth of the films affords fast preparation of nanostructured materials which is essential for multiple practical applications ranging from optical devices to ultrastrong composites.

# 611c Influence of the Functional Group, Type of Filler and Processing on the Mechanical Properties of Swcnt/nylon and Graphene Sheet/nylon Composites

**Arun K. Kota**[1], Mohammad Moniruzzaman[1], Arnab Mukherjee[2], Jayanta Chattopadhyay[2], W. Edward Billups[2], and Karen I. Winey[1]. (1) University of Pennsylvania, Philadelphia, PA 19104, (2) Rice University

We have recently prepared nylon 6,10 nanocomposites using functionalized single wall carbon nanotubes (SWCNTs) and our interfacial in situ polycondensation method. The specific functional groups $-(CH2)_4COCl$ and $-(CH2)_9COCl$ on the sidewalls of SWCNT were designed to covalently link nanotubes to the nylon matrix via alkyl segments. The composites with functionalized SWCNT show significant improvements in tensile modulus, strength and toughness relative to nylon and nylon modified with non-functionalized SWCNT. The alkyl linkages at the SWCNT/nylon 6,10 interface contribute significantly to improving the toughness of the composites.

We have now extended this effort to nylon 6,10 nanocomposites with functionalized graphene sheets. The graphene sheets were functionalized with $-COCl$, $-(CH2)_4COCl$, and $-(CH2)_9COCl$ using the Birch type reaction protocol and incorporated into nylon 6,10 using our in situ polycondensation method. The composites were melt spun into fibers or hot pressed into films for further characterization. The influence of the functional groups on the polymer-particle interface is being examined by characterizing the morphology and the tensile properties. Subsequently, influence of the type of filler on the mechanical properties will be studied by comparing the tensile properties of the graphene composites with those of the SWCNT composites. Finally, the processing effects on the tensile properties are investigated by comparing the tensile properties of the films with those of the fibers. The broad implications of this work include making an informed decision in the selection of the functional group, type of filler and the type of processing to tune the mechanical properties of polymer nanocomposites.

# 611d Deconvoluting Nanofiller Reinforcement and Polymer Morphology Via Model Polymer Nanocomposites

Gowri Dorairaju[1], Saurabh Toshniwal[1], Kunal Tulsyan[1], **Daniel F. Schmidt**[1], and Emmanuelle Reynaud[2]. (1) Plastics Engineering, University of Massachusetts Lowell, 1 University Avenue, Lowell, MA 01854, (2) Mechanical Engineering, University of Massachusetts Lowell, 1 University Avenue, Lowell, MA 01854

In spite of substantial work in the field of polymer nanocomposites, generalized predictions of their mechanical behavior have not yet been achieved. In thermoplastic systems based on nanoclays in particular, substantial mechanical properties enhancements have been observed primarily in polar, semi-crystalline systems such as polyamide-6 [1,2] and poly(vinylidene fluoride) [3,4] – but always accompanied by substantial changes in the morphology of the semi-crystalline polymer. The convolution of these effects – the presence of stiff clay layers and the remodeling of the microstructure of the polymer – represents a significant obstacle in the path to understanding the mechanical properties of these materials.

In order to address this issue, we have chosen a fully amorphous polyamide as a matrix for the production of model nanoclay-based polymer nanocomposites that lack the morphological complexities associated with their semi-crystalline analogs. In particular, we report on the successful twin-screw extrusion and injection molding of systems based on Grilamid TR 90 in combination with Cloisite 30B using realistic, industrial-scale processes and equipment.

Structural analysis of these systems (SAXS, WAXS, TEM) confirms high levels of clay dispersion, consistent with the known compatibility of this nanoclay with polyamides in general. Thermal analyses (TGA, DSC) are used to confirm the clay content and the amorphous character of these materials. Finally, we present DMA data, simultaneous tensile testing / thermal imaging results, and SEM-EDS analyses of the rupture faces in order to shed light on the reinforcement mechanisms active in these systems, and draw comparisons with their semi-crystalline counterparts.

1. 'Mechanical properties of nylon 6-clay hybrid', Y. Kojima, A. Usuki, M. Kawasumi, A. Okada, Y. Fukushima, T. Kurauchi, O. Kamigaito, Journal of Materials Research, 8, 1993, 1185-1189.

2. 'The chemistry of polymer-clay hybrids', A. Okada, A. Usuki, Materials Science and Engineering C, 3, 2, 1995, 109-115.

3. 'Dramatic Enhancements in Toughness of Polyvinylidene Fluoride Nanocomposites via Nanoclay-Directed Crystal Structure and Morphology', D. Shah, P. Maiti, E. Gunn, D. F. Schmidt, D. D. Jiang, C. A. Batt, E. P. Giannelis, Advanced Materials, 16, 14, 2004, 1173-1177.

4. 'Effect of Nanoparticle Mobility on Toughness of Polymer Nanocomposites', D. Shah, P. Maiti, D. D. Jiang, C. A. Batt, E. P. Giannelis, Advanced Materials, 17, 5, 2005, 525-528.

# 611e Influence of Nanoparticle Filler on the Dynamic Relaxation Characteristics of Polymer Nanocomposites

Anthony C. Comer[1], Victor A. Kusuma[2], Benny D. Freeman[2], and **Douglass S. Kalika**[1]. (1) Chemical and Materials Engineering, University of Kentucky, 177 Anderson Hall, Lexington, KY 40506-0046, (2) Chemical Engineering, The University of Texas at Austin, 10100 Burnet Rd., Bldg 133, Center for Energy and Environmental Resources, Austin, TX 78758

The formulation of nanocomposites by the introduction of nanoscale filler particles into a polymeric matrix can produce dramatic enhancements in material performance. Particle-polymer interactions, as well as physical confinement, lead to local conformational changes in the polymer chains and the influence of these changes on the structure and dynamics of the polymer matrix are often manifested in key performance properties, creating opportunities for new and unique material applications.

The extensive particle-polymer surface area and local confinement effects in polymer nanocomposites can have a strong influence on the segmental mobility of the polymer chains as reflected in the characteristics of the glass-rubber relaxation and corresponding glass transition temperature, Tg. Over the last several decades, there have been a number of reports examining the glass transition characteristics of filled polymers reinforced with materials such as carbon black or silica; more recently, such systems have been studied within the context of nanoscale phenomena. In filled polymers, the inclusion of inorganic particles often leads to an increase in the glass transition temperature due to favorable interactions between the particle surface and the polymer chains that limit chain mobility. Investigators have reported positive offsets in Tg, as well as the emergence of a second (higher) Tg corresponding to the presence of a distinct, constrained population of chain segments in the vicinity of the particle surface [1]. For systems that exhibit poor wetting, Tg reductions have also been encountered. Recent studies have explored the potential agreement in Tg trends between bulk nanocomposites and polymer thin films wherein both the nature of the polymer-solid interface and the degree of geometric confinement can be carefully controlled [2-3].

In this paper, we examine the influence of nanoscale particle loading on the dynamic relaxation characteristics of two model polymer nanocomposite systems based on rubbery and glassy polymers, respectively. Both systems are of interest with respect to potential improvements in gas separation performance upon nanoparticle addition. Specifically, dynamic mechanical analysis and broadband dielectric spectroscopy techniques have been employed to elucidate the characteristics of the glass-rubber and sub-glass relaxations in these systems as a function of particle surface chemistry, particle loading and nanocomposite processing.

Rubbery nanocomposite networks were prepared by UV photopolymerization of poly(ethylene glycol) diacrylate [PEGDA] in the presence of varying loadings of MgO (~ 3 nm) or SiO2 (10 nm) nanoparticles. For the nanocomposites, the liquid pre-polymerization mixture was prepared by conventional mixing methods, as well as by planetary mixing; at higher loadings, particle dispersion was facilitated by the introduction of an appropriate diluent, and the influence of the diluent on the characteristics of the crosslinked network and resulting composites was evaluated. Glassy nanocomposites were produced via solution casting of mixtures comprised of polyetherimide [PEI] and SiO2 nanoparticles according to the method described by Takahashi and Paul [4]; nanocomposites based on the inclusion of native SiO2 as well as commercially-available surface-modified silica were examined. Characterization of the static and dynamic properties of the composites included bulk density measurement and dynamic mechanical and dielectric studies. In addition, pure gas permeability measurements were performed for the PEGDA/MgO samples.

Dynamic mechanical studies on the PEGDA/MgO and PEGDA/SiO2 networks revealed shifts in the glass transition processes of the composites that correlated with the nature and dispersion of the respective metal oxide fillers. Across the PEGDA/SiO2 series, a progressive increase in Tg on the order of 15°C was observed with increasing SiO2 loading (0 to 30 wt%). For the PEGDA/MgO nanocomposites, the bulk PEGDA glass transition temperature (Tg = -35°C at 1 Hz) was largely unchanged, but a second, distinct relaxation event emerged approximately 40°C above Tg suggesting the presence of a sub-population of chain segments subject to a higher degree of motional constraint. The characteristics of this second relaxation were found to be dependent upon filler loading and the processing conditions used for preparation of the pre-polymerization mixture. Dielectric data for the PEGDA nanocomposites were consistent with the dynamic mechanical results and provided additional insight as to the influence of the nanoparticles on the individual sub-glass relaxations. Dielectric response as a function of particle loading was analyzed according to established descriptions of inhomogeneous media [5].

In the case of the PEI nanocomposites, the introduction of native SiO2 (up to 25 wt%) appeared to have little influence on the glass transition temperature of the composite as compared to the unfilled polymer (Tg ~ 225°C at 1 Hz). However, the introduction of surface-modified (i.e., hydrophobic) fumed silica led to improved particle dispersion [4] and the appearance of a second, higher-temperature Tg in the dynamic mechanical sweeps. This second Tg was offset by approximately 60°C as compared to the bulk PEI relaxation and appears to correspond to a separate glass transition emanating from regions of reduced mobility in the vicinity of the particle-polymer interface. The properties of the dual-Tg relaxation were assessed as a function of particle loading, surface characteristics, and the resulting nanocomposite morphology.

REFERENCES:

1.) G. Tsagaropoulos and A. Eisenberg, Macromolecules 1995, 28, 6067.

2.) A. Bansal, H. Yang, C. Li, K. Cho, B.C. Benicewicz, S.K. Kumar and L.S. Schadler, Nature Materials 2005, 4, 693.

3.) P. Rittigstein, R.D. Priestley, L.J. Broadbelt and J.M. Torkelson, Nature Materials 2007, 6, 278.

4.) S. Takahashi and D.R. Paul, Polymer 2006, 47, 7519.

5.) P.A.M. Steeman and J. van Turnhout, in: Broadband Dielectric Spectroscopy; Kremer F, Schonhals A, editors. Berlin: Springer-Verlag, 2003.

# 611f Nanoparticle Incorporation and Aggregation In Block Copolymer Worm-Like Micelles

**David A. Christian**[1], Karthikan Rajagopal[1], Goundla Srinivas[2], Michael L. Klein[2], and Dennis E. Discher[1]. (1) Chemical and Biomolecular Engineering, University of Pennsylvania, Room 129 Towne Bldg., 220 South 33rd Street, Philadelphia, PA 19104-6393, (2) Chemistry, University of Pennsylvania, Philadelphia, PA 19104

Top-down lithographic approaches have significant limitations in either feature size (optical lithography) or cost (electron beam lithography), whereas self-assembling polymer systems can in principle provide robust scaffolds for positioning. Block copolymers are composed of chemically distinct polymer chains joined by a covalent bond at the interface, and the distinct chemistries not only drive self-assembly in the bulk phase to form periodic nanostructures over supramolecular length scales but they might also be used to precisely order nanoparticles embedded in the polymer. By changing the surface chemistry of the particles, for example, their location can be predictably changed to different domains in the bulk diblock copolymer assembly. Factors such as nanoparticle composition, volume fraction, and concentration play crucial roles in determining the final morphology and the properties of such nanoparticle-copolymer composite materials. How these nanoparticles interact and aggregate within these block copolymer assemblies is therefore important to understand.

Amphiphilic diblock copolymers are particularly well known to self assemble in dilute solutions, forming supramolecular aggregates with a hydrophobic core surrounded by a dense, hydrophilic brush. By changing the lengths of the different polymer chains, the morphology of these aggregates can be tuned to form worm-like micelles that are nanometers in diameter and microns in length. Recent advances in assembly of worm-like micelles have demonstrated their potential to serve as linear nano-templates for ordering nanoparticles over several microns. Here, we investigate – by experiment and by coarse-grain simulation – the nanoparticle-copolymer (N-C) composite self-assembly and aggregation.

Coarse grain modeling of the self-assembly of hydrophobic nanoparticles and diblock copolymers dispersed randomly in an aqueous solution shows that nanoparticle-copolymer interactions drive the self-assembly process to produce worm-like micelles faster than when the copolymer alone is in solution. Coarse grain modeling was also used to study the aggregation of nanoparticles integrated within the core of pre-assembled worm-like micelles. In the model, nanoparticles were placed 3 nm apart in the core of a worm-like micelle. The kinetics for nanoparticle aggregation was an order of magnitude slower compared to the case with a random initial configuration. No change in the distance between the nanoparticles was observed until the density of copolymer molecules around the nanoparticles changed. It can therefore be concluded that the local density of the copolymers in the core of the worm-like micelle controls the aggregation of nanoparticles.

Integration and aggregation of hydrophobic nanoparticles – core-shell quantum dots (CSQD) – in the cores of pre-assembled worm-like micelles was investigated experimentally. Worm-like micelles composed of the amphiphilic diblock poly(ethylene oxide)-polybutadiene (PEO-PBD) were assembled by the solvent evaporation in water. CSQDs dissolved in chloroform were then dispersed into the worm-like micelle solution and were observed by fluorescence microscopy to incorporate into the hydrophobic core of worm-like micelles. Worm-like micelles were homogeneously labeled with a fluorescent hydrophobic dye, and CSQDs could be imaged as punctates along the worm-like micelle contour length. Image analysis of CSQD fluorescence intensity indicated a distribution of CSQD aggregate number. Ongoing experiments aim to incorporate CSQDs more efficiently into the worm-like micelles and also to quantify the number of CSQDs per aggregate.

## 612a Designing All-Organic MRI Contrast Particles Which Produce Contrast through Chemical Exchange Saturation Transfer

**Michael T. McMahon**[1], Yah-el Har-el[2], Guanshu Liu[1], Christopher Long[3], Assaf A. Gilad[4], Marco A. Deliso[1], George Sgouros[4], Jeff W.M. Bulte[4], and Peter C.M. van Zijl[1]. (1) F.M. Kirby Research Center for Functional Brain Imaging, Kennedy Krieger Institute, Baltimore, MD 21205, (2) Radiology and Radiological Science, Johns Hopkins University, 1150 Orleans St., Cancer Research Building 2, Room 4M62, Baltimore, MD 21231, (3) Biomedical Engineering, Johns Hopkins School of Medicine, Baltimore, MD 21205, (4) Russell H. Morgan Dept. of Radiology and Radiological Science, Johns Hopkins University, Baltimore, MD 21205

Paramagnetic and Super-Paramagnetic metal-based nanoparticles have been developed previously as MR contrast agents, and are in use for the imaging of drug delivery, gene therapy as well as to highlight processes such as plaque formation in blood vessels. We are now augmenting this library of agents with new liposome based MR particles consisting of lipids, cholesterol and either amino acids or carbohydrates and without metals. These particles can be selectively activated by frequency-specific RF irradiation, with the contrast produced due to chemical exchange saturation transfer (CEST) from exchangeable protons to water. In addition, these particles can be distinguished by the saturation frequency dependence of their sugar or amino acid content. We have injected these particles into the feet of mice, and imaged their trafficking to the popliteal lymph nodes. This trafficking was confirmed by incorporation of 111In-Cl into the liposomal lumen and obtaining SPECT-CT images.

# 612b Properties of Protein Polymer-Based MRI Contrast Agents and Their Use for Tracking Biomaterial Degradation In Vivo

**Lindsay Karfeld-Sulzer**[1], Hermann Kissler[2], Nicolynn Davis[1], Dixon B. Kaufman[2], Thomas Meade[3], and Annelise Barron[4]. (1) Chemical and Biological Engineering, Northwestern University, Tech E136, 2145 Sheridan Road, Evanston, IL 60208, (2) Surgery, Transplant Division, Northwestern University Feinberg School of Medicine, Chicago, IL 60611, (3) Chemistry; Biochemistry, Molecular Biology, and Cell Biology; Radiology, Northwestern University, 2145 Sheridan Road, Evanston, IL 60208, (4) Stanford University, 318 Campus Drive, Stanford, CA 94305

One significant barrier to the development of new generations of biocompatible materials, particularly tissue engineering hydrogels, is the difficulty of non-invasively evaluating the properties and performance of the biomaterial over time in vivo (1,2). Magnetic Resonance Imaging (MRI) is capable of whole-animal or -human imaging at high spatial and temporal resolution, and is an ideal modality for assessing hydrogel scaffolds in vivo (3). Although MRI possesses these advantages, it generally has low image contrast, making it difficult to distinguish the area of interest. Exogenous contrast agents (CAs) increase the relaxation rate (1/T1) of water protons and therefore improve image contrast. However, current clinically used CAs have low relaxivities and cannot be incorporated into biomaterials (4). Recently, researchers have been attaching small-molecule CAs to various macromolecules via either covalent or non-covalent interactions as a strategy to increase relaxivity (4). We have created a novel family of macromolecular CAs based on protein polymers to enable MRI tracking of biomaterials in vivo.

A family of protein polymer contrast agents was created by conjugating Gd(III) chelators onto evenly spaced lysines on recombinantly expressed protein polymers. Controlled cloning (5) and recombinant protein expression in E. coli were used to create the protein polymer backbone for the CAs. Inherent to this process, protein polymers are completely monodisperse and the amino acid sequence is exactly specified, allowing control over the density of reactive sites and length of the protein. Additionally, bioactive moieties, such as enyzymatic cleavage sites, could be included in the protein using controlled cloning. The chelators were conjugated onto the protein polymer in aqueous solution using peptide bond-forming agents (EDC and Sulfo-NHS). Due to the monodispersity of the protein polymers, the extent of conjugation can be characterized completely. We have demonstrated that we can modulate relaxivity through changing protein polymer length and lysine spacing. These CAs have been shown to be biodegradable by incubation with plasmin and virtually non-toxic to cells based on a viability assay. These protein polymer CAs have been crosslinked into tissue engineering hydrogels and shown dramatic contrast improvement. Since the enhanced contrast pixel volume due to the CAs can be tracked and should decrease as the material degrades, these CAs can be used to monitor in vivo scaffold degradation. In vivo experiments are in process to quantitatively analyze the degradation rate of protein polymer-based hydrogels.

References

1. Langer R., Tirrell D.A. Designing materials for biology and medicine. Nature, 428, 487, 2004.

2. Neves A.A., Medcalf N., Smith M., Brindle K.M. Evaluation of engineered meniscal cartilage constructs based on different scaffold geometries using magnetic resonance imaging and spectroscopy. Tissue Engr., 12, 53, 2006.

3. Bull S.R., Guler M.O., Bras R.E., Venkatasubramanian P.N., Stupp S.I., Meade T.J. Magnetic resonance imaging of self-assembled biomaterial scaffolds. Bioconjugate Chemistry, 16, 1343, 2005.

4. Artemov D., Bhujwalla Z., Bulte J. Magnetic resonance imaging of cell surface receptors using targeted contrast agents. Current Pharmaceutical Biotechnology, 5, 485, 2004.

5. Won J., Barron A. A new cloning method for the preparation of long repetitive polypeptides without a sequence requirement. Macromolecules, 35, 8281, 2002.

## 612c Imaging of Single Nanoparticles In Live Cell Cytoplasm

**Gang Ruan**, Chemical and Biomolecular Engineering, the Ohio State University, 140 W 19th Ave, Columbus, OH 43210, Amit Agrawal, Health Sciences and Technology, Massachusetts Institute of Technology, 77 Massachusetts Avenue, Cambridge, MA 02139, and Shuming Nie, Biomedical Enigneering, Georgia Institute of Technology and Emory University, 1639 Pierce Drive Suite 2001, Atlanta, GA 30322.

Single molecule studies can lead to insights not available from examining bulk behaviors due to ensemble averaging effects. Although imaging of single molecules in vitro and in live cell membrane by fluorescence microscopy has been extensively reported, single molecule imaging in live cell cytoplasm remains technically challenging. Single quantum dots have similar sizes as those of biomolecules and have many desirable fluorescent properties, thus offering possibilities for single molecule imaging (if quantum dots are linked with biomolecules). To the best of our knowledge, imaging of single quantum dots in the cytoplasm of live cells has not been reported prior to the present work, due to a number of challenges such as cellular delivery and nanoparticle aggregation.

We have obtained movies of single quantum dots in live cell cytoplasm by combining toxin delivery technique, optimized nanoparticle coating and dynamic confocal imaging. The status of single quantum dots is confirmed by the fluorescence intermittency of these nanoparticles. Interestingly, the aggregation of quantum dots in cytoplasm is found to be dramatically more severe than that in an extracellular environment. Aggregation of nanoparticles in the cytoplasm is greatly reduced by adjusting the surface chemistry of nanoparticles. In addition, the cell type used is also found to affect the nanoparticle aggregation behavior. We have further used thus-obtained single quantum dots to track the dynamics of motor proteins in live cells.

# 612d Quantum Dots-Capped Viruses for Antiviral Drug Screening

**Ching-An Peng**, Morris Hsu, and Chung-Hao Wang. Chemical Engineering, National Taiwan University, No. 1, Sec. 4, Roosevelt Road, Taipei, Taiwan

The increasing threats of viral diseases have gained worldwide attention in recent years. Quite a few infectious diseases are still lack of effective prevention or treatment. The pace of developing antiviral agents could be expedited by the availability of quick and efficient drug screening platforms. In this study, quantum dot (QD) was employed to form complexes with virus and used as fluorescent imaging probes for exploring potential antiviral therapeutics. Inorganic CdSe/ZnS QDs synthesized in organic phase were encapsulated by amphiphilic alginate to attain biocompatible water-soluble QDs via phase transfer. Electrostatic repulsion force generated from both negatively-charged dengue virus (or enterovirus) and QDs were neutralized by various concentration of cationic polybrene, forming colloidal complexes of QD-virus. Results showed that susceptible cells infected with QD-tagged viruses exhibited bright fluorescence intracellularly. To demonstrate the potency of QD-tagged viruses for screening antiviral agents, susceptible cells were incubated for one hour with allophycocyanin and then infected with QD-tagged viruses. Based on the developed cell-based imaging assay, allophycocyanin with concentration of 2 mg/mL led to extremely weak intracellular fluorescence post-infection. That is, the efficacy of anti-viral activity of allophycocyanin was clearly illustrated by the inorganic-organic hybrid platform constructed in current study.

## 612e Cytotoxicity of Luminescent Silicon Quantum Dots Engineered for Biological Applications

**Folarin Erogbogbo**[1], Ken-Tye Yong[2], Hong Ding[2], Paras N. Prasad[2], and Mark Swihart[1]. (1) Chemical and Biological Engineering, The University at Buffalo (SUNY), 303 Furnas Hall, Buffalo, NY 14260, (2) Institute for Lasers, Photonics, and Biophotonics, The University at Buffalo (SUNY), Buffalo, NY 14260-4200

Silicon quantum dots (QDs) have been proposed for use as luminescent biological probes. Along with the general advantages of quantum dots over traditional fluorescent dyes, such as brighter emission, photostability, and the ability to excite dots with multiple emission wavelengths using a single excitation source, interest in silicon quantum dots for biological applications is driven by their perceived low toxicity. Given the low toxicity of bulk silicon and porous silicon, this is a reasonable expectation. However, the biocompatibility of free silicon nanoparticles remains to be investigated. We have recently demonstrated in vitro fluorescence imaging using Si quantum dots in live cells. While we did not observe any acute toxicity in that study, no thorough investigation of cytotoxicity of silicon nanoparticles, or the dependence of cytotoxicity on the method of synthesis or surface modification, has been presented. Thus, in the work to be described here, we have measured the cytotoxic effect of luminescent silicon quantum dots. Bare (as prepared), organic ligand functionalized, micelle encapsulated, protein conjugated, and antibody conjugated silicon quantum dots were incubated with pancreatic cancer cells. Dose dependent cytotoxicity was evaluated using an MTS viability assay, for a wide range of Si QD concentrations. Our results confirm that silicon nanoparticles synthesized by laser pyrolysis are relatively non-toxic to cells. However, depending on the method of surface functionalization of the particles, some apoptosis may occur. In the range typically used for biological imaging, all of the silicon nanoparticle preparations were relatively non-toxic, confirming their potential for use as biocompatible imaging agents.

## 612f Gold Nanoroses for Optical Biomedical Imaging and Photothermal Therapy In Atherosclerosis and Cancer

**Li Ma**[1], Kiran Cheruku[2], Amit Paranjape[3], Vidia Paramita[1], Timothy A. Larson[3], Jignesh Shah[3], Marc D. Feldman[2], Thomas E. Milner[3], Bysani Chandrasekar[2], Konstantin Sokolov[3], Stanislav Emelianov[3], and Keith P. Johnston[1]. (1) Department of Chemical Engineering, University of Texas at Austin, 1 University Station C0400, Austin, TX 78712, (2) Division of Cardiology, Department of Medicine, University of Texas Health Science Center at San Antonio, Mail Code 7872, 7703 Floyd Curl Drive, San Antonio, TX 78229, (3) Department of Biomedical Engineering, University of Texas at Austin, 1 University Station C0800, Austin, TX 78712

Novel nanoscaled composites of iron oxide and gold, coated with hydrophilic polymers, have been designed to absorb light in the near infrared (NIR) spectral region for optical imaging in atherosclerosis and cancer. These "nanoroses" form a stable colloidal dispersion in aqueous media. The geometry of the gold domains in the nanorose particle leads to a red-shift in absorbance into the NIR region, where light scattering of blood is less. The hydrophilic coatings are designed to control the synthesis of the nanoroses and to enhance the circulation time in the bloodstream. The nanoroses are compared with four previously reported types of NIR contrast enhancement agents: nanospheres, nanorods, nanoshells, and nanocages. The nanoroses have high optical sensitivity (absorption and scattering) in the NIR and visible region (500 nm -1100 nm), and thus provide high contrast enhancement for various optical biomedical imaging techniques including optical coherence tomography (OCT), photoacoustic imaging, and optical microscopy. The nanoroses are coated with dextran to target the macrophage cells. Uptake of nanoroses into macrophage cells associated with atherosclerotic plaque is investigated with dark field and phase contrast microscopy, and the nanoroses optical properties within the cells are measured with hyperspectral microscopy. Experiments were carried out ex-vivo on animal tissue injected with the broadly absorbing gold nanocomposites. Photoacoustic imaging identified the presence of gold nanocomposites in tissue. In addition, a localized temperature increase, obtained during therapy, was monitored using photoacoustic and ultrasound imaging.

# 612g Multi-Stage Delivery of Gold Nanoparticles for Detection of Early Stage Oral Cancer Using Optical Coherence Tomography

**Chang Soo Kim**[1], Yeh-Chan Ahn[2], Lih-Huei Liaw[2], Hilari Kawakami-Wong[2], Petra Wilder-Smith[2], Matthew Brenner[2], Zhongping Chen[3], and Young Jik Kwon[4]. (1) Chemical Engineering and Materials Science, and Beckman Laser Institute, University of California, Irvine, CA 92697, (2) Beckman Laser Institute, University of California, Irvine, CA 92697, (3) Beckman Laser Institute and Biomedical Engineering, University of California, Irvine, CA 92697, (4) Chemical Engineering and Materials Science, and Pharmaceutical Sciences, University of California, Irvine, CA 92697

Early detection is an inarguably indispensable requirement for achieving successful cancer therapy. Among many clinically available imaging modalities, Optical Coherence Tomography (OCT), a technology visualizing heterogeneous structures of a tissue by quantifying local mismatches in refractive indices, offers noninvasive and fast visualization of tumor tissues at micro-resolution with several millimeter depths. However, early stages of cancer do not induce remarkable changes of tissue structures, which necessitates use of contrast agents to detect premalignancy at a molecular level. Among a number of nanomaterials considered as OCT contrast agents, gold nanoparticles offer high optical scattering properties, clinically acceptable biocompatibility, and ease of bioconjugation with additional functional modalities. In this study, engineering questions of how to overcome highly limited penetration and distribution of gold nanoparticles in an early oral tumor, where efficient transport via vasculature is not available, were explored.

Tumors in hamster cheek pouch were induced by treatment with DMBA (9,10 dimethyl-1,2 benzanthracene) for 5 months. Spherical gold nanoparticles (70 nm in diameter) conjugated with anti-epidermal growth factor receptor (EGFR) monoclonal antibodies were administrated to the cheek pouch of an anesthetized hamster. Before topical administration of gold nanoparticles, microneedles (isosceles shape with 170 μm base and 300 μm height) were used to create passages for gold nanoparticles to penetrate through epithelial layers. To mildly disturb cellular junctions in order to enhance distribution of gold nanoparticles in the tissue, ultrasound waves with the clinically accepted frequency of 1MHz and the density of 0.3 W/cm2 were applied to the area where gold nanoparticles had been administered. Results obtained by 3D spectral domain OCT (SD-OCT) showed remarkably enhanced OCT images with clearly differentiated epithelial layers in the tissue. Histological analysis was also used to confirm and correlate distributions of gold nanoparticles in the tissue with the OCT images. Parameters involved in the penetration and distribution of gold nanoparticles using this multi-stage delivery approach will also be discussed with theoretical and experimental evidence.

A fractal analysis is presented for different examples wherein biosensors have been involved in drug discovery. This is an important area of investigation wherein there is a continually increasing application of biosensors, and where biosensors are making important contributions. This is particularly so for cases wherein biosensors may be used as a HTS (high-throughput screening) method to quickly narrow down the possible candidates from a wide spectrum of potential candidates. The examples analyzed include (a) inhibitors of protein kinases (Viht et al., 2007) wherein the interactions of adenosine-oligoarginine congujates (ARC) and alpha isoform of the catalytic subunit (Calpha) of CAPK are examined, (b) the binding of phosphate ion (Pi) to a rhodamine-PBP (phosphate binding protein) fluorescence-based phosphate biosensor (Okoh et al., 2006), and (c) the binding of different concentrations (in mU) of methionine-7-amido-4-methylcoumarin (MET-AMC) and methionine in solution in the cSPA (competitive aminoacyl-tRNA synthetase charging assay) (Forbes et al., 2007). The binding kinetics is described by either a single- or a dual-fractal analysis. A dual-fractal analysis is only used when a single-fractal analysis did not provide an adequate fit. This was done using Corel Quattro Pro 8.0. The fractal dimension provides a quantitative measure of the degree of heterogeneity present on the biosensor chip surface. An increase in the fractal dimension value or the degree of heterogeneity on the surface leads, in general, to an increase in the binding and in the dissociation rate coefficient. For example, for the binding of Calpha in solution to ARC-704 immobilized on a SPR biosensor chip surface (Viht et al., 2007), and for a dual-fractal analysis, the binding rate coefficient, k1 exhibits an order of dependence between seven and seven and one-half (equal to 7.351) on the fractal dimension, Df1 or the degree of heterogeneity on the SPR biosensor chip surface. This indicates that the binding rate coefficient, k1 is very sensitive to the fractal dimension or the degree of heterogeneity present on the sensor chip surface. The three examples to be presented emphasize that the degree of heterogeneity that exists on the biosensor surface does significantly affect, in general, the rate coefficient and affinity values, and subsequently the kinetics in general.

# 655b Controlled Assembly and Disassembly of Biotinylated Au Nanopatricles on Metallic Nanowires for Nanobiosensing

**Ramazan Kizil**, Chemical Engineering, Istanbul Technical University, School of Chemical & Metallurgical Engineering, B-410 Chemical Engineering Department, Istanbul, 34469, Turkey

The reversible binding of biotin capped Au nanoparticles to the streptavidin coated surfaces can be useful for nano-electronic and nano-sensor applications. Au nanoparticles derivitized with ethylene glycol and biotin together exhibit reversible aggregation characteristics. Streptavidin functionalized metallic nanowires were electrofluidically aligned between micron-size separated planar electrodes under an AC bias and were interacted with the biotinylated Au nanoparticles dispersed in water. The biotinylated nanospheres were assembled on the nanowire surface because of the biorecognition and disassociated by adding free biotin, removing the Au nanoparticles from the surface.

The current electrical nanobiosensing platforms based on FET, conductance measurements of single molecules and molecular electronics applications suffer from the stability and reproducibility and scaling due to the local environment of the molecules being characterized, interaction with vibrational modes and microscopic coupling of the molecule to the measurement electrodes. For this reason, electrical measurements undergo large variations and drawing a good conclusion about the conductance of molecules necessitates averaging of multiple measurements. Nanoparticle based reversible binding of an analyte or a single molecule approach can be used to validate the electrical response of the nanobiosensor and molecular electronics systems.

# Development of a new generation of bone cement using nanotechnology

S. M. Khaled[1], Amin S. Rizkalla*[1, 2] and Paul A. Charpentier*[1]
1. Department of Chemical and Biochemical Engineering, Faculty of Engineering
[2] Division of Biomaterial Science, Schulich School of Medicine and Dentistry
University of Western Ontario
London, Ontario, Canada N6A 5B9

**Abstract**

A new generation of bone cement with mechanical properties significantly higher than that of commercially available acrylic bone cements are strongly desired in order to ensure the long term clinical performance of the cemented arthroplasty. Currently commercial acrylic bone cement formulations contain micron-sized particles of either $BaSO_4$ or $ZrO_2$ to provide radiopacity, however these particles lead to a deterioration in mechanical properties. In this work, we have introduced nanosized fibrous and tubular titania ($TiO_2$) particles with high aspect ratio, in the acrylic polymer matrix which not only provides radiopacity, but also acts as a reinforcing agent for the cement. The challenge for preparing nanocomposite is the agglomeration of the nano fillers in the polymer matrix which leads to poor performance of the composite. We have presented a novel method of functionalizing titania nanostructure in order to avoid the nanophase agglomeration when it is blended with the polymer matrix.

Methacrylic acid (MA), a functionalization agent that can chemically link $TiO_2$ nanomaterials and polymer matrix, was used to modify the surface of titania using a Ti-carboxylic coordination bond. Then, the double bond in MA was copolymerized with methyl methacrylate (MMA) to form a $TiO_2$-PMMA nanocomposite. In order to enhance the adsorption of carboxylic group of methacrylic acid on titania surface, the pH of the reaction medium has also been adjusted taking into account the amphoteric nature of the naturally occurred hydrated titania surface. A control material was also synthesized consisting of 6% commercial $BaSO_4$ embedded in a PMMA matrix. Fracture toughness and dynamic and static elastic moduli of the nanocomposites were evaluated in the wet condition. Experimental composites containing commercial barium sulfate particles have also been subjected to all these characterization for a comparative study with the experimental composites.

Composites with functionalized titania nanotubes and nanofibers were shown to possess significantly higher fracture toughness and tensile moduli, compared to the composites containing commercial $BaSO_4$ at $p<0.05$. High interfacial area of nanotubes and fibers promote the adhesion between the functionalized nanofillers and surrounding polymer matrix in allowing an external load to be effectively transferred to the nanofillers from the polymer matrix resulting in producing tougher and stiffer cements. More specifically, composites with nanotube exhibits stronger mechanical properties than that of composites with nanofibers. The improved mechanical properties are due to the higher aspect ratio and ability of the hollow nanotubes to mechanically interlock with the polymer matrix.

## Introduction

Acrylic bone-cement is currently the only material used for anchoring cemented arthroplasties to the contiguous bones.[1] In this application the main functions of the cement are to transfer body weight and service loads from the prosthesis to the bone and /or increase the load-carrying capacity of the prosthesis-bone cement-bone system.[1] In a joint replacement surgery it is vital for the bone cement to be radiologically detectable. Most acrylic bone cements are rendered radiopaque by the addition of heavy metal salts of barium or zirconia as a contrast medium.[2] A major disadvantage associated with the addition of heavy metal salts is the phase separation due to the incompatibility of highly polar, ionic radiopaque substances with low polarity resin, which in turn lead frequently to degradation of physical and mechanical properties of the cement.[3] Radiopacifying agents that are compatible with the polymer matrix is an alternative route for achieving radiopacity in organic polymer without compromising its mechanical properties.[4]

In this work we have used nano-sized $TiO_2$ fibers and tubes as reinforcing agent in the poly(methyl methacrylate) (PMMA) matrix which are biocompatible and possess a certain level of radiopacity.[2] In order to compatibilize them with polymer matrix we have functionalized them by using methacrylic acid (MA), which has both a carboxyl group for coordination to the nano-titania,[5, 6] and a vinyl group for subsequent polymerization.[7] In addition, the functionalization has been enhanced by varying the pH of the reaction environment. It has been found that the extent of functionalization varies with pH of the reaction medium with a maximum of 10wt% at pH 5.5 and a minimum of 2.9wt% at pH 3.85 for the case of $TiO_2$ nanofibers, while the $TiO_2$ nanotubes yield 10wt% functionalization at pH 5.0.

The purpose of this research was to evaluate the mechanical properties of PMMA based composites reinforced with $TiO_2$ nanostructures having fibrous and tubular morphologies. Moreover, the effect of extent of functionalization of nanofibers on the mechanical properties of the composites as well as a comparative study of the mechanical properties of the experimental composites with nano-titania and commercial barium sulfate particles have also been conducted in this work.

## Experimental Section

**Materials.** $TiO_2$ nanofibers were synthesized by a direct sol-gel technique in supercritical carbon dioxide in accordance with the procedure explained in the previous work of our group.[8] $TiO_2$ nanotubes were synthesized by an alkaline hydrothermal method in an autoclave with a Teflon lined high-pressure stainless steel vessel equipped with electrical resistance and temperature control.[9]

2-propanol, Methacrylic Acid (99.99%, Sigma-Aldrich) and MMA monomer (95.9 wt %) containing a butanediol dimethacrylate cross-linker (4 wt %) and activator (0.1 wt %) were used as received (Ivoclar Vivadent). PMMA beads (99.2 wt %, Ivoclar Vivadent) having an average particle diameter of 57.5 μm (as measured by laser diffraction using a Malvern Mastersizer 2000) with a free radical initiator BPO (0.8 wt %) were used as received.

**Functionalization of Titania Nanostructures.** The as-prepared $TiO_2$ nanofibers and nanotubes were calcined at 600 and 400 °C respectively to impart crystallinity and thermodynamic stability.[10] In order to functionalize the surface of the n-$TiO_2$, 0.1 g of calcined titania powder was dispersed in 35 mL of 2-propanol with the aid of ultrasonic agitation, followed by reacting with 3 mL of methacrylic acid at 80-85 °C for 24 hours with constant stirring. During the reaction the pH of the medium was measured to be 3.85 in the case of nanofibers and 4.5 in the case of nanotubes. In order to increase the extent of functionalization, the same reaction has been carried out with the medium having pH 5.5 and 5.0 for nanofibers and tubes respectively adjusted by adding few drops of potassium hydroxide solution in the reaction mixture prior to the reaction.

**Preparation of $TiO_2$-PMMA Nanocomposite.** $TiO_2$-PMMA nanocomposites were prepared by bulk polymerization free-radical technique as mentioned in our previous work.[7] Nanocomposite specimens with 2 and 6 wt% titania nanofillers were prepared and kept in distilled water at 37 °C for seven days before evaluating the mechanical properties.

**Characterization.**
The FTIR spectra were collected using KBr pellet form with a Bruker IFS 55 FTIR instrument with an attached MCT detector. Scanning electron microscopy (SEM) were recorded using a LEO 1530 instrument without gold coating. Transmission electron microscopy (TEM) was studied using a Philip CM10 operated at 80 KV. Thermogravimetric analysis (TGA) was performed using a Q500 TA instrument from 25 to 600 °C at a heating rate of 20 °C/min under nitrogen atmosphere. The NTP fracture toughness was determined using the method developed by Ruse et al.[11] Triangular prism specimens (n=12) (6mmX6mmX12mm) were produced. Testing was conducted using an Instron 2519-107 at a crosshead speed of 0.1 mm/min. The tensile moduli of the nanocomposite specimens was determined by the diametral tensile strength (DTS) method.[12, 13] Disc specimens (n=12), 4.6 mm in diameter and 2 mm in thickness, were produced and testing was conducted using an Instron 2519-107 at a crosshead speed of 2 mm/min. The tensile modulus was determined from the slope of the elastic portion of the stress-strain curve.

**Results and Discussion**

**FTIR.** The FTIR spectra of both the calcined and the functionalized $TiO_2$ nanofillers (fibers and tubes) are shown in Figure 1. In Figure 1(a) the absence of any peak from 1300-1800 $cm^{-1}$ indicates the absence of any organic groups on the titania surface. In Figure 1(b) the very strong bands at 1546, 1444 and 1414 $cm^{-1}$ are due to the bidentate coordination between titanium atoms and the carboxylic groups of methacrylic acid.[14]

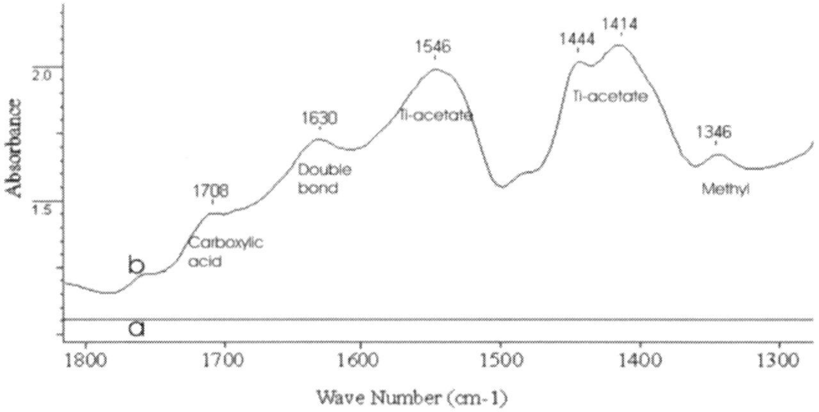


**Figure 1. FT-IR Spectra: the non-functionalized (a) and functionalized (b) titania nanofillers.**

**TGA.** The extent of functionalization of the $TiO_2$ nanofibers and nanotubes have been evaluated by thermogravimetric analysis(TGA). Figure 2 shows the thermal weight loss of nonfunctionalized and functionalized titania nanofibers. The nonfunctionalized titania does not exhibit any weight loss while the functionalized titania show 2.9 and 10 % weight loss which correspond to their extent of functionalization of 2.9 and 10 %. Similar results were obtained for functionalized $TiO_2$ nanotubes.

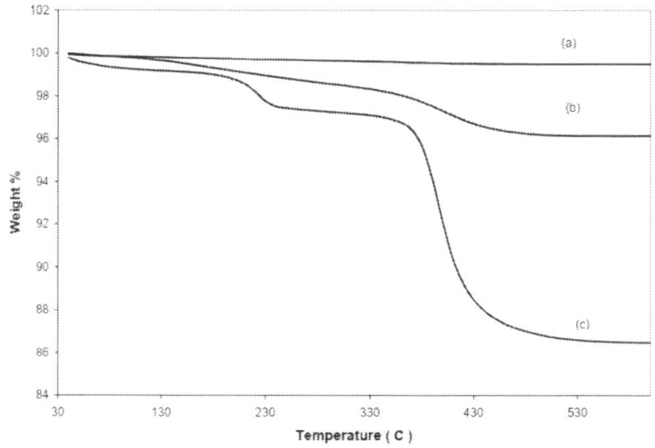


**Figure 2. TGA: extent of functionalization. (a) non-functionalized $TiO_2$ nanofibers; (b) 2.9 wt% functionalized $TiO_2$ nanofibers and (c) 10 wt% functionalized $TiO_2$ nanofibers.**

**Electron Microscopy.** The physical morphology of the functionalized $TiO_2$ nanofibers, nanotubes and commercial $BaSO_4$ particles used in this work are shown in Figure 3 a, b and c. These micrographs show that $TiO_2$ nanofibers and tubes, following functionalization, preserve their fibrous and tubular shape respectively. (Figure 3 a, b), while commercial $BaSO_4$, as received, exhibits irregular shape morphology, (Figure 3 c).

The SEM images of the fracture surfaces of the nanocomposites loaded with 6 wt% functionalized $TiO_2$ nanofibers and nanotubes are shown in Figure 3 d and e while Figure 3 f shows the PMMA composite loaded with 6wt% $BaSO_4$. The functionalized $TiO_2$ nanofibers and tubes show strong adhesion with the PMMA matrix, while the $BaSO_4$ particles are loosely embedded in the matrix and surrounded by a number of pores.

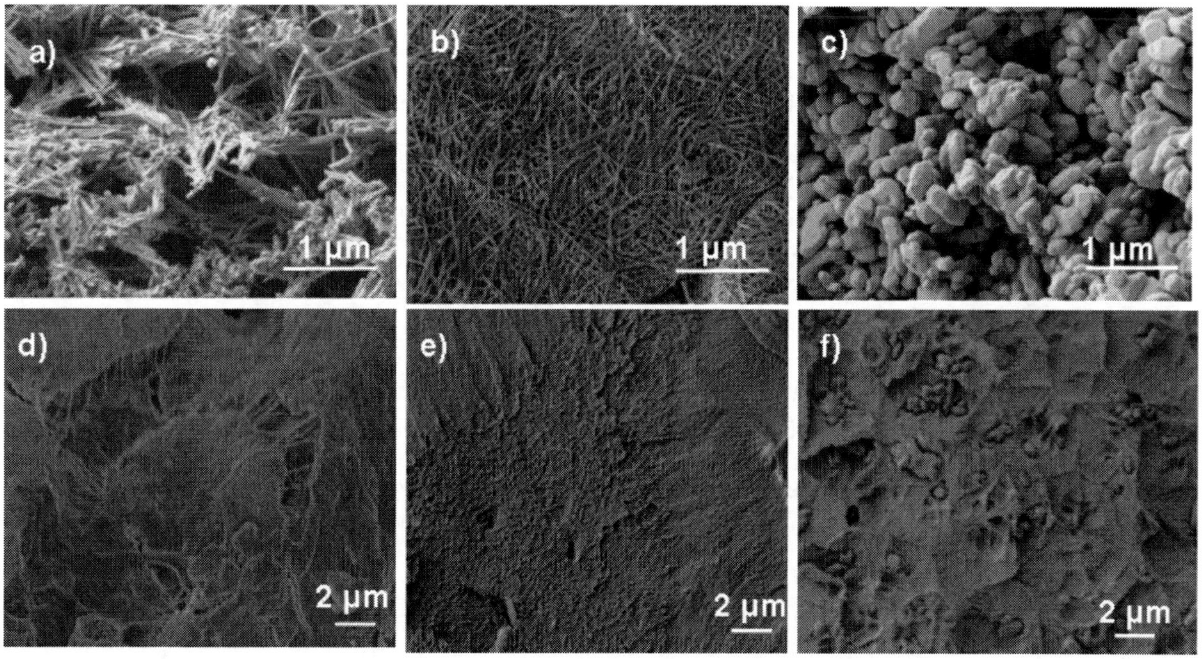


**Figure 3. SEM: (a) Functionalized $TiO_2$ nanofibers; (b) functionalized $TiO_2$ nanotubes; (c) as received commercial $BaSO_4$ particles; (d) Fracture surface of 6 wt% functionalized $TiO_2$ nanofibers-PMMA composite; (e) Fracture surface of 6 wt% functionalized $TiO_2$ nanotubes-PMMA composite (f) Fracture surface of 6 wt% $BaSO_4$-PMMA composite.**

Functionalization provides organophilicity to the $TiO_2$ nanofibers and tubes which resulted in strong adhesion of the nanofibers with the PMMA matrix (Figure 3 d and e), while $BaSO_4$ particles exhibited only a weak adhesion with the matrix (Figure 3 f).

In order to examine the microstructures and nanofiller distribution within the nanocomposites TEM analysis were conducted. Figure 4 shows TEM images of PMMA composites with 3 wt% functionalized nanofibers (a) and nanotubes (b).

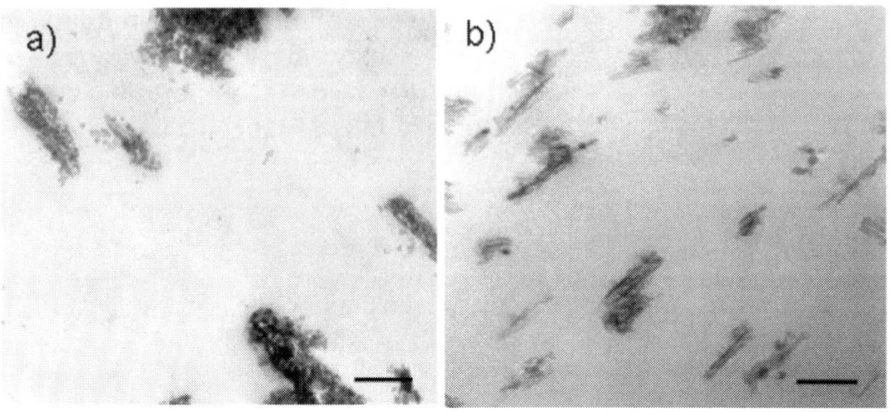


**Figure 4. TEM:(a) 3wt% $TiO_2$ functionalized nanofiber-PMMA composites; (b) 3wt% $TiO_2$ functionalized nanotubes-PMMA composites. (scale bars are in 100 nm)**

From Figure 4 a it is observed that while there are several individual particles distributed throughout the polymer matrix, most are clumped together in bundles. The bundles are in the nanosize range, and are uniformly dispersed throughout the polymer matrix. Fig 4 b shows the functionalized titania nanotubes in PMMA matrix where the 10 nm sized tubes are uniformly dispersed.

**Tensile Modulus.**The tensile modulus of the composites is mainly affected by the interfacial adhesion between the inclusions, i.e. $TiO_2$ nanofibers, nanotubes or $BaSO_4$ particles, and the polymer matrix.[7] Figure 5 shows the comparison of the tensile modulus values between nanocomposites reinforced with commercial $BaSO_4$ and experimental $TiO_2$ nanofibers having 0, 2.9 and 10% functionalization and nanotubes having 10% functionalization in the wet condition. As some commercial bone cements contain 6wt% radiopacifiers, we introduced 6wt% commercial $BaSO_4$ fillers into the PMMA matrix to prepare the composite. The nanocomposite reinforced with 2wt% $TiO_2$ nanofibers and nanotubes having 10wt% functionalization exhibited the highest value of tensile modulus, which was significantly higher than that of unfilled PMMA matrix, and the composites reinforced with $BaSO_4$ and $TiO_2$ nanofibers having 2.9wt% functionalization at $p=0.05$.

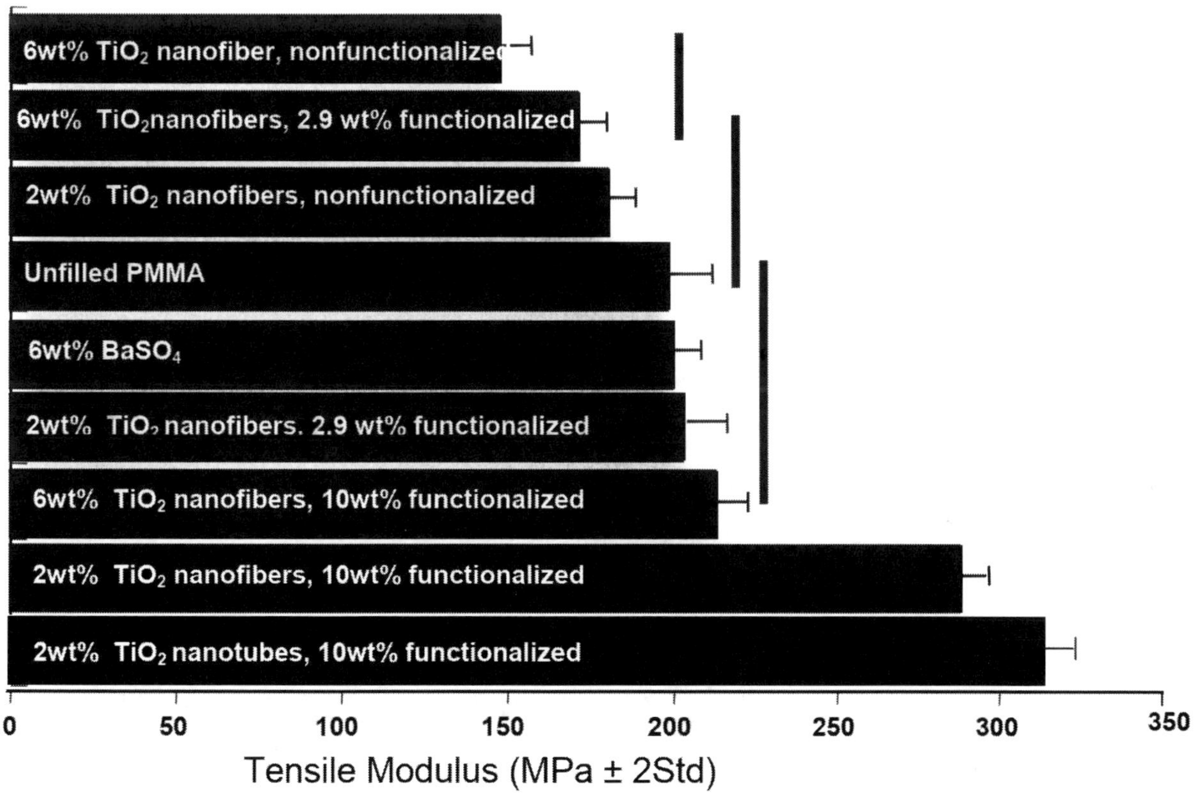


Figure 5: Tensile modulus of composites reinforced with $TiO_2$ nanofibers, nanotubes and $BaSO_4$ in wet condition (P=0.05). (A solid line connects those values which are not significantly different at *P = 0.05*)

The increase in functionalization enhances the organophilicity of the hydrophilic $TiO_2$ nanofibers and nanotubes, improving the dispersion of the nanofillers in the organophilic PMMA polymer matrix. Hence, the nanocomposites reinforced with 2wt% nanotubes and 10 wt% functionalization exhibited the highest values of tensile modulus which is 30% higher than the nanocomposite reinforced with 2% $TiO_2$ nanofibers having 10% functionalization. High aspect ratio and hollow internal structure of the nanotubes not only allows high interfacial area for chemical adhesion with the matrix through its outer surface but also mechanical interlocking with the polymer chain through its inner surface. Furthermore, the incorporation of 6wt% $BaSO_4$ did not have any significant effect on the tensile modulus of the composite, as this value is not significantly different from that of the unfilled PMMA matrix or non-functionalized $TiO_2$ nanocomposite, p=0.05. It should be noted that when comparing the nanocomposites reinforced with 6wt% $TiO_2$ nanofibers to those with 2wt% nanofibers (at either % functionalization), the tensile modulus was significantly lower, p=0.05. This may likely be due to agglomeration of the $TiO_2$ nanofibers at the higher % loading.

**Fracture Toughness.** Fracture toughness is defined as the energy absorbed by the material prior to crack propagation.[1, 15] Hence, fracture toughness is a very important material property for experimental bone cement which is also affected by the morphology, dispersion and extent of functionalization of its radiopacifying fillers. Figure 6 shows the fracture toughness of the composites reinforced with 6wt% $BaSO_4$ nanocomposites with 2 and 6 wt% of $TiO_2$ nanofibers having different extents of functionalization, nanocomposites with 2 wt% $TiO_2$ nanotubes having 10% functionalization, as well as the unfilled PMMA matrix in the wet condition. It can be seen that the nanocomposite reinforced with 2wt% $TiO_2$ nanotubes having 10wt% functionalization exhibited significantly higher fracture toughness values amongst all other composites and unfilled PMMA at p=0.05 which is obviously due to the stronger and greater adhesion of the polymer matrix with the functionalized nanotubes than any other nanofillers and particles used in this work.

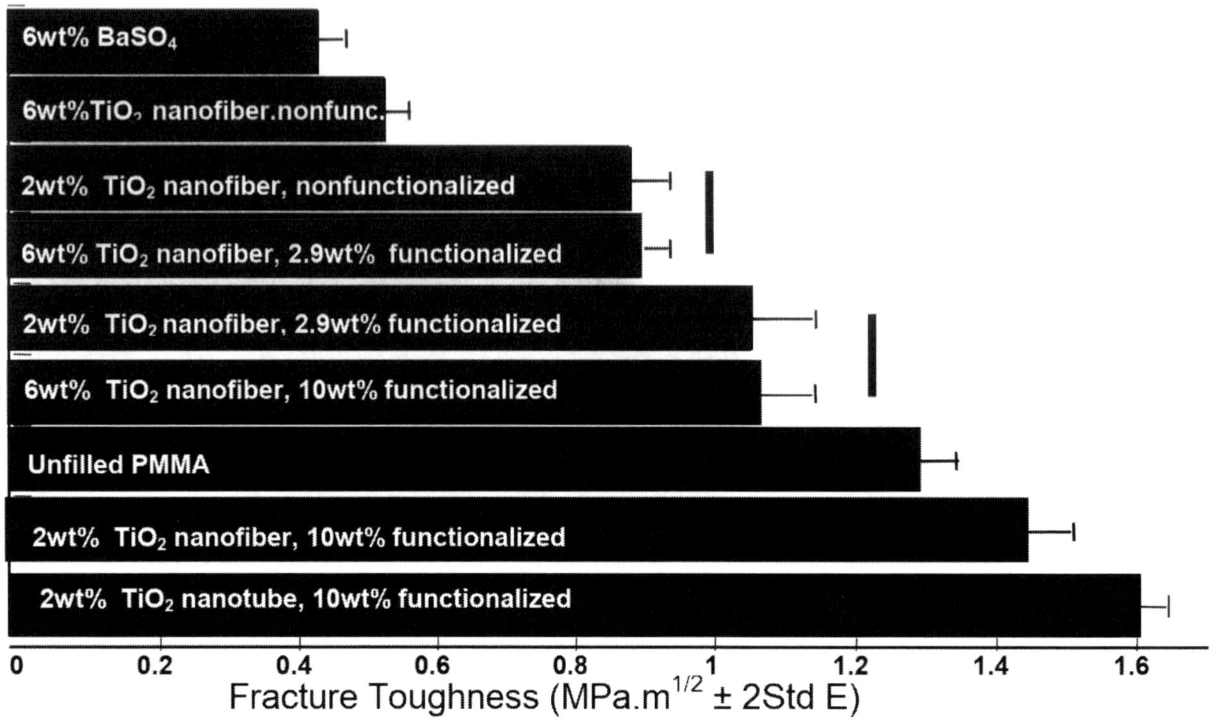


Figure 6: Fracture toughness of composites reinforced with $TiO_2$ nanofibers, nanotubes and $BaSO_4$ in Wet condition ($p=0.05$).

As shown in Figures 6, the composite with 6wt% $BaSO_4$ exhibited a significantly lower value of fracture toughness compared to the $TiO_2$ nanocomposites, as well as the unfilled PMMA matrix. Being incompatible with the organophilic PMMA matrix, the clumps of $BaSO_4$ particles resulted in the generation of pores which acted as stress concentrators and sites for initiating crack propagation during tensile loading on the composite. This is also reflected by the SEM image of the fracture surface of the composite as shown in Figure 3 (f).

**Conclusions.**

$TiO_2$-PMMA nanocomposites with chemical bonding between the filler and polymer matrix were synthesized using the coupling agent methacrylic acid. The coordination of methacrylic acid to the $TiO_2$ nanofiller was confirmed by IR spectra. An excellent distribution of the filler in the polymer matrix was observed from electron microscopy. It has also been observed that with the extent of functionalization the physical and mechanical properties of the composites also increased. After analyzing the mechanical properties, it has been found that the composites with functionalized $TiO_2$ nanotubes exhibited significantly higher elastic moduli and fracture toughness than those with functionalized nanofibers and commercial $BaSO_4$ particles at $p<0.001$. In order to evaluate the efficacy of the experimental nanocomposites as bone cement, radiopacity, rheology and osteoblast cellular response will be studied in our future work.

## References

1. Lewis, G., Properties of Acrylic Bone Cement: State of the Art Review. *Journal of Biomedical Materials Research (Applied Biomaterials)* **1997,** 38, 155-182.
2. Goto, K.; Tamura, J.; Shinzato, S.; Fujibayashi, S.; Hashimoto, M.; Kawashita, M.; Kokubo, T.; Nakamura, T., Bioactive bone cements containing nano-sized titania particles for use as bone substitutes. *Biomaterials* **2005,** 26, (33), 6496-6505.
3. Goto, K.; Hashimoto, M.; Fujibayashi, S.; Kokubo, T.; Nakamura, T., New bioactive bone cement containing nano-sized titania particles. *Key Engineering Materials* **2005,** 284-286, 97-100.
4. S.Deb; S.Abdulghani; Behiri, J. C., Radiopacity in bone cements using an organo-bismuth compound. *Biomaterials* **2002,** 23, 3387-3393.
5. Nakamoto, K., *Infrared and Raman Spectra of Inorganic and Coordination Compounds.* A Wiley-Interscience Publication: New York, 1997; Vol. Part B, p 59-62.
6. Gratzel, M., Solar Energy Conversion by Dye-Sensitized Photovoltaic Cells. *Inorg Chem* **2005,** 44, (20), 6841-6851.
7. Khaled, S. M.; Sui, R.; Charpentier, P. A.; Rizkalla, A. S., Synthesis of $TiO_2$-PMMA Nanocomposite: Using Methacrylic Acid as a Coupling Agent. *Langmuir* **2007,** 23, (7), 3988-3995.
8. Sui, R.; Rizkalla, A. S.; Charpentier, P. A., Formation of Titania Nanofibers: A Direct Sol-Gel Route in Supercritical $CO_2$. *Langmuir* **2005,** 21, (14), 6150-6153.
9. Kasuga, T.; Hiramatsu, M.; Hoson, A.; Sekino, T.; Niihara, K., Formation of Titanium Oxide Nanotube. *Langmuir* **1998,** 14, (12), 3160-3163.
10. Zhang, H.; Banfield, J. F., Thermodynamic Analysis of Phase Stability of Nanocrystalline Titania. *J Mater Chem* **1998,** 8, (9), 2073-2076.
11. Ruse, N. D.; Troczynski, T.; MacEntee, M. I.; Feduik, D., Novel fracture toughness test using a notchless triangular prism (NTP) specimen. *Journal of Biomedical Materials Research* **1996,** 31, (4), 457-463.
12. Carneiro, F.; Barcellos, A., Tensile Strength of Concretes. In *Rilem Bulletin*, 1953; Vol. 13, pp 103-107.
13. Medeiros, I. S.; Gome, M. N.; Loguercio, A. D.; Filho, L. E. R., Diametral tensile strength and Vickers hardness of a composite after storage in different solution. *Journal of Oral Science* **2007,** 49, (1), 61-66.
14. Sui, R.; Rizkalla, A. S.; Charpentier, P. A., FTIR Study on the Formation of $TiO_2$ Nanostructures in Supercritical $CO_2$. *J Phys Chem B* **2006,** 110, (33), 16212-16218.
15. Anderson, T., *Fracture Mechanics: Fundamentals and Applications.* CRC Press: Boston, 1995.

## 655d Self-Assembly of Polymer Grafted Nanoparticles In Polymer Melts

**Pinar Akcora**[1], Sanat Kumar[1], Brian C. Benicewicz Sr.[2], Yu Li[2], and Linda S. Schadler[3]. (1) Chemical Engineering, Columbia University, 500 W. 120th St. Mudd 801, New York, NY 10027, (2) Chemistry Department, Rensselaer Polytechnic Institute, New York State Center for Polymer Synthesis, 110 8th St., Troy, NY 12180-3590, (3) Department of Materials Science and Engineering, and Rensselaer Nanotechnology Center, Rensselaer Polytechnic Institute, 110 8th Street, Troy, NY 12180

Particle-polymer and particle-particle interactions determine the state of dispersion of particles and mechanical behavior of polymer nanocomposites. I will present the unique self-assembly of polymer grafted nanoparticles in polymer melts and will couple the statics results with the microscopic particle dynamics measurements which show that particles form gels through different mechanisms. Gel-like behavior of particle dynamics underpins the mechanical reinforcement measured in these gel former nanocomposites. Additionally, local dynamics in the network, formed by the strong matrix-particle interactions, show that segmental grafted chain dynamics move slower than that of matrix as the composite forms a gel.

# Biodegradable Polymer /Clay Nanocomposites Based on Poly(Butylene Adipate-co-Terephthalate) and Poly(Lactic Acid)

Mahin Shahlari and Sunggyu Lee
Department of Chemical and Biological Engineering
Missouri University of Science and Technology, Rolla, MO 65409, U.S.A.

## Abstract

Poly(butylene adipate-co-terphthalate) (PBAT) is a biodegradable polymer with high ultimate elongation but low modulus. This paper demonstrates that the addition of nano-sized clay can improve the modulus without significantly decreasing the tensile strength and biodegradability of the PBAT , while offering enhanced blendability with a rigid biodegradable polymer as a conceivable means to improve the PBAT's modulus.

Poly(butylene adipate-co-terephthalate) was melt mixed with 3 wt.% and 5 wt.% Cloisite 15A which is organically modified Montmorillonite. This process resulted in nanocomposites with higher moduli than those of the neat PBAT samples. A further increase in the modulus of the PBAT was achieved by blending it with poly(lactic acid) (PLA) in addition to Cloisite 15A. Samples containing 10, 20, or 50% PLA and 5 % clay were melt compounded with a twin screw extruder. The samples were then prepared as films and injection molded parts. The morphologies of the blends were examined using scanning electron microscopy (SEM), and the thermal properties of the blends were tested using Differential Scanning Calorimetry (DSC) and Thermal Gravimetric Analyzer (TGA). X-Ray Diffraction (XRD) was used to investigate the intercalation and dispersion of clay layers in the polymer matrix.

The addition of 5 wt. % and 3 wt. % Cloisite 15A increased the Young's modulus and flexural modulus of PBAT by 57.5% and 36.5%, respectively. DSC results revealed slight changes in the melting temperature of PBAT nanocomposites but significant changes in PLA melting and cold crystallization upon presence of clay particles in the blend. According to the XRD results, the clay layers were intercalated in both the PBAT and the PBAT and PLA blend. Finally, the addition of clay to the PBAT and PLA blend decreased the PLA phase domain size significantly, thus providing morphological evidence for enhanced compatibility.

## Introduction

Methods for modifying biodegradable plastics and expanding their application without adversely affecting the biodegradability of the material are increasingly important in the world intent on the development of environmentally friendly polymeric products. Advancements in processing and modification make biodegradable plastics more capable of competing with non-biodegradable materials.

PBAT is a random copolymer of butylene adipate and Terphthalate which has shown to be fully biodegradable when composted. This polymer owes its biodegradability to butylene adiapte group and owes its stability and mechanical properties to the therphthalate part.[1] Due to the random structure of the copolymer, PBAT generally cannot crystallize to a significant degree, hence it has a wide melting peak unlike that of either the aliphatic or the aromatic homopolymers, poly(butylenb adipate) and poly(butylene terphthalate), respectively.[2]

Due to its low crystallization, PBAT has low modulus and stiffness. In order to modify these properties, this polymer has been compounded with inorganic compounds such as talc[1] and montmorillonites (MMTs) organically modified with the protonated ammonium cations of dodecylamine, octadecylamine, and N-laurydiethanolamine[3].

Recently, many researchers in both acadamia and industry have studied polymer nanocomposites containing clay. Incorporation of layered organically modified clay into thermoplastic polymer composites has caused significant improvements over neat polymers and conventional composites. Incorporation of layered silicates of clay into thermoplastic polymers could enhance mechanical properties. This process could also change barrier properties, reduce flammability and modify polymer's thermal stability.[4] Since all these unique improvements require only a very low percentage of clay (due to its high aspect ratio) the density of products made with such polymeric nanocomposites would be dramatically lower than that of products made with traditional composites which generally use more than 30% percent of inorganic materials.

MMT is one the most commonly used clays for the production of commercial clay polymer nanocomposites, An MMT with the chemical formula of $M_x(Al_4\text{-}Mg_x)Si_8O_{20}(OH)_4$ is first purified, then chemically treated. To increase the d-spacing between the layers, ion exchanging reactions are produced, and hydrocarbon groups of varying sizes are grafted on the clay layers.[5]

In this study, PBAT was compounded with silicate layers organically modified with dimethyl dehydrogenated tallow quaternary ammonium and was melt blended with polylactide (PLA). PLA is the most commonly studied[6,7,8,9,10,11,12] and relatively inexpensive bio-based, biodegradable polymer which can be produced from renewable resources such as corn starch, sugar and wheat. At room temperature, PLA is a rigid thermoplastic with very low strain at break (3.8%) and high modulus (3.4 GPa), which makes this polymer an appealing choice to increase the overall modulus of the polymer PLA/PBAT blend. Since PBAT is produced from petroleum resources, addition of PLA which is synthesized form renewable resources enhances the renewability of the final compound.

Jiang *et al.* have studied the PBAT and PLA blend, with PLA as the dominant phase.[6] Their rheological studies show that PBAT has higher elasticity and viscosity than PLA, and a shorter Newtonian region. To the best of the authers' knowledge, there has not been any published work on PLA and PBAT blend with PBAT being the dominant phase nor has any work been published on the effect of clay on this blend.

## Experimental

### *Materials*

The PLA 4042D used in this study was a product by NatureWorks LLC (United States) with 1.24 g/cm$^3$ as density. The PBAT was a DuPont product marketed as Ecoflex® with mass density of 1.25g/cm$^3$ and melt flow rate of 2.5 to 4.9 grams per 10 min at 190 °C Cloisite® 15A were supplied by Southern Clay Products and is an MMT modified with dimethyl, dehydrogenated tallow, quaternary ammonium.

***Sample preparation***

Twin screw extruder was used for compounding PBAT with clay and blending it with PLA. The twin screw extruder is a Brabender model D6/2 with an L/D ratio of 7:1 and 42 diameter counter-rotating screw. The extruder temperature was controlled at 3 heating zones and the die and the temperature was at 180, 180, 190 and 190 respectively. A gravity hopper was used to feed the extruder and the feed throat was equipped with cooling channel which enabled water circulation to avoid premature melting of the pellets in the feed throat of the extruder.

***Characterization***

Crystallization and melting behaviors of the samples were studied using TA Instruments DSC 2010 Differential scanning calorimeter. The DSC experiments were conducted in a nitrogen atmosphere with a nitrogen flow rate of 10mL/min. The samples were initially heated to 190°C to remove any thermal history, then cooled to 35°C then reheated to 190°C with a heating rate of 10°C/min. The melting and crystallization temperature ($T_m$ and $T_c$), along with heat of fusion and crystallization ($\Delta H_m$ and $\Delta H_c$), were determined using TA universal analysis software. The samples and their replications were tested in a random order to avoid any systematic error in the results.

Scanning electron microscopy (SEM) was done using a Hitachi S 4700 FESEM at an accelerating voltage of 5KV, ultra-high resolution mode and low working distance of 3-5 mm. The samples were merged in liquid nitrogen and then fractured. Subsequently the samples were sputter-coated with gold-palladium. Although the samples were coated, low accelerating voltage was chosen to avoid charging the sample and a short working distance was adopted to assist acquiring vivid images.

For measuring tensile properties, dumbbell and rectangular samples were prepared using MI Morgan-press laboratory size injection molder. Tensile and flexural tests were conducted according to ASTM D6338 standards using an Instron 4469 Universal Testing Machine.

All XRD measurements were conducted using a Philips X-Pert Diffractometer at ambient temperature, 45 KV and 40 mA. The scanning step was 0.03 degree and time per each step was 10 seconds.

## Results and Discussion

***PBAT nanocomposites***

X-ray diffractometry was used to investigate the state of organoclay dispersion within the polymer matrix. The XRD results are shown in figure 1. The low-angle peaks correspond to the {001} basal reflection of the silicate layers. From the angular location of the peaks and the Bragg's relation[4], the interlayer spacing, d001, of each sample were determined. These are shown in table 1. Larger d-spacing for the particles compounded with polymer, indicated that silicate layers had been intercalated by the PBAT chains.

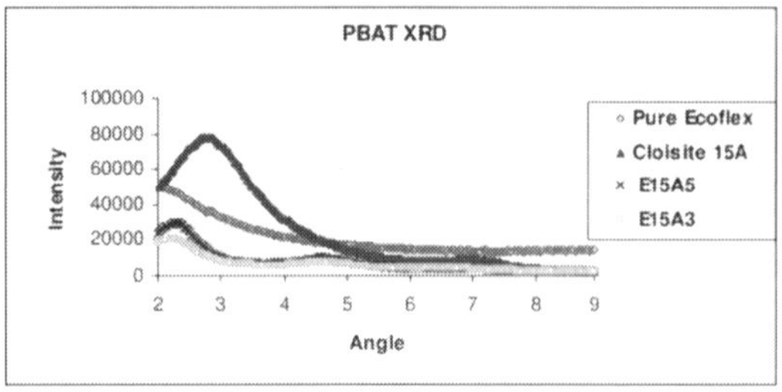


**Figure 1-X-ray differaction patterns of Cloisite 15A, PBAT-3% Cloisite 15A and PBAT 5% Cloisite 15A**

**Table 1- XRD peak positions and interlayer spacing of Cloisite 15A and PBAT with 35 and 5%**

| | 2Θ | d-spacing | 2Θ | d-spacing |
|---|---|---|---|---|
| PBAT5%15A | 2.245 | 3.932 | 4.65 | 1.901 |
| Cloisite 15A | 2.83 | 3.125 | 6.97 | 1.268 |
| PBAT3%15A | 2.275 | 3.880 | 4.672 | 1.890 |

The tensile properties of the PBAT samples containing 3% and 5% Cloisite 15A are shown in table 2. The addition of 3% Cloisite 15A to PBAT increased the elastic modulus of the samples by 24.89% and the flexural modulus by 15.56%. Adding 5% Cloisite 15A increased these two properties by 57.46% and 36.49%, respectively.

**Table 1- Modulus of the PBAT nanocomposites containing 3% and 5% Cloisite 15A**

| | Modulus of Elasticity | Flexure Modulus |
|---|---|---|
| PBAT | 13925 | 15045.71 |
| PBAT 15A 3% | 17390.4 | 17387.29 |
| PBAT 15A 5% | 21926.5 | 20536 |

The DSC results for the PBAT samples compounded with Cloisite 15A showed no significant changes in melting temperature, crystallization temperature or degree of crystallinity of the compounds compared to the pure polymer.

***PBAT/PLA nanocomposites***

XRD results of the blend with 90% PBAT-10% PLA and 50%PBAT-50%PLA blends containing 3% and 5% Cloisite 15A are compared the Cloisite 15A powder in Figure 2 and table 2. The angular location of the peaks and the interlayer spacing indicate better dispersion of the silicate layers in the 90% PBAT blend. This result may be due either to the higher compatibility of Cloisite 15A with PBAT or the relatively lower compatibility of the clay with PLA. The better dispersion of silicate layers in the 50% blend may also be due to the tendency of the silicate layers to remain in the interface.

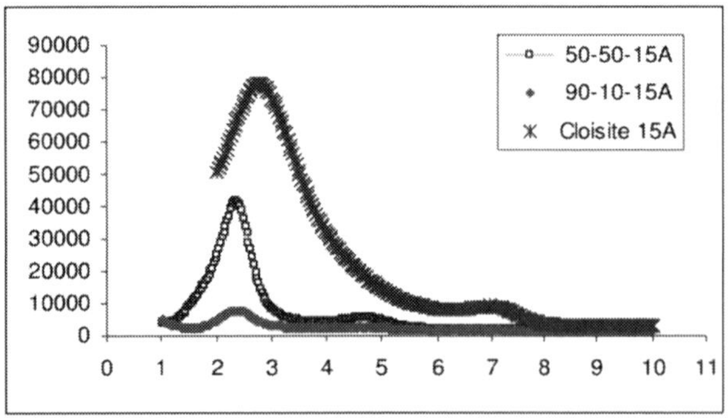


**Figure 2-Xray diffraction patterns of Cloisite 15A, PBAT-3%Cloisite 15A and PBAT-5%Cloisite 15A**

**Table 3- XRD peak positions and interlayer spacing of Cloisite 15A in PBAT 90%-PLA 10% with 5% Cloisite 15A and PBAT 50%-PLA 50% with 5% Cloisite 15A**

| | 2Θ | d-spacing | 2Θ | d-spacing |
|---|---|---|---|---|
| 90-10-15A5% | 2.4 | 3.69 | 4.62 | 1.91 |
| 50-50-15A5% | 2.37 | 3.73 | 4.67 | 1.89 |
| Cloisite 15A | 2.83 | 3.13 | 6.97 | 1.27 |

The SEM images show a distinct two-phase morphology in the absence of nano-size silicate layers. The addition of silicate layers significantly changes the morphology of the blend. As shown in figure 3, in the presence of silicate layers, the deformation of PLA phase dose not create defect cavities observable around the PLA phase in the neat blend. The stretched PLA phase in the nanocomposite case may indicate ductile fracture of the PLA in the presence of Cloisite 15A versus the brittle fracture PLA exhibited in the neat blend.

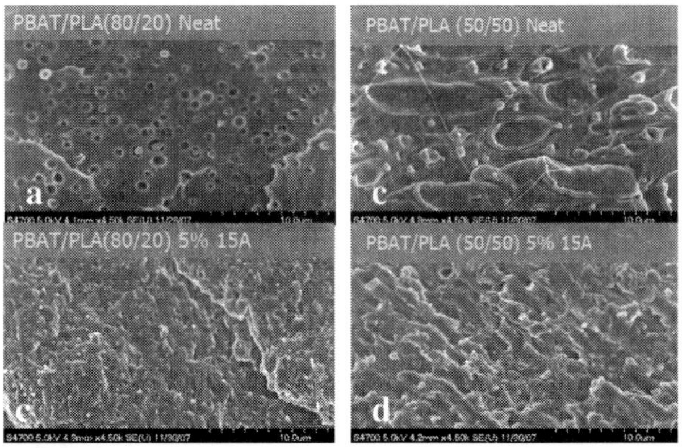


**Figure 3- SEM micrographs of PBAT/PLA blend nanocomposites; (a) PBAT80/PLA20, (b) PBAT80/PLA20/5% C15A,( c) PBAT50/PLA50, d) PBAT50/PLA50/5%C15A**

The DSC graph for pure PLA in the second heating did not show any peak beside glass transition ($T_g$). The lack of a peak indicates that the polymer chains were not able to form any crystalline structure at the heating rate of 10°C/min. On the other hand, results for PBAT /PLA blends show a wide peak for PBAT melting and a small peak for PLA melting. Upon addition of Cloisite 15A to the blend, the melting graph for the 80/20 and 50/50 blends shows an additional peak for cold crystallization which did not appear for the neat blends. This additional peak makes measurement of the area under the PBAT melting peak difficult; however, the melting point for the nanocomposites appear to be the same as that for the neat blends. In the nanocomposite case, the PLA shows two close melting peaks instead of a single peak that appears in the neat blend, indicating the possibility of different crystalline structures in presence of clay particles. Thermal properties of the samples are reported in table 4 and DSC graphs are depicted in figure 4.

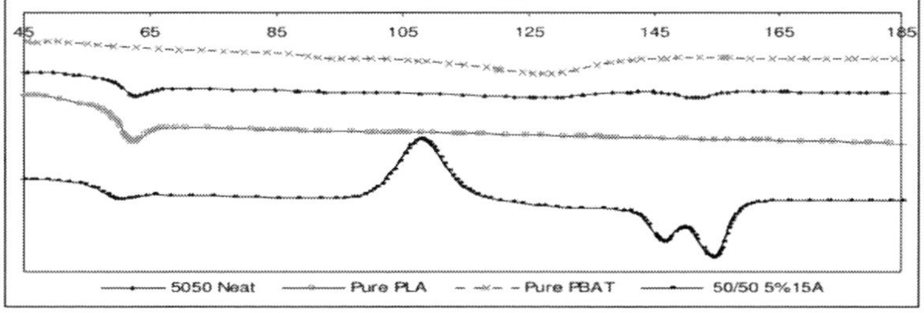


**Figure 4- DSC graphs of melting pure PLA, pure PBAT, PBAT50/PLA50 and PBAT/PLA50/5%C15A**

**Table 4-Glass transition, cold crystallization temperature, melting temperatures and the crystallization temperature of PBAT80/PLA20 and PBAT50/PLA50 with and without 5% Cloisite 15A.**

| | PBAT | 80-20 Neat | 80-20 5%15A | 50-50 Neat | 50-50 5%15A | PLA |
|---|---|---|---|---|---|---|
| Tg | - | 55.42 | 53.54 | 57.68 | 54.5 | 57.58 |
| Tcc | - | - | 106.77 | - | 107.9 | - |
| Tm 1 | 127.2 | 128.48 | 129.17 | 127.46 | 127.36 | - |
| Tm 2 | - | 149.87 | 146.52 | 152.1 | 146.06 | - |
| Tm 3 | - | - | 154.42 | - | 154.42 | - |
| Tc | 87.53 | 89.1 | 94.86 | 88.16 | 93.54 | - |

Tensile testing of the PBAT-PLA films show that the modulus of the sample, as it was predicted, improved significantly as the PLA percentage increased. The PBAT/PLA blend samples mechanical properties are shown in table 5.

**Table 5- Tensile Modulus of PBAT, PBAT80/PLA20, PBAT50/PLA50 and PLA**

| Samples | Modulus MPa/% |
|---|---|
| PBAT | 0.99 |
| 80-20 | 1.22 |
| 50-50 | 5.55 |
| PLA | 13.12 |

## Conclusion

PBAT was compounded with PLA and nano-size clay and the dispersion of clay in the pure PBAT and PBAT/blend was studied using XRD which showed that clay was intercalated in all cases and it was better dispersed when the PLA content was lower. SEM images showed that presence of clay in the PBAT/PLA blend changed the morphology of the compound. The clay presence also had significantly affected the crystallization rate of the PLA in the blend and finally, modulus of PBAT was enhanced due to compounding with clay and PLA.

## References


1- Raquez, M., Nabar, Y., Narayan, R. And Dubois, P. Macromolecular materials and engineering, (2008) 310-320

2- Shi, X.Q., Ito, H. and Kikutan, T., polymer (2005) 11442-11450

3- Someya, Y., Sugahara, Y. and Shibata, M. Applied polymer science (2005) Vol 95. 386-392

4- Chow W, Mohd. Ishak Z, Ishiaku U, Karger-Kocsis J, Apostolov A. Journal of Applied polymer Science 2004. 91, 175-189.

5- Utracki L.A. clay-containing polymeric nanocomposites, rapra technology, (2004)

6- Jiang, L.; Wolcott, M. and Zhang, J. Biomacromolecules, 2006, 199-207

7- Pluta, M. J. of Polymer Science: PArtB: Polymer Physics, 2006, 3392-3405

8- Chen, G.; Choi, J. and Yoon, J. Macromolecular Rapid Communication, 2005, 183-187

9- Bhardwaj, R. and Mohanty, A. Biomacromolecules, 2007, 183-187

10- Oksman, K. and Selin, J. Plastics and Composites from Polylactic acid 150-165

11- Ajayqn, P.; Schadler, L,; Braun, P,; *Nanocomposite Science and Technology* Wiley VCH, (2003)

12- Lee, S.; Lee, Y. and Lee, J. Macromolecular Research, 2007, 44-50

# 655f Hybrid System Consisting of Polyaniline/poly(acrylic acid) Doped with Au Nanoparticles for Biofuel Cell Electrode Platform

**Sunmook Lee**, Bokkyu Choi, Chihiro Fushimi, and Atsushi Tsutsumi. Institute of Industrial Science, The University of Tokyo, 4-6-1 Komaba, Muguro-ku, Tokyo, 153-8505, Japan

Biofuel cell electrodes were fabricated by using composite materials consisting of polyaniline/poly (acrylic acid) doped with Au nanoparticles in the form of thin films on a planar Au electrodes. The hybrid films were polymerized on the planar Au electrode surface via electropolymerization, followed by the covalent attachment of amine-terminated Au nanoparticles. The hybrid Au nanoparticles-polymer films were subjected to cyclic voltammetry for the electrochemical oxidation of ascorbic acid and the bioelectrocatalytic activation of glucose oxidase (GOx). It was observed that the incorporation of poly (acrylic acid) with polyaniline shifted its electroactivity to neutral pH, which makes it feasible to use these composite polymers in physiological conditions. Carboxyl groups of poly(acrylic acid) facilitated a facile covalent attachment of GOx and amine-terminated Au nanoparticles onto the polymer films. Furthermore, Au nanoparticles covalently attached onto the polymeric matrix significantly enhanced the charge transport compared to the polymer films without doped Au nanoparticles.

## 657a Synthesis and Characterization of Cadmium Telluride Nanowire

**Maxwell C. Kum**[1], Bongyoung Yoo[2], Youngwoo Rheem[1], Krassimir N. Bozhilov[3], Wilfred Chen[1], Ashok Mulchandani[4], and Nosang V. Myung[1]. (1) Department of Chemical and Environmental Engineering, University of California, Riverside, Riverside, CA 92521, (2) Division of Materials and Chemical Engineering, Hanyang University, Ansan, South Korea, (3) Central Facility for Advanced Microscopy and Microanalysis, University of California, Riverside, Riverside, CA 92521, (4) Department of Chemical and Environmental Engineering, University of California, Riverside, CA 92521

CdTe nanowires with controlled composition were cathodically electrodeposited using track-etched polycarbonate membrane as scaffolds and their material and electrical properties were systematically investigated. As-deposited CdTe nanowires show nanocrystalline cubic-phase structures with grain size of up to 60 nm. The dark field images of nanowires reveal that the crystallinity of nanowires was greatly improved from nanocrystalline to a few single crystals within nanowires upon annealing at 200°C for 6 hrs in a reducing environment (5 % $H_2$ + 95 % $N_2$). For electrical characterization, single CdTe nanowire was assembled across microfabricated gold electrodes using drop-casting method. In addition to increase in grain size, the electrical resistivity of an annealed single nanowire (a few $10^5$ ohm cm) was one order of magnitude greater than as-deposited nanowire indicating that crystallinity of nanowires improved and defects within nanowires were reduced during annealing. By controlling the dopants levels (e.g. Te content of nanowires), the resistivity of nanowires were varied from $10^4$ to $10^0$ ohm cm. Current-voltage (I-V)characteristics of nanowire indicated the presence of Schottky barriers at both ends of the Au/CdTe interface. Temperature dependent I-V measurements show that the electron transport mode was determined by a thermally activated component at T>-50°C and a temperature independent component below -50°C. Under optical illumination, the single CdTe nanowire exhibited enhanced conductance.

# 657b Growth of ZnO Nanowires for Dye-Sensitized Solar Cells

**Janice Boercker**[1], Yasuhide Nakamura[2], Spencer Schaber[1], Jillian Schmidt[1], and Eray S. Aydil[1]. (1) Chemical Engineering and Materials Science, University of Minnesota, 421 Washington Avenue SE, Minneapolis, MN 55455, (2) Chemical Engineering, University of California, Santa Barbara, Santa Barbara, CA 93106

Nanostructured mesoporous $TiO_2$ films are used as photoanodes in novel solar-to-electric energy conversion devices such as dye-sensitized solar cells (DSSCs) [1]. In a DSSC, a mesoporous $TiO_2$ film, made up of 5-10 nm diameter nanoparticles is deposited on a transparent conducting oxide (TCO) and a monolayer of a dye is adsorbed onto the nanoparticle surfaces to form a photosensitized anode. The porous space between the particles is filled with an electrolyte containing the $I_3^-/I^-$ redox couple and the cell is completed by sandwiching this nanostructured $TiO_2$-dye-electrolyte interface between the TCO and a photocathode. The incident photons excite electrons from the HOMO to LUMO levels in the dye and these photoexcited electrons are injected into the $TiO_2$ nanoparticles. The charged dye is reduced through an electrochemical reaction with $I^-$ in the electrolyte. The electrons hop from particle to particle, flow through the external load and reduce $I_3^-$ at the cathode to complete the circuit. Electron transport through the nanoparticle film is via random walk with multiple trapping and detrapping events. This hopping mechanism increases transport time across the mesoporous film and puts stringent limits on the electron recombination kinetics at the $TiO_2$ electrolyte interface; essentially, only electrolytes or hole conducting media with extraordinarily slow electron recombination at the $TiO_2$ surface can be used to achieve high electron collection efficiencies. Thus, high efficiency DSSCs have been limited to liquid electrolytes containing $I_3^-/I^-$, the only known hole conducting media with electron recombination much slower than electron transport. This inflexibility with the hole conducting medium has limited the DSSC open circuit voltages, which is determined by the difference between the $I_3^-/I^-$ redox level and conduction band edge of the $TiO_2$. Slow electron transport also limits the film thicknesses to less than the electron diffusion length and prevents further increase of the film's optical density in the infrared region of the spectrum where dye absorption cross sections are smaller.

Recently, there has been increased interest in replacing the nanoparticles with one dimensional nanostructures such as nanowires, nanotubes or chains of nanoparticles [2-4]. This interest is based on the conjecture that such one dimensional nanostructures would provide direct transport from the point of injection to the anode without particle-to-particle hopping. This may result in faster electron transport and/or slower recombination allowing thicker photoanodes to be assembled with hole conducting media other than the ubiquitous $I_3^-/I^-$ electrolyte.

The two candidate materials that can be used as the wide band gap semiconductor in DSSCs are ZnO and $TiO_2$. The first nanowire based DSSCs were made from ZnO nanowires due to the relative ease of their synthesis compared to $TiO_2$ nanowires. So far, the highest overall efficiency of a ZnO nanowire DSSC is ~1.7% [5]. This is a factor of six lower than that achieved using $TiO_2$ nanoparticle DSSCs. The lower overall energy conversion efficiency is due to the significantly lower light harvesting efficiency of the ZnO nanowire DSSCs as compared to $TiO_2$ nanoparticles DSSCs. The light harvesting efficiency of the ZnO nanowire DSSCs is low because the surface area of the nanowires is an order of magnitude less than that of the particle films. Improving the ZnO nanowire solution synthesis process to make nanowire arrays with surface areas similar to that of $TiO_2$ nanoparticles films would greatly improve the efficiency of ZnO nanowire DSSCs.

The solution synthesis of zinc oxide nanowires is a two step process. First, a conducting oxide substrate is coated with a ZnO seed layer and the nanowires are grown from the seed layer in an aqueous solution of zinc nitrate and methenamine. The nanowires do not all grow perpendicular to the substrate and as they grow longer they run into neighboring wires and stop growing. As a result the areal nanowire number density decreases as the height of the nanowire film is increased. This results in lower surface areas for dye adsorption and, consequently, lower light harvesting and overall energy conversion efficiencies. To make nanowire films with higher surface area one must (i) increase the density of nanowires, (ii) orient them perpendicular to the substrate and (iii) reduce their radial growth rate so that they do not grow into each other to form a continuous film. In short, one must form a very dense carpet of thin (~10-100 nm), long (10-100 microns), nanowires oriented perpendicular to the surface. Understanding the nanowire growth mechanism may help elucidate ways to engineer the solution synthesis process such that thinner, longer more dense nanowire arrays can be synthesized. Additionally, most of the reactants are lost to homogenous growth during the reaction and do not contribute to nanowire growth. Understanding of the growth mechanism may also lead to more efficient synthesis conditions and strategies.

In this presentation we describe our work exploring the growth mechanism of ZnO nanowires from zinc nitrate and methenamine. Using EDTA titration of zinc ions we found that the depletion rate of the zinc in the nanowire synthesis reaction is first order with respect to zinc ion concentration and with respect to the initial methenamine concentration. Proton NMR was used to monitor the methenamine concentration during the reaction. The methenamine concentration decreases during the first 3 hours of the reaction and then remains constant. The nanowires reach higher overall lengths when grown in solutions with lower methenamine concentrations but the growth rate slows over time.

While all hydrothermal growth processes are intended to be isothermal, in reality they rarely are and it can take many minutes or hours for the growth vessel and the growth solution to reach the steady state temperature. Indeed, in situ measurements of the solution temperature evolution with time in ZnO nanowire growth showed that the thermal inertia of the beaker and the solution determine the transient temperature rise during the synthesis reaction and it may take as long as 1.5 hours for the solution to reach the steady state growth temperature. This temporal evolution of the solution temperature strongly affects the nanowire growth. When the synthesis is done isothermally at 90°C by mixing solutions preheated to this temperature, the pH remains constant at ~5.8. If the solution is allowed to heat up to 90°C over a 1.5 hour time period, the pH starts at ~6.8 and then drops to 5.8 over the 1.5 hour time period that it takes the solution to heat to 90°C, and then remains at 5.8. Nanowires grown in an isothermal solution at 90°C are shorter than those that are grown in a solution that is gradually heated to 90°C. This difference is due to the competition between homogeneous growth which yields particles in solution and heterogeneous growth which yields nanowires on the substrate. At high temperatures the ZnO precursor supersaturation is higher causing homogenous growth to dominate and deplete the precursor too fast to allow for long periods of heterogeneous nanowire growth. In contrast, when heating slowly, the substrate is exposed to lower precursor supersaturation for longer times and heterogeneous growth can occur without significant depletion of the ZnO precursor via homogeneous growth. This information on ZnO nanowire growth may allow us to engineer the temporal evolution of the temperature and reactant concentration profiles to yield longer, thinner more dense nanowires.

References:

[1] O'Regan B, Grätzel M Nature 1991, 353, 737-40

[2] Baxter J B, Aydil E S Appl. Phys. Lett 2005, 86, 053114

[3] Zhu K, Neale N R, Miedaner A, Frank A J Nano Lett 2007, 7, 69-74

[4] Adachi M, Murata Y, Takao J, Jiu J, Sakamoto M and Wang F, J. Am. Chem. Soc. 2004, 126, 14943-49

[5] Gao Y, Nagai M, Chang T, Shyue J, Cryst. Growth Des., 2007, 7, 2467-2471

## 657c High Yield Solution Synthesis of ZnO Nanowire Arrays Using a Novel Reactor Design

**K.M. McPeak** and J.B. Baxter. Chemical & Biological Engineering, Drexel University, 3141 Chestnut St, Philadelphia, PA 19104

Chemical bath deposition is an inexpensive, low temperature, aqueous method for depositing oxide and chalcogenide thin films and nanostructures. However, such process frequently results in deposition yields of less than 5% and excessive waste solvent. Here we report the synthesis of ZnO nanowire arrays using a novel reactor which results in deposition yields of over 80% and reduces the waste solvent volume by a factor of 20. Furthermore our reactor design allows careful study of the thermodynamics and kinetics of ZnO formation. Nanowires are single crystals with strong UV photoluminescence. Dense arrays of well-aligned nanowires with diameters of 60 - 100 nm were grown on substrates pre-seeded with a thin polycrystalline ZnO film. Such nanowire arrays have applications in nanostructured solar cells, gas sensors, and photocatalysis.

# 657d Oriented Single Crystalline Titanium Dioxide Nanowires for Dye-Sensitized Solar Cells

Bin Liu, Janice E. Boercker, and **Eray S. Aydil**. Department of Chemical Engineering and Materials Science, University of Minnesota, 151 Amundson Hall, 421 Washington Avenue SE, Minneapolis, MN 55455

Nanowires of wide band gap semiconductors such as zinc oxide and titanium dioxide are emerging as alternatives to mesoporous nanocrystalline films in dye sensitized solar cells (DSSC). In DSSCs assembled using nanoparticles the electrons are transported through the film by hopping through the nanoparticle network. This hopping mechanism slows down the electron transport and limits the film thickness to electron diffusion length. This in turn limits the optical density of the film especially in the infrared where the dye absorption is weak. In nanowires, the hopping transport is replaced by a direct path to the anode so that such devices may show faster transport and higher charge collection efficiencies than nanoparticle DSSCs. While synthesis of single crystal oriented zinc oxide nanowires and polycrystalline titanium dioxide nanowires have been reported, growth of dense arrays of single crystalline titanium dioxide nanowires that are tens of microns long has been elusive. Crystallinity is very important because grain boundaries in polycrystalline materials scatter or trap photogenerated electrons and adversely affect the performance of solar cells. The single-crystallinity and ordered topology could provide direct electrical pathways for photogenerated electrons and enhance the electron transport rate. This in turn may benefit the performance of photocatalytic and photovoltaic devices such as DSSCs. To date, nanowires and nanotubes used to assemble $TiO_2$-based DSSCs have all been polycrystalline. In this talk, we will describe a simple and environmentally benign method for fabricating oriented single-crystalline anatase $TiO_2$ nanowire arrays with controlled aspect ratios. Vertically oriented $TiO_2$ nanowire arrays were grown on titanium foil using a three step synthesis method. The synthesis method relies on the ability to grow single crystal sodium titanate ($Na_2Ti_2O_5H_2O$) nanowires on titanium foil through a novel alkali hydrothermal growth process. Following growth, the $Na_2Ti_2O_5.H_2O$ nanowires are converted to protonated bititanate ($H_2Ti_2O_5.H_2O$) nanowires through an ion-exchange reaction without changing their morphology or crystal structure. Finally, the protonated bititanate nanowires are converted to single-crystalline anatase $TiO_2$ nanowires through a topotactic transformation by calcination. These three sequential steps yield a carpet of 10-50 micron long single crystalline nanowires oriented in the [100] direction and primarily normal to the titanium foil. Preliminary DSSCs using 12 μm thick $TiO_2$ nanowire films as the photoanode had overall energy conversion efficiencies of 1.4%.

## 657e Effect of Surface Structure on Interfacial Charge Transfer In Photoelectrochemical Cells

**Sudhira Pasupuleti** and Gerold A. Willing. Chemical Engineering, University of Louisville, 106 Ernst Hall, Louisville, KY 40292

Photoelectrochemical cells (PECs) are the most promising of the solar cell types for inexpensive fabrication and high energy conversions. The PEC set up consists of an organo-metallic photosensitive dye physisorbed onto a nanowire mat which is sintered onto a sputter-coated indium tin oxide-glass substrate. This structure acts as a metal oxide framework which in turn is connected to the working electrode of an electrochemical cell. Light absorption occurs in the dye which leads to the generation of sub-bandgap excitations that are used as electronic charge carriers which must then be transferred across the dye-nanowire interface. Improving the mobility of these charges across the dye/metal oxide and metal oxide/electrode interface will directly impact the overall efficiency of the photochemical cells. The photocurrents produced by this process are measured by an atomic force microscopy using an electrochemical cell where the tip acts as the counter electrode to complete the electrical circuit for current generation. The charge collection along the length of the nanowire will be mapped and its relation on the adsorption of the dye molecules along its surface will be studied. The variations in the photovoltages both in light and dark conditions will also be observed. Emphasis will lie on investigating methods for decreasing the recombination efficiency and increasing the conversion efficiency of the photoelectrochemical cells.

## 657f Atomistic Design of Thermoelectric Properties of Si Nanowires

**Trinh Vo**[1], Andrew Williamson[2], Vincenzo Lordi[1], and Giulia Galli[3]. (1) 7000 East Ave., Lawrence Livermore National Laboratory, Livermore, CA 94550, (2) Physics Venture, San Francisco, CA 94111, (3) Chemistry, University of California at Davis, Davis, CA 95616

We present predictions of the thermoelectric figure of merit (ZT) of Si nanowires with diameter up to 3 nm, based upon the Boltzman transport equation and ab-initio electronic structure calculations. We find that ZT depends significantly on the wire growth direction and surface reconstruction and we discuss how these properties can be tuned to select silicon based nanostructures with combined n-type and p-type optimal ZT. Our calculations show that only by reducing the ionic thermal conductivity by about 2 or 3 orders of magnitudes with respect to bulk values, one may attain ZT larger than 1, for 1 or 3 nm wires, respectively. We also find that ZT of p-doped wires is considerably smaller than that of their n-doped counterparts with the same size and geometry.

## 671a pH Sensitive Fluorescent-Magnetic Nanocomposites for Intracellular Detection and Manipulation

Dhananjay Thakur[1], Lee Siers[2], Thierno Baldet[2], and **Jessica O. Winter**[3]. (1) Biophysics Program, the Ohio State University, 140 W 19th Ave, Columbus, OH 43210, (2) Chemical and Biomolecular Engineering, the Ohio State University, 140 W19th Ave, Columbus, OH 43210, (3) Chemical and Biomolecular Engineering/Biomedical Engineering/ Biophysics Program, Ohio State Unversity, 140 W 19th Ave, Columbus, OH 43210

Hybrid nanoparticles have multimodal applications in biology and medicine. Fluorescent magnetic nanoparticles could be used for mechanical manipulation of cells and subcellular features with real time monitoring, while simultaneously taking advantage of the variation of fluorescence with changes in the physical structure and local environment of the nanoparticles. Although, current high temperature synthesis methods using organic solvents produce nanoparticles with better surface passivation and particle stability than aqueous methods, the aqueous route can produce particles with interesting physical properties while retaining relative simplicity in procedure. We have attempted to synthesize fluorescent - magnetic nanocomposites by using a bifunctional linker molecule to bind quantum dots and superparamagnetic iron oxide nano-particles. Cadmium sulfide nanocrystals were synthesized through arrested precipitation using known methods [1,2] and were linked to iron oxide nanoparticles using a bifunctional ligand (MPS - 3-mercaptopropyl trimethoxy silane).

Water soluble fluorescent magnetic particles were obtained in an aqueous solution at a pH below 2.0. The particles remained stable in solution for up to 48 hrs, after which aggregates were formed that precipitated out of the solution. The fluorescence intensity showed a strong dependence on the pH of the local environment, while the magnetic properties were unaffected. Quenching of fluorescence was observed at a neutral pH. Particles formed large aggregates and precipitated almost immediately. Fluorescence intensity increased rapidly as pH was reduced, with a peak intensity at pH below 2.0. The fluorescence peak was recovered after recycling of pH in the range 1.5 to 7.5. We are currently working on stabilizing the particles across the pH range required for physiological applications and characterizing them on the basis of initial particle size, ligand concentration and temperature during synthesis. This pH dependence could be harnessed for selective targeting and identification of subcellular environments and cellular processes such as endocytosis, exocytosis and changes in the proton flux during oxidative phosphorylation.

1.Winter J. O, Gatzert S., Korgel B.A., Schmidt C.E. (2005). Variation of Cadmium Sulfide Nanoparticles Size with Altered Aqueous Synthesis Conditions. Colloids and Surfaces A, 254(1-3): 147-157

2.Chen H.M., Huang X.F., Xu L., Xu J., Chen K.J., Feng D., Superlattices Microstruct., Volume: 27, (2000), p. 1

# 671b A Quantitative Description of Absolute Endocytosis and Exocytosis Rates for Nanoparticles

**Hong Jin**, Chemical Engineering, Massachusettes Institute of Technology, Building 66-580, 25 Ames St., Cambridge, MA 02139, Daniel A. Heller, Department of Chemistry, University of Illinois at Urbana-Champaign, 104 RAL, Box 93-5, MC-712, 600 S. Mathews Ave., Urbana, IL 61801, Richa Sharma, Chemical Engineering, MIT, 66-565, 25 Ames St., Cambridge, MA 02139, and Michael S. Strano, 66-566 Department of Chemical Engineering, Massachusetts Institute of Technology, 77 Massachusetts Avenue, Cambridge, MA 02139.

The endocytosis of nanoparticles is important to drug delivery, bioimaging and toxicology. However, there is no quantitative model linking internalization rate with nanoparticle geometry. Current theory suggests that the rate is bracketed by elastic energy release as the cell membrane wraps the nanoparticle and an entropic penalty as additional cell surface receptors are recruited to continue the process. While successful in predicting a maximum uptake at a given radius, the approach requires this value be a threshold below which endocytosis is impossible and this is in direct contradiction to experiment. To address this, we develop a deterministic kinetic model and assert that nanoparticles smaller than the predicted cutoff can cluster on the cell surface after diffusion, thereby gaining the free energy necessary to surmount the entropic barrier for uptake. We focus on $d(GT)_{15}$ wrapped SWNT[1], and use its intrinsic band gap fluorescence to track interactions with 3T3 cells using a 2D InGaAs imaging array coupled to an inverted microscope with a perfusion stage[2].

The model accurately describes uptake rates from single particle tracking for single walled carbon nanotubes(SWNT)[2] of lengths from 130±18 to 660±40 nm and literature data for Au nanoparticles of diameters from 14 to 100 nm. Remarkably, both have a maximum rate near 25 nm when scaled using an interaction radius for membrane diffusion. The endocytosis rate constant of SWNT ($10^{-3}$ $min^{-1}$) is nearly 1000 times that of Au nanoparticles ($10^{-6}$ $min^{-1}$). The recycling (exocytosis) rate constants are similar in magnitude ($10^{-4}$ to $10^{-3}$ $min^{-1}$) for poly(D,L-lactide-co-glycolide), SWNT and Au nanoparticles and across cell lines and generally decrease with increasing size. The ability to understand and predict the cellular uptake of nanoparticles quantitatively should find utility in designing nano-systems with controlled toxicity, efficacy and functionality.

(1) Jin, H.; Jeng, E. S.; Heller, D. A.; Jena, P. V.; Kirmse, R.; Langowski, J.; Strano, M. S. Macromolecules 2007, 40, 6731-6739.

(2) Jin, H.; Heller, D. A.; Strano, M. S. Nano Lett. 2008, ASAP.

## 671c Tracking the Intracellular Transport of Nanoparticles by Quantum Dots

**Gang Ruan**, Chemical and Biomolecular Engineering, the Ohio State University, 140 W 19th Ave, Columbus, OH 43210, Amit Agrawal, Health Sciences and Technology, Massachusetts Institute of Technology, 77 Massachusetts Avenue, Cambridge, MA 02139, and Shuming Nie, Biomedical Enigneering, Georgia Institute of Technology and Emory University, 1639 Pierce Drive Suite 2001, Atlanta, GA 30322.

Intracellular transport of nanoparticles is an issue of considerable current interest in developing advanced nanoparticle agents for molecular imaging and therapy. It has been technically challenging to directly follow nanoparticle transport in live cells, however. We have taken advantage of the bright and stable fluorescence of quantum dots (QDs), which are used as the prototype of nanoparticles of many kinds, and the dynamic confocal imaging capability of spinning disk confocal microscope to track the cellular uptake of nanoparticles and the downstream events of nanoparticle transport. Specifically, we have examined the intracellular transport of Tat peptide conjugated quantum dots (Tat-QDs). Tat peptide is a peptide sequence playing critical role in the infection of HIV virus, and has been used as a novel delivery agent for a variety of cargos, although its delivery mechanism remains controversial.

We have found that Tat-QDs are internalized by an endocytosis process triggered by Tat-QD binding to negatively charged cell membranes. The internalized Tat-QDs are tethered to the inner vesicle surfaces and are trapped in cytoplasmic organelles. The QD loaded vesicles are actively transported by motor proteins along cytoskeleton tracks and end up being accumulated at a perinuclear region. We also find that Tat-QDs strongly bind to cellular membrane structures such as filopodia and that large QD-containing vesicles are released from the tips of filopodia by vesicle shedding. These results provide new insights into the mechanisms of Tat peptide-mediated delivery as well as toward the design of functionalized nanoparticles for molecular imaging and targeted therapy.

# 671d Rare-Earth Doped and Co-Doped $Y_2O_3$ Nanomaterials as Potential Bio-Imaging Probe

**Gautom Kumar Das** and Tan Thatt Yang Timothy. School of Chemical and Biomedical Engineering, Nanyang Technological University, 62 Nanyang Drive, Singapore, 637459, Singapore

There has been a burgeoning trend in research and development for nanoparticles for potential bio-imaging application. Several classes of nanomaterials have been investigated. They include organic, metallorganic and polymer based dye, fluorescent proteins and semiconductor nanocrystals. Organic fluorophores suffer from a number of shortcomings including photobleaching and instability. Inorganic nanophosphors such as quantum dots seem ideal due to their bright fluorescence, size-dependent tunable emission and photostability. However, they usually contain toxic elements such as cadmium. Rare-earth (RE) nanophosphors, in comparison, offer an alternative for biological labeling and medical diagnostics due to their large Stokes shift, sharp emission spectra, long lifetime, multiphoton and up-conversion excitation, low toxicity and reduced photobleaching over quantum dots and organic phosphors molecule. These materials have the ability to emit intensely at various wavelengths by an appropriate choice of color-center elements instead of varying particle size.

In the current work, synthesis and characterization of various rare-earth (RE) doped and co-doped $Y_2O_3$ nanomaterials were undertaken. RE dopants such as terbium (Tb), europium (Eu) and erbium (Er) were successful doped, and ytterbium (Yb)-erbium (Er) was co-doped into yttrium oxide ($Y_2O_3$) nanomaterials. The as-synthesized nanomaterials are highly-dispersed in organic phase. X-ray diffraction (XRD) peaks confirms the crystallinity of the as-prepared $Y_2O_3$ samples and peaks can be assigned to cubic $Y_2O_3$. Room temperature photoluminescence (PL) spectra showed characteristic emission peaks of Tb, Eu and Er ions doped, and Er-Yb co-doped $Y_2O_3$ samples. Up-conversion spectra of green and red emissions were also shown for Er-Yb co-doped $Y_2O_3$ samples and the intensities of the green and red emissions could be altered by varying the dopant concentration. The nanocrystals were surface-functionalized with amine ($NH_2$) group via a reverse microemulsion method to render them water soluble and also afford them with the possibility of further functionalization by other bio-molecules. *in vitro* cytotoxicity studies showed that the synthesized nanocrystals have no appreciable toxicity on human hepatocellular carcinoma (Hep-G2) cells. Because of the $Y_2O_3$:RE nanomaterials' well-dispersity in water, low toxicity and good photoluminescence, they can potentially be used as fluorophores in bioimaging.

## 671e Dynamic Fluorescent Nanoparticle Assemblies for Biosensing Applications

**G. Daniel Lilly** and Nicholas A. Kotov. Chemical Engineering, University of Michigan, 2300 Hayward, HH Dow Building #3074, Ann Arbor, MI 49109

The conjugation of nanomaterials to one another to create unique superstructures allows novel coupling of the electronic and optical properties of the constituent materials. This paper focuses on the unique optical properties of assemblies formed by the conjugation of Au nanoparticles (NPs) and core/shell CdSe/ZnS NPs using polyethylene glycol (PEG). Similar assemblies have been formed with Au NPs and CdTe NPs whose quantum yield is temperature dependant; however, the diameter of CdTe cannot be tuned such that its PL covers all visible wavelengths, and CdTe NPs experience contact quenching when many NPs are bound together, thus limiting their potential applications. The PL of CdSe can be tuned to any visible wavelength by changing the diameter of the NP; moreover, when a shell with a larger bandgap than CdSe is applied to a CdSe core (ZnS, ZnSe, or CdS), contact quenching can be prevented and the quantum yield further enhanced. These core shell materials are also more photochemically stable and more resistant to photobleaching than traditional NPs.

## 671f Fluorescence Emission Control by Nanometal Particle

**Jianting Wang**[1], Samuel Achilefu[2], and Kyung A. Kang[1]. (1) Chemical Engineering, University of Louisville, Ernst Hall 106, Speed School of Engineering, University of Louisville, Louisville, KY 40292, (2) Department of Radiology, Washington University in St. Louis, CB 8225, Room 1150, East Building, 4525 Scott Avenue, St. Louis, MO 63110

In fluorophore mediated biosensing and bioimaging, fluorescence emission level control can be beneficial. For example, fluorescence enhancement can improve the sensitivity, and fluorescence quenching caused by a particular molecule can be used to detect that molecule. Since fluorescence is emitted during electron energy transition of a fluorophore, it may be altered when the fluorophore is placed in a plasmon field generated by free electron oscillations. Nanometal particles (NMP) possessing strong surface plasmon field can be effectively used for this fluorescence level control. The field strength is the highest at the surface of the particle and decays exponentially with the distance from the surface and the fluorescence change by NMPs depends on the plasmon field strength where the fluorohpore is placed. In other words, the level of fluorescence may be controlled by controlling the distance between the particle surface and the fluorophore.

One way of controlling the distance is placing a molecular spacer between a particle and a fluorohpore. Here we have correlated the fluorescence emission level (i.e. enhancement or quenching) with the spacer length between a fluorophore and an NMP. The information obtained may be effectively utilized for optimal design of fluorophore mediated biosensing and bioimaging.

## 684a Phase Behavior and Rheology of Biopolymers Stabilized Carbon Nanotubes

**Dhriti Nepal** and Virginia A. Davis. Department of Chemical Engineering, Auburn University, 212 Ross Hall, Auburn, AL 36849

The phase behavior of carbon nanotube (CNT) dispersions is of tremendous importance for its processing into varieties of forms such as films, composites, and fibers. Ordered phases of SWNTs in an aqueous system is of great demand which offers unique platform to create novel class of biological electronics and optical devices. Here we compare the isotropic-nematic phase transition of SWNT dispersed in aqueous solution of two most prominent biopolymers DNA and protein (lysozyme). At very low concentration, SWNTs in DNA dissolve as individual tubes which behave as Brownian rods. At higher concentration, SWNTs form a nematic phase consisting of spaghetti-like self-assembled supermolecular strands. At even higher concentration, the spaghetti strands self-assemble into polydomain nematic liquid crystal. The phase behavior of lysozyme-SWNT is different from that of DNA-SWNT confirming uniqueness on interaction of each of these biopolymers with the SWNTs. Details investigation using optical microscopy, atomic force microscopy, scanning electron microscopy, rheology and Raman spectroscopy will be discussed.

## 684b Conical Carbon Nanotube Arrays: Large Area Synthesis, Electrochemical and Field Emission Characteristics

**Santoshrupa Dumpala**[1], David Mudd[2], Gamini U. Sumanasekera[2], and Mahendra K. Sunkara[3]. (1) Chemical Engineering, University of Louisville, 106, Ernst hall, University of louisville,, Louisville, KY 40292, (2) Physics, University of louisville, 2210 south brook st, room 211, Louisville, KY 40208, (3) Department of Chemical Engineering, University of Louisville, 106 Ernst Hall, Department of Chemical Engineering,, University of Louisville, Louisville, KY 40292

Conical carbon nanotubes refer to structures that taper from micron-sized base to nanometer scale tips with central nanotube as hollow core1. The surfaces of these conical structures are composed of high density of graphene edge planes.2 Also, due to the conical geometry and high aspect ratio, the arrays of the conical carbon nanotubes contain their tips spaced from each other. So, these conical carbon tube arrays are expected to yield dramatically different behavior in terms of both electrochemical and field emission characteristics. However, for all the applications, large area (> 1 cm2 area) synthesis of these arrays on flat substrates has been challenging.

Here, we report our recent success with synthesizing these conical carbon tubular arrays over graphite and other metallic foils using a 3 kW microwave plasma enhanced chemical vapor deposition (MWCVD) reactor. Cylindrically rolled foils immersed vertically into the plasma, resulted in conical carbon tubes up to as long as 30 microns, at increased temperatures. The as-synthesized arrays show reversible behavior with electrochemical redox reactions involving neurotransmitters like Levodopa and Apomorphine. In addition, the field emission characteristics of conical carbon nanotubes show better field emission characteristics compared to multiwalled carbon nanotube arrays.

References:

1. R. C. Mani, X. Li, M. K. Sunkara, K. Rajan, Nano Lett. 2003, 3, 671

2. B. Chernomordik, S. Dumpala, Z. Chen and M.K. Sunkara, Chemical Vapor Deposition, Accepted for special issue on nanodiamond, (2008).

## 684c Anisotropic Thermal Properties of Carbon Nanotubes: Reaction Rate Enhancements and Effects of Covalent Functionalization

**Joel T. Abrahamson**[1], Nitish Nair[1], and Michael S. Strano[2]. (1) Department of Chemical Engineering, Massachusetts Institute of Technology, Building 66-153, 25 Ames St, Cambridge, MA 02139, (2) 66-566 Department of Chemical Engineering, Massachusetts Institute of Technology, 77 Massachusetts Avenue, Cambridge, MA 02139

We have numerically solved the coupled energy balances for a single-walled carbon nanotube (SWNT) with an annular coating of reactive Zr metal and oxide. In our model, thermal transport in the nanotube accelerates reaction front velocity along the length of the annulus by as much as a factor of 10. This offers a proof-of-concept for one-dimensional anisotropic energetic materials, which could find new applications in inorganic synthesis and novel propellants. Interestingly, interfacial resistance is not the most significant factor limiting the coupling due to the large temperature differences between nanotube and annulus at the reaction front. Transient thermoreflectance (TTR) spectroscopy offers a way to measure thermal conductivity and interfacial resistance in this and other novel nanotube-based materials, leading towards experimental demonstration of this effect. We also use TTR to assess the effects of covalent functionalization on the thermal properties of SWNT materials, arising from changes in the SWNT phonon band structure.

[1] J. T. Abrahamson, N. Nair, M. S. Strano. Nanotechnology. 19, (2008), 195701

[2] P. M. Norris et al. Rev. Sci. Instr. 74, (2003), 400

# 684d Voltage-Controlled Purification of DNA-Wrapped Carbon Nanotubes

**Steve R. Lustig**[1], Rick Rajter[2], and Yet-Ming Chiang[2]. (1) Central Research & Development, The DuPont Company, Experimental Station, Route 141, Wilmington, DE 19880-0356, (2) Department of Materials Science & Engineering, Massachusetts Institute of Technology, 77 Massachusetts Avenue, Cambridge, MA 02139

We report a novel method to trap preferentially and release DNA-wrapped single wall carbon nanotube (DNA/SWNT) hybrids both as a whole population and by chirality of the SWNT. The preferential trapping allows for enrichment of the initially polydisperse chiral population into its constituents. The separation system combines an HPLC, a potentiostat and a quartz crystal microbalance flow cell that was modified to allow for a standard three electrode setup. The combination of these devices allows for real time control over the working electrode surface potential, liquid flow rate and mobile phase concentration while monitoring adsorption/desorption mass on/from the working electrode and the eluting species' UV/Vis spectra. This method also has an advantage in that we can vary the coating upon the quartz crystal microbalance working electrode with non-conductive materials. This gives us more control over both the electrostatic and van der Waals/London dispersion interactions and a larger experimental parameter space to explore. Our results show that holding the surface potential at -900mV relative to an Ag/AgCl reference electrode can trap hybrids upon the microbalance surface. These trapped hybrids can be ejected back into solution via dropping the potential, adding salt, or accelerating the mobile phase flow rate. In numerous experiments we see that the injection population and the released population vary significantly in the resolution of SWNT chirality. Separation between metallic and semiconducting classes of SWNTs has been achieved. We report our most recent results fine tuning this method to isolate purified SWNT chiralities.

## 684e Photothermal Ablation of Neuroblastoma Using Carbon Nanotube Bound with GD2 Monoclonal Antibody

Chung-Hao Wang, Yao-Jhang Huang, Chia-Wei Chang, and **Ching-An Peng**. Chemical Engineering, National Taiwan University, No. 1, Sec. 4, Roosevelt Road, Taipei, Taiwan

Neuroblastoma (NB) is the most common solid tumor of children, derived from the sympathetic nervous system. Since intensive therapeutic intervention does not prolong the overall disease-free survival rate for this tumor, novel therapeutic strategies are required. It has been reported that disialoganglioside (GD2) is abundantly expressed on the surface of NB cells and minimally in normal tissues. Hence, GD2 is an ideal targeting ligand for NB cells. On the other hand, delivery of carbon nanotubes (CNTs) into mammalian cells via endocytic pathway has recently gained increasing interests because of its potential therapeutic applications. In this study, we examined the specific binding of anti-GD2-tagged multi-walled CNTs against GD2-positive NB cell lines in vitro and the ensuing ingestion of fluorescent labeled anti-GD2-CNTs were investigated. Since optically stimulated electronic excitation of CNTs could be transferred to molecular vibration energy and causing heating, the study of increasing necrosis of malignant NB cells specifically ingested anti-GD2-CNTs is worth of exploring as a promising treatment. Our results showed that after CNTs were internalized via targeting onto cell surface, washed and irradiated with near-infrared (808 nm) laser for 15 min under intensity of 6 W/cm2, large fraction of GD2-expressing NB cells were non-viable as detected by calcein AM dye. It is expected that site-targeting of NB cells with anti-GD2 conjugated CNTs and further ablation of CNT-laden malignant cells via near-infrared laser induced over-physiological temperature is an alternative NB therapy.

## 684f Aggregation Kinetics of Carbon Nanotubes In the Presence of Biomacromolecules

**Navid Saleh**[1], Lisa D. Pfefferle[1], and Menachem Elimelech[2]. (1) Chemical Engineering, Yale University, 9 Hillhouse Ave, Room 300, New Haven, CT 06511, (2) Department of Chemical Engineering, Yale University, Mason Lab, 313A, 9 Hillhouse Avenue, New Haven, CT 06520

Carbon nanotubes (CNTs) have a widespread potential for application due to their unique physicochemical properties. There has been lack of studies aimed at understanding the aggregation behavior of CNTs in the presence of macromolecules of environmental and biomedical relevance. A systematic study of CNT aggregation kinetics is essential in order to predict their fate and interaction in environmental and biological systems. Effects of macromolecules such as natural organic matter, polysaccharides, proteins, and lipids on the aggregation kinetics of single-walled and multi-walled carbon nanotubes (SWNTs and MWNTs) were evaluated in this study. CNTs were characterized using Raman scattering (for state of defect), thermo gravimetric analysis (for metal impurities), transmission electron microscopy (for length and diameter distribution), Fourier transformed infrared spectroscopy (for functional groups), and electrophoretic mobility (for surface charge). The aggregation kinetics of CNTs were slowed down considerably in the presence of the macromolecules. Calcium ions showed specific interaction with alginate (polysaccharide) and resulted in enhanced aggregation. Steric repulsion is proposed to be the mechanism of enhanced stability in presence of the macromolecules.

# 705a Nanotechnology for Regenerating Tissues: Hype or Reality ?

**Thomas J. Webster**, Engineering and Orthopaedics, Brown University, 182 Hope Street, Box D, Brown University, Providence, RI 02912

Nanotechnology is being used to mimic structural components of our tissues in synthetic materials intended for various implant applications. Nanomaterials have increased surface area, surface roughness, and energy which optimally interact with proteins that control cell functions to regenerate tissues. Coupled with such unique surface properties that promote cell functions, nanomaterials can now be fabricated to possesses structural properties necessary for supporting physiological loads (as in the case of orthopedic implants) and/or viscoelastic properties (as in the case of vascular grafts). Novel fabrication methods are being used (such as 3D printing) to create hierarchical tissue substitutes which mimic the micron and nano structure of our tissues, especially bone. Recent studies have highlighted that when compared to currently implanted materials, nanomaterials with random and oriented nanofeatures promote optimal initial protein interactions necessary to mediate cell adhesion and subsequent tissue growth. This has been demonstrated for a wide range of implant chemistries (from ceramics to metals to polymers) and for a wide range of tissues (including skin, bone, vascular, cartilage, bladder, and the central and peripheral nervous system). Recent results have even incorporated the use of stem cells with carbon nanotubes to heal stroke damage in rats. Importantly, these results have been seen at the in vitro and in vivo level. Lastly, through the use of existing antimicrobial chemistries, decreased bacteria infection has been observed through the use of nanomaterials. Since implant infection is a major cause of biomaterial failure, such promoted functions of cells to regenerate tissues coupled with decreased bacteria infection highlights a promising future for nanomaterials in tissue engineering applications. This talk will cover some of the more significant advancements in creating better tissue engineering materials through nanotechnology.

# 705b Production of Virus-Mimetic Mucus-Penetrating Particles for Drug and Gene Delivery In Mucosal Tissues

**Ming Yang**[1], Samuel K. Lai[2], Ying-Ying Wang[1], Christina Happe[2], Michael Zhang[2], and Justin Hanes[2]. (1) Department of Biomedical Engineering, The Johns Hopkins University, 221 Maryland Hall, 3400 N. Charles Street, Baltimore, MD 21218, (2) Department of Chemical and Biomolecular Engineering, The Johns Hopkins University, 221 Maryland Hall, 3400 N. Charles Street, Baltimore, MD 21218

Therapeutic nanoparticles must rapidly penetrate human mucus to avoid clearance in the lungs, GI tract, female reproductive tract, eye, and more. For example, the gene that could cure Cystic Fibrosis (CF) has been available since 1989, but gene therapy has yet to cure a single person of CF. Efficient gene delivery through CF sputum, which possesses a bulk viscosity roughly 10,000-100,000 fold higher than water at low shear rates, has been thought to be implausible or impossible since diffusion rates of even small gene-carrier nanoparticles through such a high viscosity fluid with physiological thicknesses found in the lungs would be much too slow. The same is true for nanoparticulate drug delivery through healthy human mucus, which is ~2,000-fold more viscous than water at low shear rates.

Looking to nature for guidance, we noted that pathogens that infect mucosal surfaces have evolved net neutral and hydrophilic surfaces to minimize their adhesion to mucus. We then discovered that even large polymer nanoparticles (up to 500 nm) can rapidly penetrate human mucus through low viscosity "pores" in the human mucus gel, but only if their surfaces are rendered completely non-adhesive to mucus. A critical next step is the identification of safe coating materials that can allow a variety of promising polymer particles to penetrate human mucus.

Here, we describe a simple chemical method, adsorption of Pluronic to either hydrophobic particles or to hydrophobic adhesive regions of human mucus, that readily confers virus-mimetic muco-resistant properties to standard (mucoadhesive) particles. Pluronic is used safely in humans in many products.

Pluronic, an amphiphilic triblock polymer, contains a core polypropylene glycol segment that anchors to the adhesive core of polymeric nanoparticles, coupled with two flanking polyethylene glycol chains that present an uncharged, hydrophilic surface. Pluronic may also shield the adhesive hydrophobic patches along mucin fibers. Both particle coating with Pluronic and simple addition of Pluronic to mucus markedly improves the transport of various polymeric particles in fresh, undiluted human cervicovaginal mucus, with effective diffusivities (Deff) up to only 3-fold reduced compared to those in water. In contrast, the Deff for uncoated particles was ~4000-fold lower in mucus than in water. Much larger fractions of the uncoated particles were immobilized or otherwise hindered by mucus than the Pluronic-coated particles or uncoated particles in Pluronic-treated mucus. We show that this simple chemical method can enhance the transport of a variety of synthetic polymeric particles, offering the prospect that a wide diversity of biomaterials can be adopted for transmucosal therapeutics delivery.

## 705c Influenza Virus-Like Particles as a Platform for Novel Bionanomaterials

**Harvinder S. Gill**, Fu-Shi Quan, Sang-Moo Kang, and Richard W. Compans. Department of Microbiology and Immunology, Emory University, Atlanta, GA 30332

Virus-like particles (VLPs) are virus bodies without the genome. VLPs mimic viruses in structure and morphology but are safe because they lack natural viral infectivity. The first commercial use of VLPs was as a recombinant hepatitis B vaccine in 1986. Recently, much attention has been focused on VLPs as a source of novel bionanomaterials. VLPs can be fabricated utilizing their natural ability to self-assemble into nanometer-sized biocomplexes, can be genetically tailored to incorporate new traits and can be targeted to specific cells using the natural infective mechanism of viruses, without the genome to actually cause infection. Many VLPs are constructed based on non-enveloped viruses (i.e., viruses that naturally do not possess an outer lipid bilayer) by self-assembling protein subunits comprising the viral protective protein cage. In this study we have fabricated influenza VLPs, which in contrast contain a lipid envelope and offer the advantage of anchoring membrane bound proteins with different functions for enhanced versatility.

Using recombinant molecular biology, baculovirus expression vectors were first synthesized to express hemagglutinin (HA type H1) and matrix protein (type M1) of the influenza virus A PR/8. Insect cell cultures (SF9 cells) were then used as an expression system for baculovirus vectors to produce VLPs. VLPs were separated from cell culture, concentrated and purified over a discontinuous sucrose gradient to yield 8 mg VLPs from 1L cell culture stock. VLPs were characterized through transmission electron microscopy and were found to be morphologically similar to influenza virus. Purity of VLPs was assessed through SDS-page electrophoresis and directly compared to influenza A PR/8 virus. The cleavage pattern of HA protein incorporated into VLPs suggests that protein translation and post-translational modification in insect cells yielded HA molecule similar to the native HA protein. As an in vivo test, VLPs were injected into mice to study the immune response and subsequent protection from exposure to influenza virus. VLP immunized mice were protected from lethal challenge while naïve non-immunized mice succumbed to infection. This shows that functionally HA protein of VLPs is similar to native HA protein and VLPs can induce antibodies against influenza virus.

In conclusion, we fabricated influenza VLPs that mimic influenza A PR/8 virus, using an insect cell culture system. Influenza VLPs have the potential to encapsulate DNA for combined DNA/protein vaccination and for gene therapy. As in the native virus, hemagglutinin molecule should also enable influenza VLPs to fuse with host cell membrane to deliver encapsulated DNA. Furthermore, influenza VLPs can also be used for imaging and targeted delivery of therapeutics, for example in cancer medications.

## 705d Design of Caged Protein Scaffold for Molecular Encapsulation

Mercè Dalmau[1], Sierin Lim[2], Helen C. Chen[1], and **Szu-Wen Wang**[1]. (1) Department of Chemical Engineering and Materials Science, University of California, Irvine, 916 Engineering Tower, Irvine, CA 92697-2575, (2) Division of Bioengineering, School of Chemical and Biomedical Engineering, Nanyang Technological University, Singapore, 637457

Caged protein systems have been proposed as general drug carriers since they form highly symmetric nanoscale architectures that offer the potential to be tailored according to the desired application. The E2 subunit of the pyruvate dehydrogenase protein from *Bacillus stearothermophilus* forms a 25-nm nanocapsule structure with a hollow cavity. We produced a variant of this protein consisting only of the structural core, and found the thermostability of this self-assembled scaffold to be unusually high, with an onset unfolding temperature of 81.1°C and an apparent midpoint unfolding temperature of 91.4°C. To evaluate the potential of this scaffold for encapsulation of guest molecules in the internal cavity, we made variants at residues 381 and 239 which altered the physicochemical properties of the hollow internal cavity. These mutants, yielding 60 and 120 mutations within this cavity, assembled into the correct architecture and exhibited high thermostability that was also comparable to the wild-type scaffold. To show the applicability of this scaffold for small molecule encapsulation, fluorescein-5-maleimide and Alexa Fluor 532 C5-maleimide were coupled to the internal cavity. We demonstrate that these mutations can introduce non-native functionality and enable molecular encapsulation within the cavity while still retaining the dodecahedral structure. The unusually robust nature of this scaffold and its amenability to internal changes reveal its potential for nanoscale applications.

## 705e Self-Assembly of Biomimetic Peptide Nanocages with Enzymatic Control

**Ying Chau** and Weiping Wang. Department of Chemical Engineering, The Hong Kong University of Science and Technology, Clear Water Bay, Kowloon, Hong Kong

We have designed clathrin-like triskelion molecules with enzyme-responsive peptide units, Fmoc-Phe-Tyr-phosphate. At physiological pH, these three-legged molecules can self-assemble into spherical nanocages in the presence of phosphatase. The dimension and hollow spherical structure are characterized by dynamic light scattering and transmission electron microscopy. The ability to use enzymatic trigger for smart release is assessed. The potential to develop the new nanocages as drug delivery vehicles are further evaluated by measuring the encapsulation efficiency and release profile of model drug compounds.

## 705f Formation and Characterization of Virus-Nanoparticle Clusters

**Leonard F. Pease III**[1], De-Hao Tsai[2], Michael Zachariah[2], and Michael J. Tarlov[1]. (1) Process Measurements Division, National Institute of Standards and Technology, 100 Bureau Drive, Gaithersburg, MD 20899, (2) University of Maryland and NIST, College Park, MD 20742

Here we controllably generate and rapidly characterize hybrid structures composed of inorganic particles and viruses. Nano and mesostructured materials and modular assemblies promise to improve interfaces between biomolecular and inorganic components for tissue engineering, novel therapeutics, cancer treatments, and microelectronics. Realization of these goals requires reliable methods to produce, characterize, and tune the component building blocks. Here we advance towards this vision by combining a gold nanoparticle and a viral particle into a chemically specific heterodimer, which we characterize using electrospray differential mobility analysis (ES-DMA). The advantages of ES-DMA are that it requires no labeling, provides a direct read-out of particle size distributions, rapidly measures statistically significant populations of viruses, allows the extraction of kinetic rate constants from temporal variation, and detects changes in virus size as little as 0.2 nm. The use of phage display libraries may open this technique to an entire family of nanoparticle-virus clusters. This talk will describe the theory of operation of ES-DMA and its use to size viruses, inorganic nanoparticles and their composites.

## 705g Molecular Assemblies of Nanoparticles: Surprising Analogy with Proteins and New Optical Properties

**Nicholas A. Kotov**, Chemical Engineering Department, Materials Science Department, Biomedical Engineering Department, University of Michigan, 2300 Hayward, Ann Arbor, MI 48109

Physical properties of nanoparticles and nanowires are currently well understood. The next frontier is conceptualization of larger structures, such as nanoscale assemblies. This talk will discuss the two paradigms in this field: (1) spontaneous assemblies of nanoparticles and (2) quantum mechanical interactions of metallic and semiconductor building blocks in nanoscale assemblies. Anisotropic forces arising between nanocrystalline particles drive the self-assembly behavior of these colloidal particles. Interaction anisotropy between CdTe nanoparticles in solution leads to their spontaneous, template-free organization into free-floating sheets. Electrostatic interactions arising from a dipole moment and a small positive charge combined with directional hydrophobic attraction between the nanoparticles are the driving forces for the self-assembly, which we demonstrate by computer simulation. We found that nanoparticles show conceptual similarities with assembly of proteins. This supposition was recently confirmed by assembly of nanoparticles into spiral systems similar to those found in many biological systems.

Electronic interactions in nanoparticle assemblies represent one of the fundamental problems of nanotechnology. Excitons and plasmons are the two most typical excited states of nanostructures, which were shown to produce coupled electronic systems. The concept of these interactions between the Au and CdTe nanoparticles and nanowires will be discussed in terms of quantum mechanical coupling of excited states and unusual optical effects. As such, in presence of dynamic component for excitons theory predicts that emission of coupled excitations in nanowires with variable electronic confinement is stronger, shorter, and blue-shifted. These predictions were confirmed with high degree of accuracy in molecular spring assemblies, where one can reversibly change the distance between the exciton and plasmon. The prepared systems were made protein-sensitive by incorporating antibodies in molecular springs. Modulation of exciton-plasmon interactions can serve as wavelength-based biodetection tool, which can resolve difficulties of quantification of luminescence intensity for complex media and optical pathways.

## 706a An Imaging Nanoprobe Predicts the Outcome of Chemotherapy

**Efstathios Karathanasis**[1], Ioannis Sechopoulos[2], Ananth Annapragada[3], and Ravi Bellamkonda[1]. (1) W.H. Coulter Department of Biomedical Engineering, Georgia Institute of Technology, 313 Ferst Dr NW, Atlanta, GA 30332, (2) Radiology, Emory University, 1701 Uppergate Drive NE Suite 5018, Atlanta, GA 30322, (3) Health Information Sciences, University of Texas Health Science Center at Houston, 7000 Fannin St Suite 600, Houston, TX 77030

**Introduction:** The success of nanoparticle chemotherapy is primarily dependent on the access these agents have to tumors via the leaky vasculature, the so called Enhanced Permeation and Retention (EPR) effect. Yet, the extent of EPR in individual tumors varies widely, resulting in a correspondingly wide range of responses to the therapy. However, there exist no tools currently to rationally determine whether tumor blood vessels are amenable to nanocarrier mediated therapy in a patient-specific manner today. To address this need, we developed a long-circulating liposomal contrast nanoprobe for x-ray imaging based probing of tumor EPR status. In this work, the intratumoral uptake of the nanoprobe was imaged using a clinical digital mammography unit, and subsequently the animals were treated with liposomal doxorubicin of similar composition and particle size as the clinically used liposomal chemotherapy. One expects that the nanoprobe will accumulate in tumors in the same way that liposomal doxorubicin does by the EPR effect.

**Methods:** Using a clinical digital mammography unit, preliminary studies were performed to identify the agent dose that would result in adequate tumor enhancement without enhancement of the normal vasculature. This dose was used to image a rat breast tumor (n=14) over a period of three days using mammography, and subsequently the animals were treated with a single dose of liposomal doxorubicin (10 mg/kg doxorubicin) of similar composition and particle size as the probe. The predictive capability of the probe was characterized by analyzing the correlation between the tumor enhancement found during imaging and the tumor growth rate after treatment.

**Results:** The nanoprobe had a diameter of 103 nm and contained 155 mg/mL iodine (all encapsulated). A dose was identified that produced undetectable signal from the blood while the intratumoral accumulation of the probe produced adequate signal for detection. The high iodine content and the method of use allowed for quantification of the intratumoral, extravascular accumulation of the nanoprobe. The imaging measurements enabled identification of two subgroups prior to treatment: a 'good prognosis' and a 'bad prognosis' subgroup and indeed these demonstrated differential tumor growth rates. An inverse linear relationship between the contrast enhancement rate constant during imaging and the tumor growth rate constant during chemotherapy treatment was found.

**Conclusions:** To our knowledge, this is the first demonstration of a contrast agent capable of predicting the therapeutic efficacy of a clinically approved chemotherapeutic. Such an a priori determination would therefore facilitate personalized therapy, and spare potential non-responders from the rigors of a chemotherapy regimen.

## 706b Targeted Radionuclide Therapy of Metastatic Ovarian Cancer

**Yah-el Har-el** and George Sgouros. Russell H. Morgan Dept. of Radiology and Radiological Science, Johns Hopkins University, 1150 Orleans St., Cancer Research Building 2, Room 4M62, Balitmore, MD 21231

Metastatic ovarian cancer is initially confined to the intraperitoneal (i.p.) space and can be targeted using liposomes immunolabeled with the humanized monoclonal antibody, trastuzumab (Herceptin) that are injected directly into the i.p. cavity. The alpha-particle ($\alpha$)-emitting radionuclide, actinium-225 (Ac-225) was encapsulated within these immunoliposomes as a therapeutic in an animal model of ovarian cancer metastases. This approach is more effective than the traditional delivery of chemotherapeutics by i.v. injection for a number of reasons. By using an i.p. injection, drug delivery targets tumor nodules that are not vascularized and therefore would not be reached by i.v. treatment. Using an $\alpha$-emitting radionuclide as the drug of choice delivers high local energy deposition which is effective at sterilizing targeted cells, but only has a short path length, which would spare adjacent normal tissue. By encapsulating the Ac-225 within immunoliposomes, specific targeting is achieved, and it effectively decreases the toxicity associated with Ac-225's free daughters.

Previously, it has been shown that the process of actively loading Ac-225 within the liposomes can achieve high loading of the radionuclide (~80-95%) as well as stable retention (80-90% over 4 weeks). Prior to beginning therapy, fluorescent and $\mu$ SPECT/CT imaging analysis demonstrated that liposomes were selectively targeted to the sites of tumor nodules. Immunoliposomes containing Ac-225 or trastuzumab directly conjugated to Ac-225 were injected into tumor-free mice in order to determine the maximal tolerated dose (MTD) prior to treatment. It was found that immunoliposomal encapsulation and delivery of the Ac-225 resulted in a 5-fold higher MTD than antibody-mediated delivery of Ac-225 alone (1300nCi vs. 250nCi). Biodistribution analysis of a sub-therapeutic dose in tumor bearing mice suggests that the tumor targeting effect of the two formulations is very similar, but there is a much higher concentration of the free daughter, Bismuth-213 in the liver and kidneys when the antibody alone is injected. This suggests that the higher MTD of the Ac-225 loaded liposomes may be due to its efficient retention of Ac-225 and its daughters at the target site, thereby decreasing their systemic toxicity. Currently underway are therapy studies in which tumor bearing mice are injected with radionuclide loaded immunoliposomes or antibody at their respective MTDs to determine the therapeutic efficacy of this approach.

## 706c Composite Nanoparticles (CNPs) Containing Upconverting Nanophosphors for Photodynamic Therapy

**Stephanie J. Budijono**[1], Robert K. Prud'homme[1], Yiguang Ju[2], Jingning Shan[2], Robert Austin[3], Baris Ungun[1], Josh Collins[4], and Joseph Friedberg[4]. (1) Chemical Engineering, Princeton University, Princeton, NJ 08540, (2) Mechanical and Aerospace Engineering, Princeton University, Princeton, NJ 08540, (3) Physics, Princeton University, Princeton, NJ 08540, (4) University of Pennsylvania, Philadelphia, PA

Photodynamic therapy is an approved treatment for pulmonary and pleural mesothelial cancers. The therapy relies on cytotoxic singlet oxygen production due to excitation of a photosensitizer by a visible light. The current method has significant limitations because visible light has a limited penetration distance in tissue and the mode of singlet oxygen delivery is not targeted specifically to cancer cells.

We present the design and proof of concept studies of composite nanoparticles containing upconverting nanophosphors (UCNPs) and a photosensitizer. The nanophosphors, which absorb infra red light and emit in the visible region, remedy the penetration depth problem. Utilization of infra red light in place of visible light enhances light penetration depth five folds, thus significantly improves therapeutic accessibility. Both nanophosphors and the photosensitizer (meso tetraphenyl porphine) are co-localized within biocompatible, water soluble poly(ethylene glycol)-block-poly(caprolactone) nanoparticles. The composite nanoparticles (CNPs) containing inorganic UCNPs, organic photosensitizers, and biocompatible polymer coatings are readily produced by Flash NanoPrecipitation technology. Work on targeting of the CNPs with antibody ligands is the next extension of the technology.

## 706d Catalase Polymer Nanofilaments of Tunable Stiffness and Morphology

**Eric A. Simone**, Bioengineering, University of Pennsylvania, 3620 Hamilton Walk, 1 John Morgan Bldg, IFEM, Philadelphia, PA 19104, Thomas D. Dziubla, Chemical and Materials Engineering, University of Kentucky, 177 F. Paul Anderson Tower, Lexington, KY 40506, and Vladimir Muzykantov, Institute for Environmental Medicine, Department of Pharmacology, University of Pennsylvania School of Medicine, 1 J Morgan Building, 3620 Hamilton Walk, Philadelphia, PA 19104.

Filamentous polymer nanocarriers (*f-PNC*) can be used to encapsulate the highly potent antioxidant enzyme catalase, protecting this therapeutic from proteases and inhibitors found in many biological settings [1]. The primary factor that controls morphology of these *f-PNC* is the degree of amphiphilicity of the constituent poly(ethylene glycol)-b-poly(lactic acid) (PEG-PLA) diblock copolymers. While maintaining the same degree of amphiphilicity through ratio of the molecular weights (MW) of PEG to PLA (i.e. 80% PLA, a *f-PNC* producing polymer), different absolute MW diblocks were synthesized. A ring opening polymerization of lactide initiated by different MW PEG polymers resulted in diblock MWs that ranged from 12 kDa to 100 kDa, as determined by $^1$H-NMR and GPC. Thermal properties such as glass transition temperature ($T_g$) and melt temperature ($T_m$) were determined via differential scanning calorimetry (DSC). These polymers were subsequently used to encapsulate catalase by previously established methods [1, 2]. Mass and enzymatic activity loading of catalase, as well as the *f-PNC* protection of the cargo enzyme from proteolysis, were determined. While we found these measures to be relatively similar between different MW *f-PNC*, it is known that characteristics such as carrier flexibility and length can play a major role in *f-PNC* circulation profiles *in vivo* [3]. Therefore, morphological analyses were performed by transmission electron microscopy (TEM) as well as fluorescence microscopy. From fluorescence time-lapse microscopy, *f-PNC* persistence lengths ($l_p$), which describe nanocarrier flexibility, as well as contour lengths, were measured. Lower MW *f-PNC* had relatively smaller $l_p$'s and are thus more flexible at room temperature. Since we found that the $T_g$ was around 25°C for the lowest MW *f-PNC*, and increased to 30°C for the highest MW *f-PNC*, we also determined the $l_p$'s at physiologic temperature, 37°C, well above the highest MW polymer $T_g$. Heating increased flexibility (decreased $l_p$) of *f-PNC* overall, however the highest MW *f-PNC* still had larger $l_p$'s than lower MW carriers. Contour lengths also significantly increased with increasing polymer MW. Thus it would seem that polymer MW remains the most important factor in determining nanocarrier morphology and stiffness. In conclusion, several new preparations of catalase-loaded *f-PNC* are reported here, with polymer MW tunable flexibility, offering several potential therapeutic applications.

**References:**

[1] E.A. Simone, T.D. Dziubla, F. Colon-Gonzalez, D.E. Discher, V.R. Muzykantov, Effect of polymer amphiphilicity on loading of a therapeutic enzyme into protective filamentous and spherical polymer nanocarriers. Biomacromolecules 8(12) (2007) 3914-3921.

[2] T.D. Dziubla, A. Karim, V.R. Muzykantov, Polymer nanocarriers protecting active enzyme cargo against proteolysis. J Control Release 102(2) (2005) 427-439.

[3] Y. Geng, P. Dalhaimer, S. Cai, R. Tsai, M. Tewari, T. Minko, D.E. Discher, Shape effects of filaments versus spherical particles in flow and drug delivery. Nat Nano 2(4) (2007) 249-255.

## 706e Effect of Solvent Properties on Antioxidant Enzyme Polymer Nanocarriers Formed with a Modified Double Emulsion

**Elizabeth Hood**, Institute for Environmental Medicine, University of Pennsylvania, 3620 Hamilton Walk, 1 John Morgan Bldg, Philadelphia, PA 19107, Eric A. Simone, Bioengineering, University of Pennsylvania, 3620 Hamilton Walk, 1 John Morgan Bldg, IFEM, Philadelphia, PA 19104, and Vladimir Muzykantov, Institute for Environmental Medicine, Department of Pharmacology, University of Pennsylvania School of Medicine, 1 J Morgan Building, 3620 Hamilton Walk, Philadelphia, PA 19104.

We evaluated a battery of different solvents used during the formation of polymer nanocarriers (PNCs) that may improve loading and protection of the antioxidant enzyme catalase. The polymers used for carrier formulation were poly(ethylene glycol)-poly(lactic acid) diblock copolymers (PEG-PLA), synthesized by ring opening polymerization of lactide. The polymer number averaged molecular weight was determined via proton nuclear magnetic resonance ($^1$H-NMR), and the weight averaged MW and corresponding polydispersity index were measured with gel permeation chromatography (GPC). PNCs are formed from these polymers by a freeze-thaw modified water/oil/water double emulsion technique. The aqueous/oil interface has been shown to affect enzyme loading and denaturation, therefore we examined the effects of different solvents on those parameters in formed polymer nanocarriers. Candidate solvents were selected by desirable freezing points, water miscibility, and vapor pressures. Tetrahydrofuran (THF), in particular, was compared directly to dichloromethane (DCM), our established solvent used in previous PNC formulations. THF has a higher miscibility and a lower vapor pressure than DCM. We found that nanocarriers formulated with THF showed a higher percentage of entrapped protein by weight loading when measured using radio-isotope tracing of $^{125}$I labeled-catalase. Concentration changes in hydrogen peroxide, catalase's substrate, were determined by UV absorbance of the substrate and provided a direct measure of catalase activity. The THF-formed PNCs showed a moderate improvement in activity as a percentage of the initial enzyme activity versus those made using DCM. Size distribution of the particles formed as measured by dynamic light scattering was largely unchanged between the two solvents. Protection of catalase by polymer carriers from proteolysis via the nonspecific protease, pronase, appeared relatively independent of solvent selection and was significantly enhanced over that of non-loaded catalase. We have concluded that solvent selection has a significant impact on formulation of PNC loaded with active enzymes. Specifically, THF results in protease-resistant PNC with a significant increase in encapsulation of the potent therapeutic enzyme, catalase.

## 706f Formation of Magnetite-Containing Nanoparticles Using a Confined Impinging Jet Mixer

Raquel Mejia-Ariza[1], Jonathan Goff[2], John S. Boyd[3], Judy S. Riffle[3], and **Richey M. Davis**[4]. (1) Macromolecular Science & Engineering Program, Virginia Tech, Blacksburg, VA 24061, (2) Macromolecular Science and Engineering, Virginia Tech, Blacksburg, VA 24061, (3) Chemistry, Virginia Tech, Blacksburg, VA 24061, (4) Virginia Tech, Department of Chemical Engineering, Blacksburg, VA 24061

The formation of nanospheres comprised of polylactic acid and magnetite particles was studied using a Confined Impinging Jet (CIJ) mixer system, a continuous precipitation process that yields particles with relatively narrow and controlled size distributions. Control of nanoparticle size distribution is very important in biological systems for cellular uptake and targeted delivery. In the CIJ mixer, two liquid streams collide under turbulent conditions in a confined mixing chamber where particle nucleation and precipitation occur within milliseconds. The precipitation of polylactides and magnetite particles was studied in the presence of different amphiphilic block copolymers containing poly(ethylene oxide) to provide steric stabilization and control of size distributions. After the nanoparticles were formed, they were dialyzed to remove organic solvent and unimer chains and were then characterized by dynamic light scattering and electron microscopy. Particles in the size range of 60–200 nm were formed.

## 707a Nanowire Chem-Bio Sensors

Stergios Papadakis, JHU-Applied Physics Laboratory, 11100 Johns Hopkins Rd, Laurel, MD 20723 and **David H. Gracias**, Department of Chemical and Biomolecular Engineering, Johns Hopkins University, 125 Maryland Hall, 3400 N Charles Street, Baltimore, MD 21218.

We describe a chem-bio sensor platform in the form of arrays of nanowires for threat detection of chemical and biological compounds. The wires are assembled in arrays using dielectrophoresis. The platform is based on simultaneous optical and electronic detection based on Surface Enhanced Raman and tunneling spectroscopy.

# 707b Sensing Chemical and Biological Substances with CdS Nanowires

**Shih-Yuan Lu**, Department of Chemical Engineering, National Tsing-Hua University, Hsin-Chu, 30013, Taiwan, Yi-Feng Lin, Dept. of Chemical Engineering, National Tsing-Hua University, Hsin-Chu, 30013, Taiwan, and Yung-Jung Hsu, Dept. of Materials Science and Engineering, National Chiao-Tung University, Hsin-Chu, 30010, Taiwan.

Sensing of amino acids and metal ions with CdS nanowires was discussed. First, differential sensing of tyrosine and serine achieved with well aligned CdS nanowire arrays by exploring the pH-dependent photoluminescence behavior of the nanowire arrays toward exposure to the two amino acid solutions was successfully demonstrated and discussed. The contrasting trend in photoluminescence intensity with respect to variations in analyte concentration observed at pH 11 served as the check point for the present differential sensing. Second, sensing of Ag+ and Cu2+ with ethylenediamine (EN)-capped CdS nanowires (NWs) was successfully demonstrated and the relevant mechanism was revealed with a direct observation of the side product of the sensing event. The binding of Ag+ and Cu2+ at the surfaces of the EN-capped CdS NWs triggered an in-situ growth of Ag and Cu nanocrystals, respectively, causing decay in the photoluminescence of the CdS NWs and rapid color change of the solution recognizable to naked eyes.

# 707c Probing the "Quality" of Intrinsic Semiconducting –O–Ti–O–Ti– Chains In Titanosilicate ETS-10

**Zhaoxia Ji**[1], Morag Murdoch[2], Dennis M. Callahan Jr.[1], Juliusz Warzywoda[1], Russell F. Howe[2], and Albert Sacco Jr.[1]. (1) Department of Chemical Engineering, Center for Advanced Microgravity Materials Processing, Northeastern University, 360 Huntington Ave, 147 Snell Engineering Center, Boston, MA 02115, (2) Chemistry Department, University of Aberdeen, Aberdeen, AB24 3UE, United Kingdom

Titanosilicate ETS-10 crystals contain monatomic (diameter 0.67 nm) semiconducting …Ti–O–Ti–O–Ti… chains, which show quantum confinement effects and are regarded as quantum wires [1]. "Quality" of these crystals/quantum wires is expected to affect the performance of electronic devices. It has been hypothesized that larger ETS-10 crystals may have fewer discontinuities in the …Ti–O–Ti–O–Ti… chains, which was indicated by the red shift of the UV absorption edge. Here, Raman and UV-vis spectroscopy were used to probe the "quality" and examine the optical properties of large ETS-10 crystals.

ETS-10 crystals with the average size ($L_{ave}$) of 35, 14, 6, and 2 $\mu$m were synthesized hydrothermally at 473 K [2, 3]. Raman spectra of these crystals showed all the bands characteristic for ETS-10 [4, 5]. However, the full width at half maximum (FWHM) and the position of the most intense Raman band that was assigned to the symmetric Ti – O stretching vibrations in the …Ti–O–Ti–O–Ti… chains varied from sample to sample. The FWHM of this band increased from 19-21 $cm^{-1}$ for the larger ($L_{ave} \geq 14$ μm) crystals to 38-39 $cm^{-1}$ for the smaller ($L_{ave} \leq 6$ μm) crystals. In addition, the band position shifted to higher frequency, i.e., from 725 $cm^{-1}$ for the crystals with $L_{ave} \geq 14$ μm to 733 $cm^{-1}$ for the crystals with $L_{ave} \leq 6$ μm. The band broadening and shift to higher frequency observed for the smaller crystals suggest more stacking defects in these crystals (i.e., wider length distribution of the …Ti–O–Ti–O–Ti… chains). Diffuse reflectance UV-vis spectroscopic analysis showed that the absorption edge of the spectra for the larger ETS-10 crystals appeared progressively more red-shifted from the spectra of the smallest ETS-10 crystals. This red shift may be attributed to the fewer defects present along the …Ti–O–Ti–O–Ti… chains in the larger ETS-10 crystals. Consistent with Raman spectroscopic analysis, this result suggests increased average length of the …Ti–O–Ti–O–Ti… chains embedded in increasingly larger ETS-10 crystals.

[1] C. Lamberti, Micropor. Mesopor. Mater., 30 (1999) 155.

[2] Z. Ji, J. Warzywoda, A. Sacco Jr., Micropor. Mesopor. Mater., 81 (2005) 1.

[3] Z. Ji, J. Warzywoda, A. Sacco Jr., Micropor. Mesopor. Mater., 109 (2008) 1.

[4] N. C. Jeong, M. H. Lee and K. B. Yoon, Angew. Chem., Int. Ed., 46 (2007) 5868.

[5] P.D. Southon and R.F. Howe, Chem. Mater., 14 (2002) 4209.

Web Page: nuweb.neu.edu/cammp/naturalquantum-wires.htm

## 707d Nanowire Based Hybrid Materials as Anodes for Lithium Ion Batteries

**Praveen Meduri**[1], Chandrashekhar Pendyala[2], Vivekanand Kumar[3], Gamini U. Sumanasekera[4], and Mahendra K. Sunkara[3]. (1) Chemical Engineering, The University of Louisville, 106 ERNST HALL, Louisville, KY 40217, (2) Chemical Engineering, University of Louisville, Louisville, KY 40292, (3) Department of Chemical Engineering, University of Louisville, 106 Ernst Hall, Department of Chemical Engineering,, University of Louisville, Louisville, KY 40292, (4) Physics, University of Louisville, 2210 south brook st, room 211, Louisville, KY 40208

Lithium ion batteries (LIBs) have considerable advantages over conventional Ni-Cd batteries in terms of energy to weight ratios and battery life. Present energy demands have increased interest to develop new class of high capacity LIBs. Constant search for a stable high capacity anode material for LIB's is underway because of the low capacity of graphite (372 mAhg-1). Sn and SnO2 have theoretical capacities three times the capacity of graphite but it has been shown that the capacity fading with cycling is a considerable problem in these materials.

Nanowires, with high surface areas and better charge transport properties have been shown to perform better than their thin film counterparts. However, pure phase SnO2 nanowires have shown severe degradation with cycling and the specific capacity fades to 170 mAhg-1 over 40 cycles. The post-lithiated samples show a distorted nanowire structure with big agglomerates. To overcome the problem of capacity loss with cycling, we present a simple and generic concept of nanowire based hybrid materials as high capacity anodes for Li-ion batteries. A hybrid system with Sn nanoclusters/ SnO2 nanowires is shown to retain high capacity over 100 charge-discharge cycles. The capacity fading is low and the structural integrity of the nanowire architecture is confirmed by scanning electron microscopy. This hybrid structure concept is being extended to other high capacity materials like silicon and germanium which can offer significant enhancements in portable electronics.

## 707e Interface Thermal Transport Measurements across Nano-Junctions

**Monalisa Mazumder**, Department of Chemical & Biological Engineering, Rensselaer Polytechnic Institute, 110 8th Street, Troy, NY 12180 and Theodorian Borca-Tasciuc, Department of Mechanical, Aerospace & Nuclear Engineering, Rensselaer Polytechnic Institute, 110 8th Street, Troy, NY 12180.

With increased miniaturization of electronic devices and consequently increased integration density, effective dissipation of heat stands out as one of the most important factors dictating performance efficiency and sustainability of the devices. Thermal management and fundamental research on thermal transport properties in low-dimensional nano-structures is thus crucial for engineering high performance devices. Moreover, nanotechnology based devices contain a large number of interfaces and junctions, and transport across the junctions plays a major role in device operation. The present paper aims at reporting proof-of-concept demonstration of measurement of interfacial thermal resistance (ITR) by an AFM based technique. Topographic imaging is used to locate the nano-structures and then electric current is passed through the circuit formed by the nano-structure patterned on a dielectric layer, two current-carrying electrodes on the substrate, a known macroscopic resistor and an external power supply. A set of two electrodes taps the voltage drop across the nano-structure on the dielectric. Due to passive heating across the nano-structure, the resistance changes resulting in a dc voltage drop across the wire from which the rise in temperature due to ITR is computed by fitting the experimental data to the theoretical model, knowing temperature coefficient of resistance. The rise in temperature across the nanowire-substrate interface, in turn, gives the ITR. Alternatively, the 3ω method has been used to monitor the temperature when an AC current replaced the DC. This work presents the technique along with the measurements of nano-interface thermal resistance of nanowire-substrate junction using the aforesaid technique.

# 707f Materials, Electrical, and Sensing Properties of ZnO Nanowires

**Sandra Catalina Hernández**[1], Min Lai[1], Ashok Mulchandani[2], Marc A. Deshusses[1], and Nosang V. Myung[1]. (1) Department of Chemcial and Environmental Engineering, University of California, Riverside, Riverside, CA 92521, (2) Department of Chemical and Environmental Engineering, University of California, Riverside, CA 92521

ZnO is an interesting and versatile material due to its ample material properties and diverse applications. ZnO is an n-type direct and wide band gap semiconductor with band gap of 3.35eV, it has excellent chemical and thermal stability, photonic properties, with strong piezoelectric effect, and it is sensitive to some volatile organic compounds (VOCs) and can be fabricated in diverse morphologies. Additionally, improvement of electronic properties and enhancement of selectivity towards VOCs can be achieved with selective doping. This opens up the possibility of multiple gaseous analyte sensing with real time detection by using ZnO nanostructures as the sensing element. Important parameters to consider are the geometry of the nanostructures, for example nanowires vs. nanoparticles, the dimensions of the nanostructures such the diameter of the wires, the material quality (i.e. crystallinity), and the selection and quantity of the dopant material. In this work we have explored an electrosynthesis approach towards fabricating ZnO nanowires and investigated optimal deposition parameters such as solution composition, applied voltage and temperature. We have been able to obtain a voltage deposition window spanning from -0.6 to -1.0 V vs. Ag/AgCl where ZnO nanowires can be fabricated which allows for further implementation of simultaneous electrodeposition of dopants with ZnO. Deposition temperature analysis demonstrated that amorphous nanowires can be obtained at 25°C while single crystal nanowires can be acquired at 70°C. Furthermore, the diameter of the nanowires can be manipulated by controlling the pore size of the polycarbonate membrane used as the deposition templates. The resultant ZnO nanowires were electrophoretically aligned onto prefabricated micro electrodes and gas sensing performance as a function of crystallinity and nanowires diameter were investigated.

# 714a Self-Assembling Bifunctional Proteins for Bioelectrocatalytic Hydrogels: A Protein Engineering Approach to Advanced Materials Design

**Ian R. Wheeldon**[1], Joshua Gallaway[1], Scott Calabrese Barton[2], and Scott Banta[1]. (1) Chemical Engineering, Columbia University, 500 West 120th Street, New York, NY 10027, (2) Chemical Engineering and Materials Science, Michigan State University, East Lansing, MI 48824-1226

Protein engineering provides the tools to develop novel biomaterials endowed with multiple functionalities. The focus of much of the work in this area has been the development of bioactive hydrogels for drug delivery and tissue engineering applications. The fields of biosensing and bioelectrocatalysis will benefit from the advantages of protein-based hydrogels provided that electron transport and catalytic functionalities can be incorporated into the protein structures. Here we present two bifunctional protein building blocks that co-assemble into a bioelectrocatalytic hydrogel. One building block, a metallo-polypeptide, supports electron transfer, and a second, an engineered chimeric poly-phenol oxidase, provides enzymatic activity towards the reduction of dioxygen to water. Both proteins are bifunctional in that, in addition to electron transport and catalytic functionalities, they contain physical cross-linking domains that promote the self-assembly of a supramolecular hydrogel. The metallo-polypeptide is based on a previously designed tri-block polypeptide comprised of a soluble region flanked by two α-helical leucine zipper domains. The helical domains self-assemble into physically cross-linked junctions through coiled-coil formation. Compatible helices genetically fused to the poly-phenol oxidase allow mixtures of the two building blocks to co-assemble into bioelectrocatalytic hydrogel.

Electron transport functionality of the metallo-polypeptide is derived from the divalent attachment of an osmium redox moiety to the tri-block polypeptide. Attachment of the osmium moiety is demonstrated by mass spectroscopy (MS-MALDI-TOF) and cyclic voltammetry. Physical cross-linking functionality of α-helical domains in both building blocks is confirmed by circular dichroism spectroscopy and by rheological measurements. Catalytic activity of the chimeric poly-phenol oxidase and electron conduction of the metallo-polypeptide are evaluated by dilute solution kinetic assays and cyclic voltammetry, respectively. Mixtures of the two building blocks co-assemble into a bioelectrocatalytic hydrogel that generates a catalytic current in the presence of oxygen.

This bioelectrocatalytic system is the first example of an all protein-based electrode surface modification. With further optimization, the system could readily be used as the cathode to a biofuel cell or as an oxygen sensor. The design is also an example of a general method of multi-functional materials design, the co-assembly of multi-functional protein building blocks. In addition, the modularity of the system allows for the independent tuning of the physical properties of the hydrogel and its bulk function, an aspect of hydrogel design that will prove beneficial to a wide range of applications.

# 714b Morphology and Size-Controlled Self-Assembling of Enzyme-Responsive Peptides

**Weiping Wang** and Ying Chau. Department of Chemical Engineering, The Hong Kong University of Science and Technology, Clear Water Bay, Kowloon, Hong Kong

We have derived a novel reverse microemulsion process to impose temporal and spatial control of self-assembly. The ability of this process to transform morphology and size has been demonstrated using an enzyme-responsive dipeptide, Fmoc-Phe-Tyr-phosphate. In the presence of phosphatase, these molecules self-assemble into nanofibers of about 28nm in aqueous solution. By controlling the self-assembly process with an enzymatic trigger and confining it within nanoscale reverse micelles, the same molecules are fabricated into nanoparticles of around 6.5nm. We further investigate the morphology and size control of peptide assemblies by changing the process parameters, including the ratio of various reagents and incubation time. The resulting nanostructures are characterized by transmission electron microscopy and dynamic light scattering. The possible process for peptide nanoparticle formation inside reverse micelle is proposed according to the results. This method is applicable to other stimuli-responsive self-assembling blocks.

## 714c A DNA-Programmed Protein Nanoassembly Capable of Sensing Single Nucleotide Polymorphisms

Cho-Ah Im, Jandi Kim, and **Jongshik Shin**. Department of Biotechnology, Yonsei University, B509, Engineering Building 2, Shinchon-Dong 134, Seodaemun-Gu, Seoul, 120-749, South Korea

Autonomous bottom-up fabrication based on molecular recognition is a powerful approach for engineering structures and devices at the molecular scale. DNA is a versatile construction material that can be programmed to self-assemble into nanoscale structures based on specific complementarity between bases. Here, we aim at developing bio-inspired nanomaterials capable of detecting single nucleotide polymorphism (SNP). We developed a novel SNP biosensor made of DNA-programmed protein nanostructures of which disassembly of the whole structure is triggered by a target DNA oligonucleotide carrying a specific sequence. We designed DNA sequences incorporated in the sensor materials in a way that even a single base mutation in the target DNA (e.g. a single mutation in a 24-mer DNA oligonucleotide) effectively prevents the protein nanostructures from being disassembled. We could detect single nucleotie mutations by monitoring structural changes in the protein nanostructure. The SNP biosensor material is promising for fast identification of SNP in genes of interest.

# 714d Biological Fabrication of Photoluminescent Nanocomb Structures by Metabolic Incorporation of Germanium into the Biosilica of the Diatom Nitzschia Frustulum

Tian Qin[1], Timothy Gutu[2], Chih-hung Chang[1], Jun Jiao[2], and **Gregory L. Rorrer**[1]. (1) Oregon State University, Department of Chemical Engineering, Corvallis, OR 97331, (2) Department of Physics, Portland State University, Portland, OR 97207

Diatoms are single-celled algae that make microscale silica shells or "frustules" with intricate nanoscale features such as two-dimensional pore arrays. In this study, the metabolic insertion of low levels of germanium into the frustule biosilica of the pennate diatom Nitzschia frustulum by a two-stage cultivation process induced the formation of frustules which strongly resembled double-sided nanocomb structures. The final product from the two-stage cultivation process contained 0.41 wt% Ge in biosilica, and consisted of an equal mixture of parent frustule valves possessing a normal two-dimensional array of 200 nm pores, and daughter valves possessing the nanocomb structure. The nanocomb structures had overall length of 8 microns, rib width of 200 nm, rib length of 500 nm, and slit width of 100 nm. About 10 million of these replicate microstructures were fabricated per mL of culture. Each slit of the nanocomb was most likely formed by a directed morphology change of a row of 200 nm pores to a single open slit following Ge incorporation into the developing frustule during the final cell division. The frustules possessed blue photoluminescence at peak wavelengths between 450-480 nm, which was attributed to contributions from nanostructured biosilica in both the parent valves and in the nanocomb daughter valves. This is the first reported study of using a cell culture system to biologically fabricate a photoluminescent nanocomb structure on a massively parallel scale.

# 714e Biodegradable Nanoparticles Conjugated with ávâ3 Integrin-Binding Ligand for Targeted Tumor Delivery

**Angel Mercado**, Xuezhong He, and Esmaiel Jabbari. Chemical Engineering, University of South Carolina, Dept. of Chemical Eng., Rm. 2C11, Swearingen Engineering Center, Columbia, SC 29208

Introduction and Objective: A novel strategy for selective delivery to malignant cells is to encapsulate the antitumor drug in nanoparticles (NPs) that are grafted with ligands with very high affinity for tumor cells. Biodegradable NPs <100 nm in size and surface-modified with hydrophilic polymers can evade the mononuclear phagocytes system (MPS), overcome resistance at the tumor level (EPR effect), and can be conjugated with ligands that bind with high specificity/affinity to the tumor cell surface to trigger the receptor-mediated endocytosis to localize the NPs (hence the antitumor drug) in the cell cytoplasm. Furthermore, it is well established that ligand-conjugated NPs display a stronger multivalent interaction and a much higher apparent affinity to cell surface receptors than the free ligand. Our laboratory has developed novel biodegradable poly(lactide-co-glycolide fumarate) (PLGF) and poly(lactide-co-ethylene oxide fumarate) (PLEOF) macromers that self-assemble into nanoparticles in the aqueous medium. After self-assembly, cell-responsive ligands can be conjugated to these NPs for targeted delivery of antitumor drugs. Cyclic RGD-based biomimetic peptides have high affinity for áíâ3 integrin receptor, which is implicated in tumor-induced angiogenesis and metastasis, and is up-regulated on both cancer cells and tumor-associated blood vessels. The objective of this study was to determine the effect of conjugation of cyclic RGD peptide [c(-RGDfk-)] conjugation on the uptake of PLGF-PLEOF NPs by tumor cells.

Methods: A 10% w/w solution of the PLGF-PLEOF macromers in dimethyl sulfoxide (DMSO) was dialyzed against distilled deionized (DDI) water to form a stable colloidal suspension of NPs. The morphology of the NPs was characterized by TEM and the size distribution of NPs was measured by dynamic light scattering. Degradation was measured as a function of time in vitro in primary culture media (CM) without fetal bovine serum (FBS) at 37°C. Dextran fluorescein isothiocyanate (FITC-dextran) and Paclitaxel were used as surrogate molecules for release studies. The release characteristic of Paclitaxel from the NPs was measured by HPLC. Near-infrared imaging was used for determination of in vivo distribution of NPs in mice with intestinal tumor. The MCF-7 and U87MG cancer cell lines with low and high ávâ3 integrin expression, respectively, were used for determination of in vitro cell binding affinity. The linear D-Phe-Cys-Arg-Gly-Asp peptide was cyclized in the solid-phase on the NovaSyn® TGA resin using Fmoc- chemistry. Cyclic (GRGfC) peptide was incubated for 10 h with NPs in PBS for conjugation.

Results: PLGF-PLEOF and PLAF-PLEOF NPs degraded in 15 and 28 days, respectively, which demonstrated that the release was dominated by hydrolytic degradation and erosion of the matrix. As the concentration of Paclitaxel was increased from 0 to 10, and 40 ìg/ml, cell viability with free Paclitaxel decreased from 100 to 65 and 40%, respectively, while those encapsulated in NPs decreased from 93 to 54 and 28%. Groups with Paclitaxel loaded NPs had higher cytotoxicity compared to Paclitaxel directly added to the media at the same concentration. Infrared image of mice injected with NPs showed significantly higher concentration of NPs in the tumor tissue. When cyclic RGD conjugated NPs were incubated with MCF-7 and U87MG cells, significantly higher uptake of NPs by U87MG cells with high ávâ3 integrin expression was observed.

## 714f Inulin Surfactants for Colon-Specific Drug Delivery

**Julio C. Arboleda**, G. Toriz, S. Beltran, V. Zuñiga, and E. Delgado. Departamento de Madera Celulosa y Papel, Universidad de Guadalajara, Km 15,5 Carretera Guadalajara-Nogales, Las Agujas, Zapopan, Mexico

Inulin is a storage polysaccharide found in plants such as onion, garlic, artichoke, chicory, and tequila agave. It is interesting for biomedical applications because it is metabolized only in the human colon by enzimatic means, which allows the design of site-specific drug delivery systems. Such systems are needed for oral administration of drugs that tend to be unstable, either by being absorbed in the gastrointestinal tract or by leading to side effects. Drugs used for colorectal cancer, ulcerative colitis and inflammatory bowel disease belong to this class.

Derivatives from inulin, with emulsifying behavior, can be produced by reacting hydrophobic moieties with the abundant hydroxyl groups in the polysacharide. It is possible to produce emulsions of different hydrophobic drugs with these surfactants and then, to obtain capsules by crosslinking and drying of the formed micelles.

These capsules are distinctively structured in such a way as to confine the drug in the interior and exposing the inulin moiety to the exterior. The exterior would just be enzymatically metabolized in the colon, triggering the aperture of the capsules and the subsequent release of the drug.

In order to emulsify a drug it is necessary to know its required HLB. Therefore, the synthesis of inulin derivatives had to be optimized as a function of the required HLB, which could be done by changing the molar relationship between inulin and the hydrofobic moiety during the synthesis of the surfactant.

In this work the preparation of surfactants with stearic acid/stearyl chloride and agave inulin is shown. HLB for the emulsification of 5-ASA was determined and the synthesis of agave surfactants was carried out. Drug emulsification, crosslinking and drying of micelles as well as the characterization of the obtained microcapsules are presented.

**Aluminum and copper nanostructures for luminescence enhancement / quenching**

Sanchari Chowdhury[1], Venkat R. Bhethanabotla[1], and Rajan Sen[2]
Department of Chemical and Biomedical Engineering [(1)], Department of Civil and Environmental Engineering [(2)], University of South Florida, Tampa, Fl 33620-5350

## Abstract

Metal enhanced/quenched luminescence was explored at the vicinity of aluminum and copper nanostructures. Ordered arrays of Al nano-islands with high aspect ratio were fabricated using nanosphere lithography combined with e-beam evaporation. Smaller-sized Cu nanoparticles ($\leq 10$ nm) were deposited on glass substrates using DC magnetron sputtering. Luminescence emission from Cy3 at the vicinity of these nanostructures was studied. Quenching of fluorescence was observed at the pure copper nanoparticle platforms. Signal manipulation of luminescent dye using Al and Cu nanoparticles will extend applications in different fields ranging from optoelectronics to biological research.

## Introduction

Luminescence techniques have increasingly found promising applications in biological research including single molecule detection, cellular imaging, gene profiling, proteomics, drug discovery and disease diagnostic. Fluorescent markers are extensively being used for labeling for cell imaging. Signal weakness due to low concentration of fluorophores attached with molecules and low photostability of molecular fluorophores are two major concerns for this technique(*1*). Nearby conducting metallic particles, colloids, or surfaces are known to significantly influence the emission of vicinal luminophores [(2)]. Using nanoparticles platforms, it is possible to increase the quantum yield and photostability of weakly luminescent probes by increasing their emission efficiency, by modifying radiative decay rate or by coupling the emission with far field through scattering. Metal enhanced luminescence (MEL) has been studied mostly using silver and gold nanoparticles due to their intense, narrow SPR peaks. Gold nanoparticles are known to both quench and enhance luminescence depending on the fluorophore-particle separation distance, molecular dipole orientation with respect to particle surface and size of the nanoparticle (*3-5*). Relatively smaller (typically less than 30 nm) gold nanoparticles quench the fluorescence emission due to the non-radiative transfer from the excited states of luminophore molecules to the gold nanoparticles (*4*). Larger gold nanoparticles can enhance the luminescence due to the increased contribution of nanoparticle scattering (*3, 6*). Enhanced signal and photostability of luminophores, improved surface immunoassay and DNA detection, and enhanced wavelength-ratiometric sensing, amplified assay detection are few examples of applications of MEL. On the other hand, quenching resulted due to metallic nanoparticles are successfully utilized for the improvement of homogeneous and competitive fluorescence immunoassay (*7, 8*), optical detection of DNA hybridization (*9*) , competitive hybridization assay (*10*) and in optoelectronics (*11*). The imaginary component of dielectric constant of copper is much larger (more than twice) than that of silver in the wavelength range of 300 nm to 600 nm. So, it is expected that in this wavelength range due to higher ohmic losses, Cu nanoparticles will mostly quench the luminescence at their close proximity (*12*). On the other hand, for large nanoparticles (>100nm) with high aspect ratios, MEL occurs mainly due to increase of the coupling efficiency of the fluorescence emission to the far field through nanoparticle scattering. In this case, as effect of ohmic losses on MEL is reduced, less expensive metals such as Al and Cu, which are more lossy than Ag and Au, can be considered as promising candidates for enhancing luminescence(*13*). This work focuses on the metal enhanced or quenched luminescence observed from aluminum and copper nanostructures.

## Experimental Procedure

We created the single layer polystyrene nanosphere (Interfacial Dynamics Corporation, Portland, OR) mask by spin coating. Before coating, the glass substrates were cleaned in a plasma cleaner for 15 minutes. The nanospheres were obtained from the manufacturer as a suspension in water and were diluted using surfactant Triton X-100/methanol (1:400) before spin coating. The dilution factor for 400 nm nanospheres was 1:1. The Spin coating speed was 500 rpm for nanospheres. Al was deposited on nanospheres coated glass substrates using electron beam evaporation under high vacuum ($2\times10^{-6}$) and Al deposition was followed by 10 nm of $SiO_2$ deposition. The $SiO_2$ layer was deposited on Al nanostructures to maintain a particular fluorophore nanostructure separation distance and to protect the Al film from oxidation. The thickness of film is measured with a quartz crystal microbalance. After deposition of metal, the polystyrene nanosphere mask was dissolved in ethanol by sonicating for 2 minutes. Copper nanoparticles were deposited on $22\times22$ mm glass cover slips (Fisher finest cover glass, thickness approximately 140 microns) using DC magnetron sputtering (Plasma Sciences CRC-100 Sputter Tool).

Cy3-streptavidin was first dispersed in 0.25% Poly (vinyl alcohol) (PVA, mw 15000) aqueous solution by sonicating. This fluorophore solution was coated on substrate by spin coating (3000 rpm speed). Surface morphology of nanosphere mask and nanostructures was observed by atomic force microscope (AFM) (Digital Instruments, Nanoscope IIIa). An UV- Vis spectrometer (JASCO, V-530) was used for measuring the light absorption spectra attributed to SPR of nanoparticles. The Leica DMI 4000b inverted fluorescence microscope equipped with Leica DFC340 FX CCD camera was utilized for MEL measurements. We obtained fluorescence intensities for each sample by analyzing a 1.64 mm$\times$ 2.19 mm image-section of each substrate.

## Result & Discussion

Atomic force microscopy images of Al nanostructures indicate that it is not possible to obtain uniform nanosphere lithography over an area much larger than 10 micron using spin coating. Figure 1 (A) shows the AFM image of Al deposited on nanosphere lithographed glass slides. It can be seen that the image is comprised of two distinct areas 1) particulated thin film of Al where nanosphere coating was not present 2) isolated nanostructures of Al which was obtained from nanosphere coated area (Figure 1B). The interesting observation is that luminescence of Cy3 was observed to be enhanced significantly at the vicinity of Al nanostructured area where particulated thin film area does not show any enhancement (Figure 2). Cu nanoparticles quench (4.8 $\pm$ 2.48 times) the luminescence of Cy3.

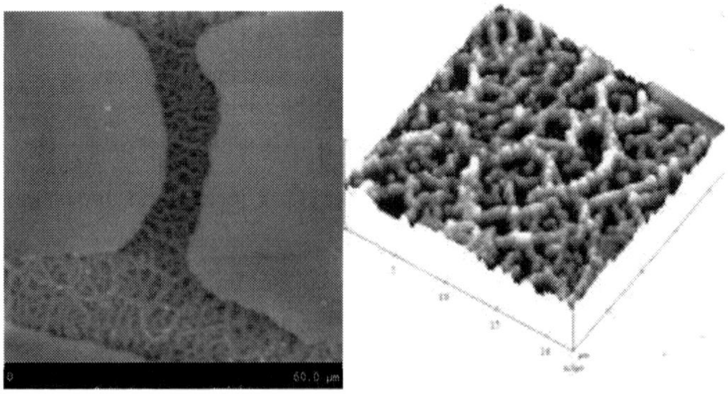


Figure 1:(A) AFM image of aluminum coated on glass substrate. (B) Isolated nanostructures of Al obtained from the nanosphere lithographed area

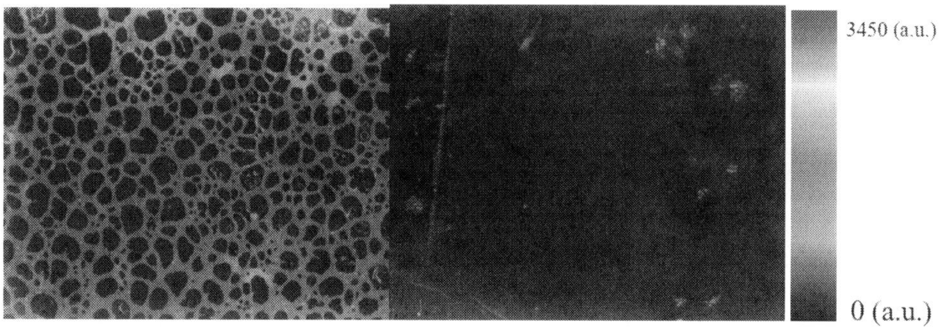

**Figure 5:** Pseudo colored image (intensity scale increases from blue to red) of Cy3 coated on (A) aluminum nanostructures (B) glass substrates.

The emission rate is increased due to the coupling of emission at far field through nanoparticle scattering. For large nanoparticles (>100 nm), effect of scattering is high and the effect of ohmic losses are not important (*15*). Hence, if scattering is high, even lossy metals such as Al and Cu can produce significant enhancements. This may be the reason why enhancement of Cy3 at the close proximity of Al nanostructures. On the other hand, since the Cu nanoparticles are of very small size (about 10 nm), we expect nanoparticle scattering to have less importance (*16*) so these nanoparticles result in quenching.

## Conclusion

Using nanosphere lithography and E-beam evaporation we fabricated single-layers of larger (>100 nm) nanostructures of aluminum. Smaller sizes of copper nanoparticles were deposited using DC sputtering. We got promising results from preliminary study of fluorescence enhancement of Cy3 by Al nanostructures and fluorescence quenching by Cu nanoparticles. This finding opens new avenues for utilization of easily fabricated Al nanoparticles in MEL-based biological applications. We expect quenching effect of copper nanoparticles will motivate the utilization of these nanoparticles as less expensive alternative to gold in biological applications such as homogeneous and competitive fluorescence immunoassay, detection of DNA hybridization, competitive hybridization assay and in optoelectronics.

## Acknowledgement

This study was funded by a grant from the National Science Foundation (No. CMS-409401). This funding is gratefully acknowledged.

## Reference


1. E. L. M. e. al., *Biophys. J* . **92**, 2150–2161 (2007).
2. C. D. Geddes, J. R. Lakowicz, *J. Fluoresc.* **12**, 121-129 (2002).
3. P. P. Pompa *et al.*, *Nat. Nanotechnol.* **1** ( 2006 ).
4. E. Dulkeith *et al.*, *Nano Lett.* **5**, 585-589 (2005).
5. K. Aslan, S. N. Malyn, C. D. Geddes, *J. Fluoresc.* **17**, 7-13 (2007).
6. K. Aslan, S. N. Malyn, C. D. Geddes, *Chem. Phys. Letts.* **453**, 222-228 (2008).
7. N. Kato, F. Caruso, *J. Phys. Chem. B* **109**, 19604-19612 (2005).
8. L. Ao, F. Gao, B. Pan, R. He, D. Cui, *Anal. Chem.* **78**, 1104-1106 (2006).
9. W. Zai-Sheng, J. Jian-Hui, L. FU, S. Guo-Li, Y. Ru-Qin, *Anal. Biochem.* **353**, 22-29 (2006).
10. B. Dubertret, M. Calame, A. J. Libchaber, *Nature Biotechnol.* **19**, 365-370 (2001).
11. H. Imahori, S. Fukuzumi, *Adv. Mater.* **13**, 1197-1199 (2001).
12. Y. Zhang, K. Aslan, M. J. R. Previte, C. D. Geddes, *Appl. Phys. Lett.* **90**, 173116-173113 (2007).
13. H. Mertens, A. Polman, in *ArXiv e-prints*. (2007), vol. 711.
14. K. B. Male, S. Hrapovic, Y. Liu, D. Wang, J. H. T. Luonga, *Anal. Chim. Acta* **516** 35–41 (2004).
15. H. Mertens, Utrecht University ( 2007).
16. J. H. Hodak, A. Henglein, G. V. Hartland, *J. Phys. Chem. B* **104**, 9954-9965 (2000).

# 715b Enhanced Cellular Activation with Single Walled Carbon Nanotube Bundles Presenting Antibody Stimuli

**Tarek Fadel**[1], Tarek Fahmy[1], Gary Haller[2], and Lisa Pfefferle[2]. (1) Biomedical Engineering, Yale University, 120 Mitchell Dr, New Haven, CT 06511, (2) Chemical Engineering, Yale University, 120 Mitchell Dr, New Haven, CT 06511

Efficient immunotherapy can be accomplished by expanding T cells outside the body using single walled carbon nanotube (SWNT) bundles presenting antibody stimulus. Owing to the large surface area of these bundles, which can reach 1560 m2/g, T cell stimulating antibodies such as anti-CD3, can be presented at high local concentrations inducing potent activation of T cells. We show that anti-CD3 adsorbed onto SWNT bundles stimulate cells more effectively than equivalent concentrations of soluble anti-CD3. Stimulation by antibody-adsorbed onto SWNT is significantly more than other high surface area materials (activated carbon, polystyrene and C60 nanoparticles), suggesting unique properties of SWNT bundles in stimuli presentation. We demonstrate the surface area tunability of these bundles by chemical treatment and its effect on antibody adsorption and subsequent T cell activation. In addition, the T cell response varied with the concentration of SWNT in a dose dependent manner. Antibody stimuli adsorbed onto SWNT bundles represent a novel paradigm for efficient activation of lymphocytes, useful for basic science applications and clinical immunotherapy.

# 715c Multiplex Analysis of Raman Nanoparticle Signatures for Protein Imaging In Tissues

**Barry R. Lutz**[1], Claire Dentinger[1], Lienchi Nguyen[1], Lei Sun[1], Jingwu Zhang[1], April Allen[2], Selena Chan[1], and Beatrice Knudsen[2]. (1) Biomedical and Life Sciences, Intel Corporation, Digital Health Group, Mission College Boulevard, SC3-41 2200, Santa Clara, CA 95054, (2) Division of Public Health Sciences, Fred Hutchinson Cancer Research Center, 1212 Aloha Street, M5-A864, Seattle, WA 98109

The ability to simultaneously detect multiple biomolecule targets in a single sample reduces demands on precious samples and allows new insight into expression colocalization in applications including cell and animal imaging, tissue histology, and flow cytometry. Nanoparticle probes based on surface-enhanced Raman scattering (SERS) are emerging as a new class of optical labels with unique spectral fingerprints that are well-suited for multiplex detection. Since the discovery of nanoparticle SERS in 1997, a variety of fabrication approaches and biomolecule conjugation methods have been developed to enable early applications in immunoassays, nucleic acid hybridization, cell and tissue staining, and in vivo imaging. Despite the significant potential of Raman probes for improved multiplex quantification, they have primarily been used for qualitative detection of single targets, due in part to the lack of analysis methods that take full advantage of the complex spectral fingerprint.

We describe application of antibody-conjugated Composite Organic-Inorganic Nanoparticles (COINs) for multiplex detection in fixed human tissue samples, and we present a simple but robust spectral fitting method that quantitatively separates overlapping probe signatures and background autofluorescence. Simplified solution and protein plate binding assays show that multiplex spectral analysis allows sensitive detection of small signals (<2% of total signal) and provides quantitative multiplex detection of at least four overlapping probe signatures. In a triplex assay on human prostate tissue, two antibody-conjugated COINs and a fluorophore are used to image expression of prostate-specific antigen, cytokeratin-18, and DNA. Spectral analysis allows point-by-point removal of unknown background autofluorescence and direct measurement of quantification error at each point in the sample (~2% of total signal). The multiplex assays are carried out by single-laser excitation using a small fraction of the visible window (~10%). This suggests that multiple-laser excitation of Raman probes and quantification by spectral analysis will allow high-order multiplexing beyond that currently possible with fluorophores.

# 715d Multifunctional Surface Ligands for Improved Stability of Quantum Dots and Gold Nanoparticles for Biological Applications

**Bing C. Mei**[1], Kimihiro Susumu[2], Igor L. Medintz[3], T. J. Mountziaris[1], and Hedi Mattoussi[2]. (1) Chemical Engineering, University of Massachusetts, Amherst, MA 01003, (2) Optical Sciences, Naval Research Laboratory, 4555 Overlook Ave SW, Washington, DC 20375, (3) Center for Bio/Molecular Sciences and Engineering, Naval Research Laboratory, 4555 Overlook Ave SW, Washington, DC 20375

Over the past decade, quantum dots (QDs) and gold nanoparticles (AuNPs) have attracted significant attention for biological sensing applications. Such interest is driven by their unique spectroscopic properties, including tunable emission and large extinction coefficients. Among the major challenges when designing QDs and AuNPs for use in biology is to design surface-functionalization strategies that provide long term stability over a broad range of biologically relevant conditions and are compatible with common bioconjugation techniques.

We have previously developed a series of compact and modular dihydrolipoic acid appended poly (ethylene glycol) (DHLA-PEG-OH) ligands, which can render QDs (via cap exchange strategy) water-soluble and functional (J. Am. Chem. Soc., 2005, 127: 3870-8). We report here the design of a new series of ligands that feature a more robust linkage group between the DHLA and PEG; the new ligands use an amide bond while the previously reported featured an ester bond. We have also developed a series of ligands with a variety of functional end groups, including biotin and carboxyl groups. Each of these ligands has a bidentate anchoring group, a variable length PEG to promote hydrophilicity, and a terminal group for bioconjugation.

Cap exchange of nanoparticles, both QDs and AuNPs, with the new ligands enabled their dispersion in aqueous buffer solutions. The dispersions were homogeneous and stable over a broad pH range (from 3 to 13 for CdSe/ZnS QDs and 2-13 for AuNPs) and in high ion concentration. Using DHLA-PEG-OCH3 either neat or mixed with amine- or carboxyl-terminated ligands (DHLA-PEG-NH2 or DHLA-PEG-COOH) allowed tuning of the surface functionalities of these nanoparticles, hence tailoring their interaction with their environment.

#759a – Commercialization of Nanotechnology (22006)

# Decontamination, Safety Applications, and Products Based on Nanomaterials

*David Jones, Olga Koper, Kyle Knappenberger, Megan Winter*
*NanoScale® Corporation, Manhattan, Kansas, USA*

NanoScale Corporation is a dynamic and innovative company focused on the development and commercialization of advanced products, including our proprietary nanocrystalline metal oxides and related technologies. NanoScale works in four primary business segments: the sale of packaged and bulk high performance materials, application services related to the materials, application development contracts, and providing analytical services.

NanoScale's business model includes pro-actively identifying and pursuing mutually rewarding joint efforts with private and government sectors to provide unique solutions. In conjunction with the Department of Defense (DoD), Department of Energy (DoE), National Institutes of Health (NIH), National Science Foundation (NSF), Environmental Protection Agency (EPA), other federal agencies, and commercial customers, NanoScale has invested over $26 million in research and application development. Additionally, independent of our joint efforts with these entities, NanoScale maintains a well funded and active internal R&D program that is dedicated to advancing our Intellectual Property portfolio, identifying new scientific applications for our advanced earth minerals, identifying new markets and applications, and providing new added value products and services for our potential and existing customers.

NanoScale has a strong and ever-growing portfolio of Intellectual Properties and trade secrets. The company has ownership and exclusive rights to twenty-four (24) issued U.S. patents and seventeen (17) issued International patents, with an additional six (6) U.S. and thirty-eight (38) International patents filed and pending.

## Core Technology Overview

NanoScale has developed a line of nanocrystalline materials (NanoActive® materials) with remarkably high chemical reactivity. These materials have been shown to be effective against a broad range of hazards, including Toxic Industrial Chemicals (TICs), Chemical Warfare Agents (CWAs)[1], as well as Biological Warfare Agents (BWAs). Reactive nanomaterials are non-flammable and stable at high and low temperatures.

[1] Wagner, G.W.; Bartram, P.W.; Koper, O.; Klabunde, K.J. "Reactions of VX, GD, and HD with Nanoscale MgO," *J. Phys. Chem. B.* **1999**, *103,* 3225; Wagner, G.W.; Koper, O.B.; Lucas, E.; Decker, S.; Klabunde, K.J. "Reactions of VX, GD, and HD with Nanosize CaO: Autocatalytic Dehydrohalogenation of HD," *J. Phys. Chem. B.* **2000**, *104*, 5118; Wagner, G.W.; Procell, L.R.; O'Connor, R.J.; Munavalli, S.; Carnes, C.L.; Kapoor, P.N.; Klabunde, K.J. "Reactions of VX, GB, GD, and HD with Nanosize $Al_2O_3$. Formation of Aluminophosphonates," *J. Am Chem. Soc.* **2001**, *123*, 1636.

Their high surface area and unique morphology give NanoActive materials very different properties compared to either bulk materials or singular atoms.[2]

Macroscopically, metal oxide nanoparticles are fine powders, but closer examination shows that nanoparticles consist of very small crystals (usually 2-10 nanometers in size) that form larger clusters. One example is NanoActive MgO Plus, which typically has a surface area of 650 - 700 $m^2/g$ and an average crystallite size of 4 nm. By way of comparison, the specific surface area of many conventional MgO products is 20-40 $m^2/g$.

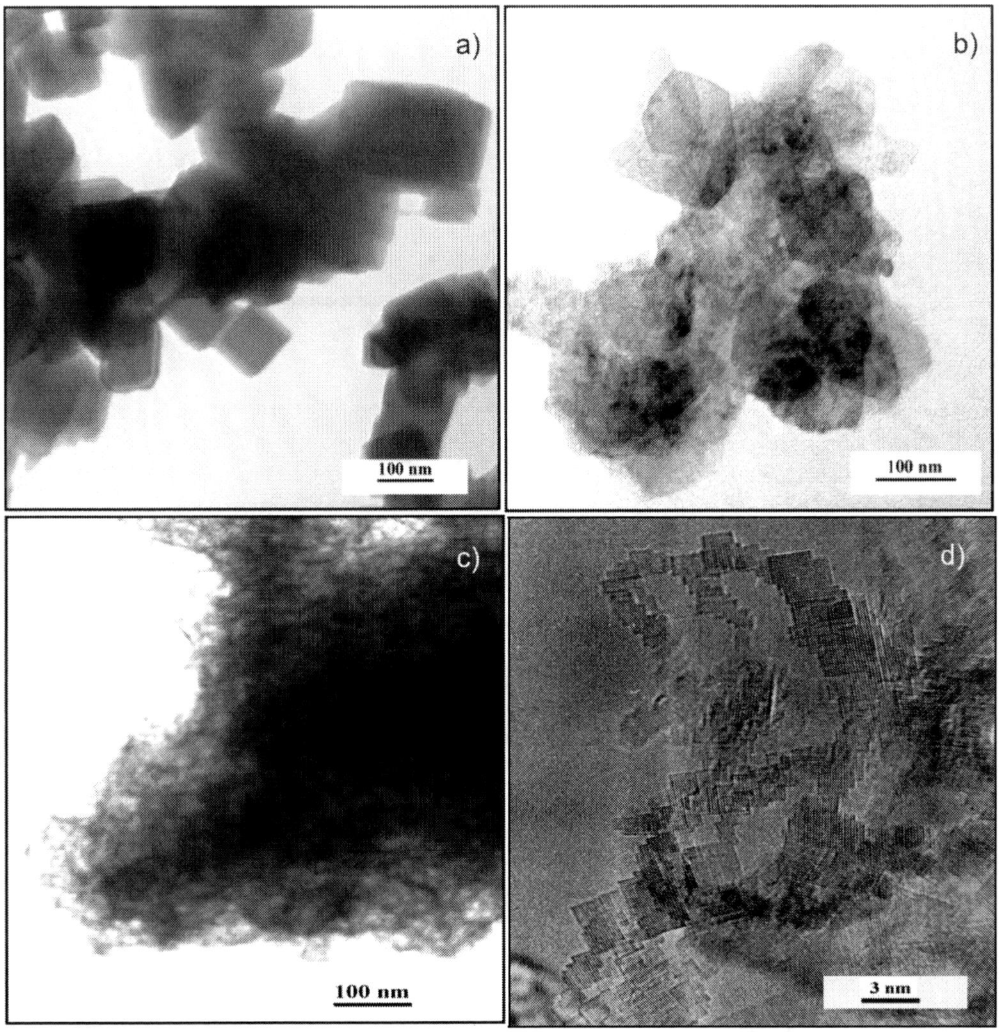


***Figure 1: TEM of Magnesium Oxides a) Conventional, b) NanoActive MgO, c) NanoActive MgO Plus & d) High-resolution NanoActive MgO Plus.***

---

[2] Klabunde, K.J.; Stark, J. V; Koper, O.; Mohs, C.; Park, D.G.; Decker, S.; Jiang, Y; Lagadic, I.; Zhang, D. "Nanocrystals as Stoichiometric Reagents with Unique Surface Chemistry," *J. Phys. Chem. B* **1996**, *100*, 12142-12153 (invited feature article).

Differences between NanoActive products and their conventional analogs are shown by the appearance of various samples of MgO as shown in Figure 1. The Conventional sample (1a) has an appearance of cubes 60-80 nm in length while NanoActive MgO (1b) is composed of hexagonal platelets ~ 80 nm in length and 10 nm thick. NanoActive MgO Plus, (1c) and (1d), is made up of a porous web-like material. The high resolution TEM, Figure 1d, shows very small 2-4 nm cubes that aggregate into interconnected polyhedral structures, forming a highly porous material with numerous, low-coordinate corners and edge sites.

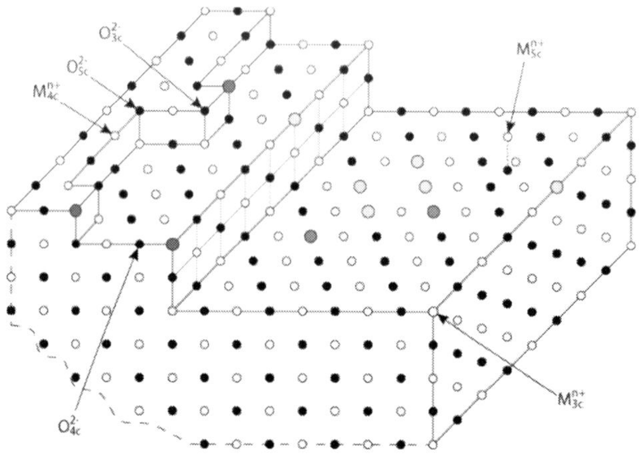


***Figure 2: Schematic of Metal Oxide Surface Illustrating Surface, Edge and Corner Ions.***

A conceptual view of a typicalmetal oxide nanocrystal is given in Figure 2. Cation ($M^{n+}$) sites in the interior of the crystal (not visible) have a coordination number of 6, surface cations (representative ions shown in blue) have a coordination number of 5, edge cations, shown in yellow, have a coordination number of 4, and corner sites, shown in red, have a coordination number of 3. A surface vacancy defect reduces the coordination number of surrounding ions; this is shown by a hole surrounded by yellow ions. The lower the coordination number of a cation site, the higher its reactivity, in general.

This unusual morphology results in enhanced chemical reactivity and suggests a two-step decomposition mechanism; the first step is adsorption of toxic agent by physisorption, followed by a second step of chemical decomposition. This two-step mechanism substantially enhances the detoxification abilities of nanocrystalline materials because it makes the decomposition less dependent on temperature. Although chemical reactivity is slowed at lower temperatures, physisorption, which is required before chemical reaction, is *enhanced* at low temperatures.

The unique characteristics of materials developed at NanoScale lend themselves to applications in the fields of chemical neutralization, odor control and water decontamination. Many materials currently used for these applications, including activated carbon, can adsorb the harmful contaminant but do not always destroy it. NanoScale's nanocrystalline materials have been proven to not only remove harmful contaminants but also chemically destroy them.

For example, testing conducted at the Aberdeen Proving Grounds elucidated the mechanism of agent destruction for chemical warfare agents. Wagner et al.[3] reported that NanoScale metal oxide nanomaterials were very effective in room temperature destructive adsorption of chemical agents, such as VX, GD and HD. The destruction products of GD (soman) are pinacolyl methylphosphonic acid (GD-acid) that converts to surface bound methylphosphonic acid (MPA), and (hydrofluoric acid) HF, which is also neutralized by metal oxides. Figure 3 illustrates this reaction.

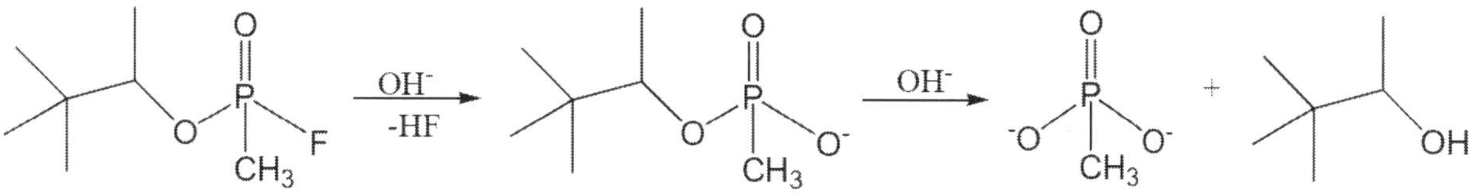


GD GD-acid MPA

***Figure 3: GD Degradation by Reactive Nanomaterials***

**Product Development and Commercialization**

NanoScale's core technology has recognized value in the areas of containment and neutralization of CWAs and TICs, odor control, antimicrobial applications, water decontamination, as well as others. In addition to extensive research and development efforts, NanoScale focuses on the transfer of laboratory developments into commercial products for the improvement of environmental safety and overall quality of life. Current commercial offerings include FAST-ACT®, NanoActive chemicals, and OdorKlenz®, which have been sold throughout the world. Other products are currently in the development phase with both government and commercial partners.

***FAST-ACT***

FAST-ACT (First Applied Sorbent Treatment Against Chemical Threats), a revolutionary chemical hazard containment and neutralization system, is NanoScale's first commercial product and was developed with the end user in mind. The FAST-ACT family of products utilizes a non-toxic, non-corrosive, non-flammable dry powder that is capable of neutralizing a wide range of toxic chemicals with the added capability to destroy CWAs. FAST-ACT is sold to first responders and HAZMAT teams around the world and used in response to liquid chemical spills and vapor releases. It has significant advantage over other chemical release mitigation measures in that it can be used on any release without the need for prior identification of the contaminant. Figure 4 shows the efficacy of FAST-ACT against vapor hazards.

---

[3] Wagner, G.W.; Bartram, P.W.; Koper, O.; Klabunde, K.J. "Reactions of VX, GD, and HD with Nanoscale MgO," *J. Phys. Chem. B.* **1999**, *103*, 3225; Wagner, G.W.; Koper, O.B.; Lucas, E.; Decker, S.; Klabunde, K.J. "Reactions of VX, GD, and HD with Nanosize CaO: Autocatalytic Dehydrohalogenation of HD," *J. Phys. Chem. B.* **2000**, *104*, 5118; Wagner, G.W.; Procell, L.R.; O'Connor, R.J.; Munavalli, S.; Carnes, C.L.; Kapoor, P.N.; Klabunde, K.J. "Reactions of VX, GB, GD, and HD with Nanosize $Al_2O_3$. Formation of Aluminophosphonates," *J. Am Chem. Soc.* **2001**, *123*, 1636.

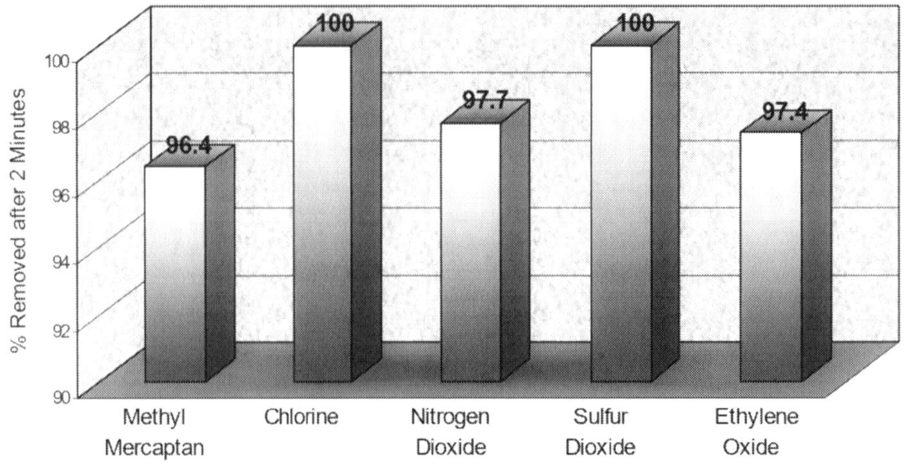


***Figure 4: FAST-ACT Efficacy Against Vapor Hazards***

The FAST-ACT material formulation was optimized for decontamination of a variety of chemical hazards (Table 1) and the product line includes pressurized cylinders, bulk pails and shakers. The variety of delivery methods allows a user to safely contain any size spill or leak.

***Table 1: Efficacy of FAST-ACT Against Hazardous Materials***

| Neutralization | | Adsorption | Containment | Not Protected |
|---|---|---|---|---|
| Corrosive Materials | | Vapor Hazards | Liquid Solvent Spills | Biologicals<br>Bacteria<br>Viruses<br>Spores<br><br>Nuclear<br><br>Radiological<br><br>Heavy Metals<br><br>Solid Waste |
| Acids<br>Inorganic and Organic<br>Hydrochloric Acid<br>Hydrofluoric Acid*<br>Hydrobromic Acid<br>Nitric Acid*<br>Phosphoric Acid<br>Sulfuric Acid*<br>Acetic Acid<br>Methanesulfonic Acid<br>Ethanesulfonic Acid<br>Benzenesulfonic Acid<br>Toluenesulfonic Acid<br><br>Phosphorus<br>Pesticides<br>Dimethylmethyl Phosphonate<br>Paraoxon<br>Parathion*<br><br>Sulfur<br>2-Chloroethyl Ethyl Sulfide<br>Methyl Mercaptan<br><br>Phenols<br>Nitrophenols<br>Chlorophenols | Carbonyl Compounds<br>Aldehydes*<br>Ketones<br>Carboxylic Acids<br><br>Nitrogen Compounds<br>Acetonitrile*<br>Sodium Cyanide (aq)<br>4-vinylpyridine<br><br>Halogens/Halides<br>Acetyl Chloride<br>Chloroacetyl Chloride<br>Chlorine<br>Chloroform<br>Hydrogen Bromide*<br>Cyanogen Chloride<br>Methylene Chloride<br>Carbon Tetrachloride<br>TCE, PCE<br><br>Bis-(2-Chloroethyl) Sulfide<br><br>Pinacolyl methylphosphonofluoridate<br><br>O-ethyl S-(2-diisopropylaminoethyl) methylphosphonothioate | Acidic and Caustic Gases<br>Hydrogen Chloride<br>Hydrogen Fluoride<br>Hydrogen Bromide*<br>NOx/N2O4*<br>Sulfur Dioxide<br>Hydrogen Sulfide*<br>Diborane*<br>Hydrogen Selenide*<br>Phosphine*<br>Ammonia<br>Carbonyl Sufide<br>Hydrogen Cyanide*<br><br>Chlorinated Organics<br>Acetyl Chloride<br>Chloroacetyl Chloride<br>Chloroform<br>Methylene Chloride<br><br>Halogens<br>Chlorine*<br>Bromine<br>Iodine<br><br>Volatile Organics<br>Methyl Mercaptan*<br>Ethylene Oxide*<br>Formaldehyde*<br>Phosgene*<br>Arsine* | Alcohols/Phenols<br>Ethanol<br>Methanol<br>Allyl Alcohol*<br>Nitrophenols<br>Chlorophenols<br><br>Caustics<br>Anhydrous<br>Ammonia*<br>Metal Hydroxides (aq)<br><br>Petrochemicals<br>Diesel<br>Gasoline<br>Oils<br><br>Others<br>Acrylonitrile*<br>Benzene<br>Hydrazine*<br>Toluene<br>Acrolein*<br>Methylhydrazine*<br>Methylisocynate* | |
| LIQUID & VAPOR CHEMICAL SPILLS AND RELEASES | | | | |

## *NanoActive Chemicals*

The NanoActive line of specialty chemicals are highly porous metal oxides with unmatched surface areas and unique nanocrystalline morphologies. The distinctive properties provide exceptional chemical reactivity for a broad range of industrial, commercial, military, and research applications. The markets include decontamination, personal protection, energy, electronics, catalysts, and other advanced materials. Despite the absence of any regulatory requirements for these materials, NanoScale has utilized independent certified laboratories to conduct evaluations of its products for health and safety risks. NanoActive materials have been proven safe through oral, pulmonary, ocular, and dermal toxicology testing. Additionally, NanoActive materials have been shown to be no more toxic than their non-nanocrystalline analogues. The materials are available to markets in a variety of forms including powders, granules, suspensions, and can be customized for market specific applications.

***Table 2: Toxicity Data for NanoActive Materials***

| | *NanoActive $TiO_2$** | *NanoActive MgO*** |
|---|---|---|
| *Acute Oral Toxicity* | $LD_{50}$ > 2 g/kg | $LD_{50}$ > 5 g/kg |
| *Acute Dermal Toxicity* | $LD_{50}$ > 5 g/kg | - |
| *Acute Dermal Irritation* | (PII) of "0" Non-irritating | EPA Category IV Non-irritating |
| *Skin Sensitization* | Non-sensitizer | Non-sensitizer |
| *Acute Eye Irritation* | Practically Non-irritating | EPA Category III<br>Slightly irritating |
| *Acute Inhalation* | Nontoxic<br>2334 mg/m$^3$ for 4 hrs | Nontoxic<br>259 mg/m$^3$ for 4 hrs |

** - Testing conducted at MPI* *** - Testing conducted at CHPPM*
***PII** – Primary Irritation Index*
*In all Oral and Dermal Toxicity studies no test animals died during the exposure time, therefore $LD_{50}$ is > test limit dose for the specific NanoActive material studied.*

## *OdorKlenz*

OdorKlenz, NanoScale's family of odor elimination products, is based on the same destructive adsorption technology. OdorKlenz has been proven commercially effective at mitigation of a wide variety of odors and is being used in areas such as the commercial and residential real estate industry, death care and medical industries, veterinary necropsy labs, and gross anatomy labs. The largest market impacted by the

OdorKlenz technology is disaster recovery. NanoScale was able to adapt the technology and configure it into a cartridge that can be used with existing industry air scrubbers for the elimination of airborne odors. The technology allows for safer and faster odor elimination.

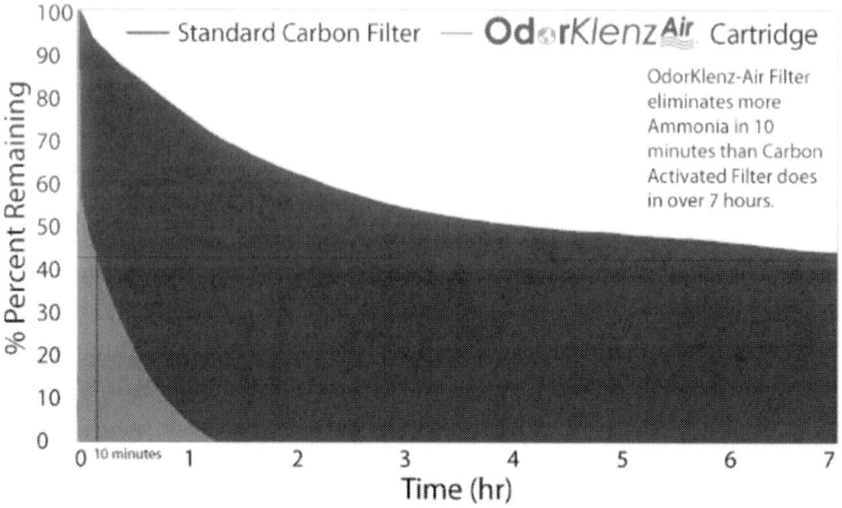


***Figure 5: Ammonia Removal from Air***

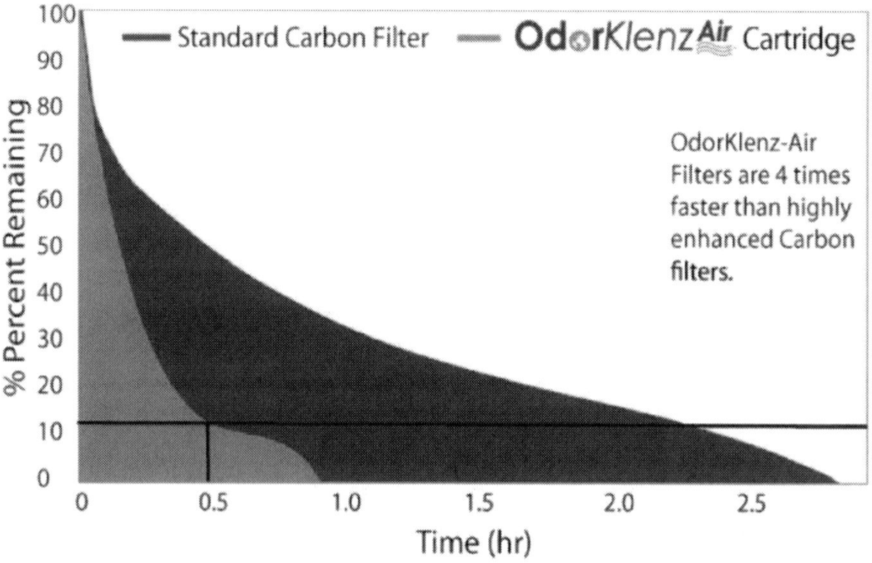


***Figure 6: Hydrogen Sulfide Removal from Air***

## 759b Report on the Nist Cross-Industry Issues in Nanomanufacturing Workshop

**Anne M. Chaka**, Physics Laboratory, National Institute of Standards and Technology, 100 Bureau Drive Stop 8443, Gaithersburg, MD 20899-8443, John Cowie, Agenda 2020 Technology Alliance, American Forest & Paper Association, 1111 19th Street, N.W., Suite 800, Washington, DC 20036, Gerard Closset, Consultant, American Forest & Paper Association, 1111 19th Street, N.W., Suite 800, Washington, DC 20036, C. Michael Garner, Intel Corporation, Santa Clara, CA 95054, Michael Postek, Manufacturing Engineering Laboratory, National Institute of Standards and Technology, 100 Bureau Drive Stop 8210, Gaithersburg, MD 20899-8210, and Dianne Poster, Chemical Sciences and Technology Laboratory, National Institute of Standards and Technology, 100 Bureau Drive Stop 8300, Gaithersburg, MD 20899-8300.

This presentation will discuss the results of a workshop held in May at the National Institute of Standards and Technology that brought together experts from across industrial sectors, national labs, and academia to identify common problems and potential solutions in the manufacture of nanomaterials and nano-containing products at a precompetitive level. The objective was to establish a framework to enable cross-fertilization and identification of best practices utilizing currently available science and technology, to define collaborative research programs, and to develop a cross-industry roadmap specific to nanomanufacturing. Opportunities for participation in the development of the roadmap and subsequent activities will be presented. The Workshop was endorsed by the Nanoscale Science, Engineering, and Technology Subcommittee of the National Nanotechnology Initiative, in collaboration with the Nanomanufacturing Industry Liason and Innovation Working Group, and sponsored by Agenda 2020 Technology Alliance with the American Forest Products and Paper Association, U.S. Forest Service, Imerys, and the University of Maryland. Industrial sectors which participated included the aerospace, automotive, chemical, food, forest products, medical, pharmaceutical, and semiconductor industries. High priority needs in nanomanufacturing science and technology were identified by industrial participants in the following three areas:

(1). Characterizing nanomaterials and enhancing their separation and fractionation to address challenges in commercial production of uniform, high quality, stable, and consistent (reproducible) nanomaterials in high volume.

(2). Understanding and controlling the surface-dependent properties of nanomaterials such as dispersion, aggregation, and adhesion at their interface with a matrix, with an emphasis on non-covalent bonding interactions.

(3) Understanding and controlling multiple properties of nanocomposites.

Web Page: www.energetics.com/nanocrosscutmay08/

# AUTHOR INDEX

# AUTHOR INDEX

# AUTHOR INDEX

# AUTHOR INDEX

# AUTHOR INDEX

# AUTHOR INDEX

# AUTHOR INDEX

# AUTHOR INDEX

# AUTHOR INDEX

# AUTHOR INDEX

# AUTHOR INDEX

# AUTHOR INDEX

# AUTHOR INDEX